Migration and Fate of Pollutants in Soils and Subsoils

NATO ASI Series

Advanced Science Institutes Series

A series presenting the results of activities sponsored by the NATO Science Committee, which aims at the dissemination of advanced scientific and technological knowledge, with a view to strengthening links between scientific communities.

The Series is published by an international board of publishers in conjunction with the NATO Scientific Affairs Division

A Life Sciences	Plenum Publishing Corporation
B Physics	London and New York
C Mathematical and Physical Sciences	Kluwer Academic Publishers Dordrecht, Boston and London
D Behavioural and Social Sciences	
E Applied Sciences	
F Computer and Systems Sciences	Springer-Verlag Berlin Heidelberg New York
G Ecological Sciences	London Paris Tokyo Hong Kong
H Cell Biology	Barcelona Budapest
I Global Environmental Change	

NATO-PCO DATABASE

The electronic index to the NATO ASI Series provides full bibliographical references (with keywords and/or abstracts) to more than 30 000 contributions from international scientists published in all sections of the NATO ASI Series. Access to the NATO-PCO DATABASE compiled by the NATO Publication Coordination Office is possible in two ways:

- via online FILE 128 (NATO-PCO DATABASE) hosted by ESRIN, Via Galileo Galilei, I-00044 Frascati, Italy.

- via CD-ROM "NATO-PCO DATABASE" with user-friendly retrieval software in English, French and German (© WTV GmbH and DATAWARE Technologies Inc. 1989).

The CD-ROM can be ordered through any member of the Board of Publishers or through NATO-PCO, Overijse, Belgium.

Series G: Ecological Sciences, Vol. 32

Migration and Fate of Pollutants in Soils and Subsoils

Edited by

Domenico Petruzzelli

Istituto Ricerca Acque
National Research Council
5, Via de Blasio
I-70123 Bari, Italy

Friedrich G. Helfferich

The Pennsylvania State University
Department Chemical Engineering
University Park, PA 16802-4400, USA

Springer-Verlag
Berlin Heidelberg New York London Paris Tokyo
Hong Kong Barcelona Budapest
Published in cooperation with NATO Scientific Affairs Division

Proceedings of the NATO Advanced Study Institute on Migration and Fate of
Pollutants in Soils and Subsoils held at Maratea, Italy, from May 24 to June 5, 1992

ISBN 3-540-56041-6 Springer-Verlag Berlin Heidelberg New York
ISBN 0-387-56041-6 Springer-Verlag New York Berlin Heidelberg

Typesetting: Camera ready by authors
31/3145 - 5 4 3 2 1 0 - Printed on acid-free paper

Preface

Mass transport phenomena in natural permeable media are of practical importance in a variety of fields. Examples from civil, hydraulic and environment engineering are groundwater pollution and seepage from landfills or other disposal operations, an example from petroleum engineering is enhanced oil recovery. Other disciplines having major impact include chemistry, biology, hydrology, geochemistry, ecology.

By their very nature, permeable media and thus the transport phenomena taking place in them are highly complex. Often, finding solutions to practical problems calls for close cooperation between experts from different fields. In recent years, highly complex theories and methods, based on novel concept of non-linear wave propagation have been developed, verified and successfully applied in chemical, petroleum and environmental engineering. Yet, a truly comprehensive understanding of migration and fate of chemicals in soils and subsoils and an ability to make reliable predictions have not been fully obtained.

The authors (from European and American countries) are experts in soil and environmental sciences as well as in theory of wave propagation and numerical modeling methods. The focus was essentially posed on the analysis of contributing phenomena and their interactions, modeling, and practical use of such knowledge and models for guidance in disposal operations, preventive measures to minimize ecological damage, prediction of consequences of seepage, and design of remedial actions.

No specialty thrives in isolation. It was from this background that the idea arose to convene qualified experts from different areas to work together in an international study meeting for the benefit of those attending as well as for mutual enrichment. This gave the meeting the spice of contribution, formal and informal, from experts of many diverse backgrounds and nationalities, and helped to secure the broad-based support from NATO, Consiglio Nazionale delle Ricerche of Italy (National Research Council), Ente Autonomo Acquedotto Pugliese (Apulian Water Authority) and industrial companies that made the Summer School possible in the first place.

It is not for the organizers and lecturers to say that the meeting was a success.   This has to come from future perspectives by those who derived benefit from attending the School or from reading the contributions presented here.

It is a pleasure, however, to express our deep appreciation for the untiring work of the other members of the Scientific Board of the meeting: Prof. W.J.Weber, Prof. R.Passino, Prof. L.Liberti, Prof. A.J.Valocchi, Dr G.Tiravanti, Prof.T.Asano, Prof.V.S.Soldatov, and for the support of all who helped to provide a stimulating climate, both technical and social, for a true meeting of minds amid beatiful surroundings, at Acquafredda di Maratea.

D.Petruzzelli   F.G.Helfferich

NATO ADVANCED STUDY INSTITUTE
MIGRATION AND FATE OF POLLUTANTS
IN SOILS AND SUBSOILS
Acquafredda di Maratea, Italy, May 24 - June 5, 1992

Director: **D. PETRUZZELLI**, Istituto Ricerca Acque, C.N.R.,
                          Bari, Italy.

Co-Director: **F.G. HELFFERICH**, The Pennsylvania State University,
                          University Park, U.S.A.

Lecturers:

ASANO T., University of California, Davis, CA, USA.
CORAPCIOGLU M.Y., Texas A&M University, College Station, TX, USA.
FORSTNER U., Technical University Hamburg-Harburg, Germany.
GAMBOLATI G., Università di Padova, Padua, Italy.
GANOULIS J., Aristotle University, Thessaloniki, Greece.
HARWELL J.H., The University of Oklahoma, Norman, OK, USA.
HELFFERICH F.G., Penn State University, University Park, PA, USA.
LOPEZ A., Istituto Ricerca Acque C.N.R., Bari, Italy.
MANSOUR M., Institut Okologische Chemie, Freising, Germany.
NERETNIEKS I., Kungl Tekniska Ogskolan, Stockholm, Sweden.
NOVAK C.F., Sandia National Laboratories, Albuquerque, NM, USA.
PETRUZZELLI D., Istituto Ricerca Acque C.N.R., Bari, Italy.
ROY W.R., Illinois State Geological Survey, Champaign, IL, USA.
SCHWEICH D., Lab.Sci. Genie Chimique, CNRS-ENSIC, Nancy, France.
SENESI N., Università di Bari, Bari, Italy.
SOLDATOV V.S., Bielorussia Academy of Science, Minsk, Bielorussia.
SPARKS D.L., University of Delaware, Newark, DE, USA.
VALOCCHI A.J., University of Illinois, Urbana, USA.
VAN DER ZEE S.E., Agricultural University, Wageningen, The Netherlands.
WEBER W.J. Jr., The University of Michigan, Ann Arbor, MI, USA.

TABLE OF CONTENTS

PART II. GLOBAL PROPAGATION PHENOMENA AND MODELING

Daniel Schweich

TRANSPORT OF LINEARLY REACTIVE SOLUTES
IN POROUS MEDIA. BASIC MODELS AND CONCEPTS.

Friedrich G. Helfferich

MULTICOMPONENT WAVE PROPAGATION: THE COHERENCE PRINCIPLE. AN INTRODUCTION

Jeffrey H. Harwell, David A. Sabatini, Thomas S. Soerens
FATE OF NON-AQUEOUS PHASE LIQUIDS:
MODELING SURFACTANTS EFFECTS. 309

Giuseppe Gambolati, Claudio Paniconi, Mario Putti
NUMERICAL MODELING OF CONTAMINANT
TRANSPORT IN GROUNDWATER. 381

PART III. SPECIFIC PROBLEMS AND APPLICATIONS

Jacques Ganoulis

RISK ANALYSIS OF GROUNDWATER CONTAMINATION. 455

Vladimir S. Soldatov

MIGRATION OF RADIONUCLIDES IN NATURAL POROUS MEDIA
THE CHERNOBYL CASE. 475

PART I. LOCAL PHENOMENA

Transport and Fate of Pollutants in Subsurface Systems: Contaminant Sorption and Retardation

Walter J. Weber, Jr.
Department of Civil and Environmental Engineering
The University of Michigan
181 Engineering - Bldg. 1A
Ann Arbor, Michigan 48109-2125
U.S.A

Sorption phenomena significantly affect the transport and ultimate fate of many organic and inorganic contaminants in soils and subsurface systems. The effects are complex, with each particular combination of contaminant, soil, and set of local conditions yielding a unique distribution of contaminant mass. Thermodynamic considerations govern ultimate contaminant distributions, but the rates at which such distributions are approached involve equally important considerations. In fact, sorption rates frequently determine the relative significance of sorption with respect to other reaction and transport processes occurring in subsurface environments.

SUBSURFACE SORPTION PROCESSES

Sorption reactions occur among all phases present in subsurface systems, and at the interfaces between phases. Two broad categories of phenomena, adsorption and absorption, can be differentiated. In adsorption, solute accumulation is generally restricted to a surface or interface between a solution and an adsorbent. In contrast, absorption is a process in which solute transferred from one phase to another interpenetrates the molecular structure of the sorbent phase. A further variation of the process occurs if a sufficiently high accumulation of solute occurs at an interface to form a precipitate, or some other type of molecular solute-solute association; e.g., a polymer or a micelle. Such processes differ from both adsorption and absorption in that they result in formation of new and distinct three-dimensional phases. Because adsorption processes generally yield surface or interface concentrations of solute greater than those in the bulk phase, it is possible for precipitation or association to occur on a surface in the absence of a solution phase reaction of the same type.

Sorption results from a variety of different types of attractive and repulsive forces between solute molecules, solvent molecules and the molecules of a sorbent. Such forces usually act in concert, but one type or another is generally more significant than the others in any particular situation. Absorption processes, such as dissolution of a relatively immiscible

NATO ASI Series, Vol. G 32
Migration and Fate of Pollutants in Soils and Subsoils
Edited by D. Petruzzelli and F. G. Helfferich
© Springer-Verlag Berlin Heidelberg 1993

phase into an aqueous phase, or accumulation of a lipophilic substance in an organic phase, involve exchanges of molecular environments. In such cases, the energy of an individual molecule is altered by its interactions with the solvent and sorbent phases. The distribution of the solute between phases results from its relative affinity for each phase, which in turn relates to the nature of the forces which exist between molecules of the sorbate and those of the solvent and sorbent phases. These forces can be likened to those associated with classical chemical reactions. Adsorption also entails intermolecular forces, but in this case, molecules at surfaces rather than bulk phase molecules are involved, and the former typically manifest a broader range of interactions. Accordingly, three loosely defined categories of adsorption -- physical, chemical and electrostatic -- are traditionally distinguished, according to the class of attractive force which predominates.

Categorization of the primary classes of interactions leading to sorption reaction provides a useful means to bridge the gap between detailed enumerations of intermolecular forces and working descriptions of observed sorption phenomena. In reality, however, such forces do not act independently; it is, rather, the effect of their action collectively and in concert which dictates the sorptive separation of substances in any system. As such, there are two predominant means by which sorption is motivated. The first derives from the net forces of affinity between the sorbate and the sorbent, here defined as "sorbent-motivated" sorption. Cation exchange reactions with clays, for example, are sorbent-motivated reactions in which electrostatic charges on the sorbent are overwhelmingly attractive to the sorbate. The second involves the sum of adverse interactions of the sorbate with the solution phase, yielding "solvent-motivated" sorptions. The repulsion by water of an oil or other hydrophobic contaminant, leading to its accumulation at a soil/water interface, constitutes a principally solvent-motivated sorption. As expected, the combined influence of all intermolecular forces in complex systems usually leads to sorption processes which are not as easily categorized as specifically or exclusively either sorbent motivated or solvent motivated, but rather which fall somewhere in the interlying continuum.

SORPTION MODELS
MECHANISTIC EQUILIBRIUM MODELS

Models providing "first principle" description of the energetics of intermolecular reactions underlying sorption have been developed. These models include mechanistic characterization of ion exchange, surface complexation, and hydrophobic sorption. Such models can often provide insights into mechanisms controlling sorption reactions in particular types of systems, and thus aid in the analysis of anticipated system responses to changes in critical conditions.

Ion Exchange and Surface Complexation

Soils materials typically contain surfaces which exhibit electrical charge characteristics, which in turn can exert influence on the sorption of ionic and polar species. Surface charges are instrumental, for example, in the sorption of metal species in subsurface systems. The charge on the surface must be counterbalanced in the aqueous phase to maintain electroneutrality. As a result, an electrical double layer exists at interfaces. This double layer consists of the charged surface sites and an equivalent aqueous-phase excess of ions of opposite charge (counter-ions) which accumulate in the water near the surface of the particle. The counter-ions are attracted electrostatically to the interfacial region, giving rise to a concentration gradient, which in turn sets up a potential for random diffusion of ions away from the surface. The competing processes of electrostatic attraction and counter- diffusion spread the charge over a diffuse layer in which the excess concentration of counter-ions is highest immediately adjacent to the surface of the particle and decreases gradually with increasing distance from the solid-water interface.

Several types of reactions can be attributed to forces associated with charged sites. Ion exchange reactions resulting from the action of electrostatic forces occur at fixed sites on soil surfaces (Stumm and Morgan, 1981; James and Parks, 1982). Fixed-charge sites, those not subject to change with changes in solution phase concentration, result from isomorphic substitution of ions in the lattice structure of clay-like minerals. A number of relationships have been developed to describe ion-exchange equilibria, including equations based on the Gouy-Chapman model for the diffuse double layer (Eriksson, 1952).

Sorption reactions which occur on variable (vis-à-vis fixed) charged surfaces, such as soil organic matter, mineral oxides (SiO_2, Al_2O_3, TiO_2 and $FeOOH$) and on the edge sites of layered silicate minerals comprise another class of electrostatic interactions. The association of ions with these surfaces is hypothesized to occur through surface complexation or ligand exchange reactions analogous to those which occur in solution. The charges on these surfaces arise most commonly through protonation and deprotonation reactions, and are thus highly pH dependent. A number of surface complexation models have been developed over the past several decades. These utilize mass law relationships and mass and charge balance equations to describe equilibria between solution species and surface complexes, and various hypotheses regarding the structure of the interfacial region to identify the location of surface complexes and describe the diffuse layer-charge potential relationship (Sposito, 1984b; Barrow and Bowden, 1987; Hayes and Leckie, 1987).

Hydrophobic Sorption

Hydrophobic interactions comprise the primary motivation for a large class of sorption reactions in the subsurface. The association of neutral, relatively nonpolar organic molecules with soils often results in quasi-linear equilibrium sorption patterns, and the magnitudes of the associated coefficients often vary with the organic carbon content of the soil. Such observations suggest that the sorption reaction may well arise from "partitioning" of the solute into an organic phase on the surface of, or within, soil particles or aggregates. The linear isotherms which result, more specifically their distribution coefficients, K_D, are often normalized by the fractional organic carbon content of the soil, (ϕ_{OC}), to give an organic carbon normalized isotherm coefficient, K_{OC}:

$$q_e = \frac{K_D}{\phi_{OC}} C_e = K_{OC}C_e \qquad (1)$$

where q_e and C_e represent the solid and solution phase equilibrium concentrations of solute, respectively, and K_{OC} represents the hypothesized distribution coefficient for a sorbent composed entirely of organic carbon. Values for K_{OC} have been found to vary by a factor of only two or so for a wide range of soils and sediments (Schwarzenbach and Westall, 1981). This variability has been linked to the nature of the organic material in the soil (Garbarini and Lion, 1986) and to concurrent mineral-site sorption (Karickhoff, 1984). The relative consistency of K_{OC} values in many instances has strengthened the notion that the sorption of hydrophobic organic compounds onto soils may be likened to partitioning or absorption into a uniform organic phase in such instances (Chiou et al., 1983).

Predictions based on traditional partitioning theory frequently suffer from their failure to account for the presence and role of macromolecular dissolved organic matter in the solvent phase. Natural "dissolved organic material" has been shown to increase the effective solubility of hydrophobic organic compounds (Carter and Suffet, 1983). This solubility enhancement has been ascribed to either alteration of the structure of the aqueous phase by the organic material or to a partitioning of solute into organic polymers (Chiou et al., 1986). This association can lead to a decrease in the extent of sorption of solutes on solid phases. Chin and Weber (1989, 1990) have developed a three-phase binding model formulated on the basis of a modified Flory-Huggins equation for a dispersed polymer phase. Their comparisons of model predictions and experimental observations demonstrate, as might be expected, that decreases in sorption attributable to dispersed polymers in solution phase are most marked for relatively hydrophobic solutes.

Hydrophobic sorption models provide a convenient basis for prediction of sorption equilibria, but only for those classes of compounds which meet the several implicitly or explicitly assumed conditions of such models; most particularly, i) relatively nonpolar, neutral solutes, and sorbents similar to those with which the correlations were developed; and ii) in general, soils and sediments with greater than 0.1% organic carbon content. The sorptive behaviors of polar and ionic organic solutes often manifest significant deviations from the correlations presented, relating to differences in the forces responsible for the sorption reactions.

PHENOMENOLOGICAL EQUILIBRIUM MODELS
Classical Models

Models for characterizing the equilibrium distributions of solute among the phases and interfaces of environmental systems typically relate the amount of solute, q_e, sorbed per unit of sorbing phase or interface to the amount of solute, C_e, retained in the solvent phase. An expression of this type evaluated at a fixed system temperature constitutes what is termed a sorption "isotherm". A number of conceptual and empirical models have been developed to describe observed isotherm patterns. The most simple is the linear model, which relates the accumulation of solute by the sorbent directly to the solution phase concentration. This relationship was given previously in Equation 1. The linear isotherm is appropriate for sorption relationships in which the energetics of sorption are uniform with increasing concentration, and the loading of the sorbent is low ("Henry's region" sorption). It accurately describes absorption, and has been found to adequately describe adsorption in certain instances, but usually only at very low solute concentrations. When justified, linear approximations to sorption equilibrium data are useful for modeling contaminant fate and transport because they substantially reduce the mathematical complexity of the effort. However, even when a particular set of data are reasonably well described by a linear model, caution should be exercised in application of that model because it may not be valid over concentration ranges beyond those represented by the data to which it is calibrated.

The Freundlich isotherm is perhaps the most widely used non-linear sorption equilibrium model. Although both its origins and applications are for the most part empirical, the model can be shown to be thermodynamically rigorous for a special case of sorption on heterogeneous surfaces. This model has the general form:

$$q_e = K_F C_e^n \tag{2}$$

The parameter K_F relates to sorption capacity and n to sorption intensity and heterogeneity. For determination of these empirically derived coefficients, data are usually fit to the logarithmic transform of Equation 2.

The Distributed Reactivity Model

Natural soils and sediments are inherently comprised by different materials and domains manifesting different affinities for sorption of organic solutes. While forces associated with the hydrophobic expulsion of organic solutes from aqueous phase may figure prominently in the sorption of many organic species, a solute will be concentrated first, and perhaps primarily, on regions of a sorbent which are energetically most favorable. As illustrated by Weber and van Vliet (1981a, 1981b) for other sorbent materials, physicochemical and structural features of soils and sediments which lead to energetic differences between or within individual particles are likely to be important determinants of sorption behavior. Heterogeneous mixtures of different constituents will exhibit heterogeneity with respect to mechanisms and reactivities, and thus likely give rise to different combinations of linear and non-linear "local" sorptions. If overall sorption on regions or components exhibiting non-linear behavior is large relative to that of linearly sorbing regions, then the observed "composite" isotherm will also be non-linear. To accommodate such sorption heterogeneity, Weber et al. (1992) introduced a composite isotherm model termed the Distributed Reactivity Model (DRM).

It is evident that the complexities of individually quantifying the parameters of each of many different individual local isotherms render full exposition of a complex model impractical for most environmental sorbent/sorbate systems. In the DRM model the linear components are grouped, and only major classes of non-linear components discriminated:

$$q_e = x_l K_{D_r} C_e + \sum_{i=1}^{m} (x_{nl})_i K_{F_i} C_e^{n_i} \tag{3}$$

where x_l is the summed mass fraction of solid phase exhibiting linear sorptions, K_{D_r} is the average partition coefficient for the summed linear components, and $(x_{nl})_i$ is the mass fraction of the ith non-linearly sorbing component and all other terms are as previously defined. This is the conceptual form of the DRM. From a practical perspective, the number of operationally distinguishable non-linear components, m, will typically be only one or two.

It is important to note that the assumptions associated with the conceptual developments of most isotherm models are rarely satisfied in natural systems. Thus the fact that any particular model may provide a phenomenological description of a sorption process in any given situation should not be taken as verification of the concept, mechanism, or assumptions upon which it is based. The ability of any phenomenological model to describe observed data may establish its utility for a specific set of conditions, but the inherent lack of mechanistic rigor associated with such models dictates against extrapolation to ranges of system conditions not experimentally quantified. Indeed, none of the isotherm models discussed here, or otherwise available, has been demonstrated to be capable of describing data over a wide range of conditions without parameter recalibration. It is not uncommon for a model to describe observed sorption behavior for a given sorbate/sorbent combination under one set of conditions, but fail to do so when system conditions change.

RATE MODELS

Equilibrium relationships comprise a set of limiting conditions for sorption processes, a set of conditions predicated on there being sufficient time for a system to achieve thermodynamic stability. In practical systems, however, the time scales required for attainment of equilibrium may approximate or exceed time scales associated with changes in solute concentrations due to macroscopic transport processes (i.e., advection and dispersion) or microscopic transport processes (i.e., diffusion, retardation). Under such conditions, the rates at which equilibrium is approached may significantly affect the process and the distribution of contaminants among the phases of a system. Whereas the extent of sorption is dependent only on the initial and final equilibrium states, rates of sorption depend on the path leading from the initial to the final state. In porous media, these paths include events that are controlled either chemically or by molecular-level mass transport. Molecular-level mass transfer refers in this context to stationary phase diffusion processes, as differentiated from the fluid-associated macro-scale transport processes of advection (convection) and hydrodynamic dispersion. The influence of these latter processes is addressed in the final section of this paper. Two molecular processes or, more realistically model representations of two molecular processes, are commonly envisaged to control rates of adsorption at the microscopic level: the first is reaction rate and the second is local mass transfer. These models lead ultimately to similar representations of the rate process for simple systems. Therefore, in that the local mass transfer model is probably a more realistic representation of actual rate controlling processes in subsurface systems, we will focus our discussion on that model.

Mass Transfer Models

In general, mass transport and transfer processes operative in subsurface environments may be categorized as either "macroscopic" or "microscopic". In the context of this discussion, macroscopic transport refers to movement of solute controlled by movement of bulk solvent, either by advection or hydrodynamic (mechanical) dispersion. By distinction, microscopic mass transfer, the focus of the discussion, refers to movement of solute under the influence of its own molecular or mass distribution or, more precisely, gradient.

One of the fundamental steps involved in characterizing and modeling microscopic mass transfer is an appropriate and accurate representation of associated resistances or impedances, including relevant distances over which solute is transferred and relevant properties of the medium through which transfer occurs. The nature and characteristics of such resistances vary with local conditions associated with particular combinations of sorbent, solute, fluid and system configuration. Differences in local conditions and associated transport phenomena are typified in subsurface systems by differences between solute transport through the interstitial cracks and crevices of rocks or soil particles, through organic polymer matrices associated with soils, and through internal fluid regions of soil aggregates. Mathematical descriptions of microscopic impedances and mass transfer processes within fluid and sorbing phases are generally structured upon one of several different types of conceptual models, tailored as necessary for a particular circumstance by appropriate assumptions and constraints regarding initial and boundary conditions and system behavior or state.

Models for describing microscopic mass transfer are generally predicated on assumptions regarding predominant or controlling transport mechanisms operating within specific types of media or domains. Microscopic mechanisms of mass transport in fluid phases include diffusion of solute molecules through elements of fluid, and solute transport facilitated by molecular-scale movement of fluid elements at or within fluid phase interfaces (surface renewal) and across microscopic velocity gradients (Taylor dispersion). The particular mass transfer mechanism which predominates in any situation depends on the properties of the solute and the medium comprising the domain, and on the microscopic hydrodynamics of the flow regime. Under fluid flow conditions typical of subsurface systems, molecular diffusion generally dominates microscopic mass transfer. Molecular diffusion can be either random ("Fickian") or constrained ("Knudsen") by the boundaries of the medium, such as surfaces bounding pore spaces. Knudsen diffusion occurs when both molecular velocities and ratios of longitudinal to radial pore lengths are high, and can be significant in gas phase mass transfer operations. Molecular diffusion in liquid phase is, however, generally controlled by random motion. In this type of diffusion, the velocity at which solute migrates along a linear path within a particular coordinate system is directly proportional to the gradient in its chemical potential, μ_i, along that path; that is, to the thermodynamic "driving force". The chemical potential of a substance is directly related to its activity and, in dilute aqueous solutions, to its concentration, C_i. It follows then that the time rate of solute mass flow by diffusion along a path, x, and across a normal (perpendicular) unit cross-sectional area (i.e., a one-dimensional flux) is directly proportional to the solute velocity, and thus to the gradient in concentration. For point-wise instantaneous diffusion, this flux, $J_{i,x}$, is given by:

$$J_{i,x} = -D_l \frac{\partial C}{\partial x} \qquad (4)$$

Equation 4 expresses Fick's first law for diffusion under non-steady state conditions. For liquid phase diffusion the constant of proportionality, D_l, is termed the free liquid diffusion coefficient, most commonly referenced to the aqueous phase.

The driving force, $\partial C/\partial x$, in Equation 4 may relate to either a constant or instantaneous difference in mass concentration across a homogeneous layer or "film" of fixed size (i.e., $\Delta x = \delta$), or to a time-variable concentration profile along a continuous path of variable length. The proportionality constant or diffusion coefficient is affected by various factors which relate to molecular interactions between the solute and the solvent, including the size, configuration and chemical structure of the diffusing molecule and the chemical structure and physical properties (e.g.,viscosity) of the liquid. Values of D_l for diffusion of typical organic contaminants in pure aqueous solutions generally fall in the range $0.5 - 5.0 \times 10^{-5}$ cm^2/sec. It should be noted, however, that the diffusion of any solute through interfacial aqueous regions between fluid and/or sorbent phases may involve resistances or impedances which differ from those of pure water. These differences arise because the properties of interfacial regions often reflect molecular interactions between adjacent bulk phases. The magnitude of the diffusion coefficient for a solute in an interfacial domain reflects these variations. For example, the accumulation of molecules other than those of the diffusing solute in an interface can increase resistance to transfer and yield a decreased diffusion coefficient, or drag forces near surfaces may effect reductions in diffusion coefficients in interfacial region between liquid and solid phases.

Models depicting microscopic molecular transport of conservative (non-reactive) substances in homogeneous or single-phase domains involve relatively straightforward applications of Equation 4. These are referred to here as Type I models. If the domain is homogeneous but also involves a reaction of the species being transported, the transport is described by a Type II model. Transport in heterogeneous domains (multi-phase) is described by either Type III or Type IV models, depending upon whether solute reactions are involved. Development of the mathematical forms associated with transport phenomena in the various types of domains which may apply in subsurface systems has been detailed by Weber et al., 1991. The Type I domain and associated model, the most simple of the foregoing models, is summarized here for illustrative purposes.

Type I Domains and Models

Models to describe mass transfer rates in any particular system typically incorporate the instantaneous point-form description of diffusion given in Equation 4 in an appropriate mass balance or continuity equation for that system. For transport of a conservative solute through a homogeneous diffusion domain, such as that depicted schematically in Figure 1, the mass continuity relationship states that the time rate of change in mass within the domain is given by the difference between the mass fluxes into and out of that domain. Thus, for a domain of volume (V), length Δx, and unit cross sectional area:

$$V\left(\frac{\Delta C}{\Delta t}\right)_V = J_{i,x}\Big|_x - J_{i,x}\Big|_{x+\Delta x} \tag{5}$$

In the limit ($\Delta x \to 0$, $\Delta t \to 0$), Equation 5, takes the form:

$$\left(\frac{\partial C}{\partial t}\right)_V = \frac{\partial}{\partial x}(-J_{i,x}) \tag{6}$$

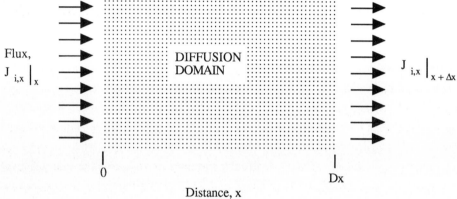

Flux, $J_{i,x}\Big|_x$ DIFFUSION DOMAIN $J_{i,x}\Big|_{x+\Delta x}$

0   Dx

Distance, x

Figure 1. Schematic depiction of a diffusion domain for application of the mass continuity equation (after Weber, et al., 1991).

In Type I models the only impedance to mass transfer arises from the uniform resistance of a homogeneous medium over a straight-line distance of travel. In this case the flux associated with the microscopic mass transfer process is defined by Fick's first law, Equation 4, and Equation 6 can be written:

$$\left(\frac{\partial C}{\partial t}\right)_V = \frac{\partial}{\partial x}\left(D_1\frac{\partial C}{\partial x}\right) \tag{7}$$

If there is no net accumulation of solute in the diffusion domain (i.e., steady-state) the left-hand side of Equation 7 is zero and integration yields a constant gradient in concentration across the domain; that is, a linear driving force for mass transfer. If in a given system the spatial integration applies to a fixed distance $\Delta x = \delta$, then a steady-state constant flux over this distance can be represented in terms of the upgradient and downgradient boundary concentrations, C_0 and C_δ respectively, and a solute velocity or mass transfer coefficient, k_f, as follows:

$$J_{i,x}\Big|_{x=0} = -D_l\frac{(C_\delta-C_0)}{\delta} = \frac{D_l}{\delta}(C_0 - C_\delta) = k_f(C_0 - C_\delta) \qquad (8)$$

It is apparent from Equation 8 that this "linear driving force" model for mass transfer is similar in form to a first-order reaction rate equation, and its solution can be approached in similar fashion.

It is imperative to note that the mass transfer coefficient and concentration relationship given in Equation 8 have been developed for, and apply strictly only to, steady-state diffusion in a homogeneous medium in which there is no impedance to diffusion other than that provided by resistance of the medium to movement of solute molecules. Other types of solute diffusion involving non-steady conditions, concurrent reactions, sorption and accumulation, and/or tortuous paths around obstacles, involve additional impedances and require different model formulations. This is emphasized here because expressions of the same general form as that given in Equation 8 are often employed as expedients to estimate mass transfer in more complex systems. When this is done, the gradient in concentration in the more complex system may in fact not remain constant, and the mass transfer coefficient may implicitly include factors other than just the free liquid diffusion coefficient and a fixed diffusion distance, factors not specifically identified and/or quantified. Under such circumstances the model, even if it can be fitted well to a particular data set, becomes a condition-specific relationship which may have limited utility for any application other than characterizing that particular data set.

Application Considerations

Subsurface systems are often comprised by multiple diffusion domains of different types and degrees of impedance. As a consequence, mass transfer processes associated with sorption reactions in such systems frequently involve two or more consecutive diffusion steps. It is generally the case, however, that one of these steps is significantly slower than the others, and is therefore, "rate determining" or "rate limiting". Identification and characterization of the step which controls overall rate in any given situation greatly facilitates the process of modeling. Indeed, because of the potential mathematical and parameter evaluation complexities otherwise involved, it is in many cases an imperative aspect of model implementation. Once a rate determining step has been identified, models of the type presented above can be adapted to the particular boundary and state conditions appropriate for the system in question.

A further complication of real systems is that solute concentrations in domains of interest in such systems are generally time dependent, thus the steady-state form of any particular model is seldom rigorously applicable. While departures from the boundary conditions associated with development of different models take various forms and yield different degrees of non-steady-state in real systems, practical approximations can frequently be made using certain quasi-steady-state approaches.

The condition of true steady state requires that the boundary concentrations of a domain, as well as the contibutions of all sink and source terms, remain constant in time. If this requirement is not met, it may still be reasonable to assume a quasi-steady condition over time periods and/or for other specific conditions for which the solute flux through the domain is large compared to the rate of change in boundary concentrations. Quasi-steady-state modeling approaches are predicated on the assumption that the concentration profile remains approximately linear throughout the domain, even though the concentration at one or both of the boundaries varies with time. The rate of change of concentration at a domain boundary is usually dependent on the rate of concentration change in the phase adjacent to the domain, which may relate either to advective flow in/out of, or accumulation/depletion within, the adjacent domain. This discussion focuses on situations involving accumulation in the down-gradient domain because that condition pertains most directly to microscopic transport processes associated with sorption reactions in subsurface systems.

A simplified representation of diffusion through a Type I domain of fixed depth δ, cross section A, and constant upgradient boundary concentration into an absorbing domain of volume $\forall$ is shown in Figure 2. The instantaneous rate of change in solute concentration, q, in an absorbing domain having a density $\rho_\forall$ is given by:

$$\frac{\partial q}{\partial t} = \frac{A}{\forall \rho_\forall} \frac{D_l}{\delta} (C_o - C_\delta) \tag{9}$$

The accumulation of solute in the absorbing domain results in a change in its concentration at the interface between the two domains; that is, the down-gradient concentration of the diffusion domain, C_δ. If the sorption process is linear and a local equilibrium state is maintained between the two domains, then the instantaneous rate of change in C_δ can be related to $\partial q/\partial t$ by a simple distribution coefficient, K_D (see Equation 1 and ensuing discussion). Thus:

$$\frac{\partial C_\delta}{\partial t} = \frac{A}{\forall} \frac{D_l}{\delta K_D \rho_\forall} (C_o - C_\delta) = \frac{k_\alpha}{K_D \rho_\forall} (C_o - C_\delta) \tag{10}$$

The effective mass transfer coefficient, k_α, in Equation 10 incorporates the cross-sectional area of the diffusion domain, the diffusivity of the solute and the volume of the absorbing domain, and has the dimensions of inverse time. If the sorption process is not specifically characterized, experimentally measured values of k_α may also incorporate, by default, the effect of solute partitioning. In any case, this coefficient, which generally must in fact be determined empirically, is highly system-specific and thus restricted in applicability.

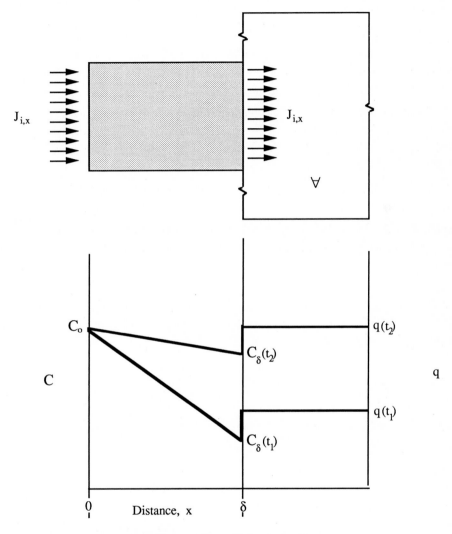

Figure 2. Schematic representation of solute migration
and concentration profiles for a Type I domain with
absorption in an adjacent region (after Weber, et al., 1991).

It is apparent from Equation 10 that a number of concentration-independent factors impact the magnitude of ($\partial C_\delta/\partial t$). If these factors are such that the change in C_δ over the period of interest is small relative to the difference in upgradient and downgradient concentrations (i.e., large K_D, small k_α), then solute migration through the diffusion domain can be approximated as a quasi-steady-state process. If this approach is applied to systems for which the assumption is inappropriate, the effective mass transfer coefficient k_α (or k_α/K_D) typically will be found to vary with time, and thus be even further restricted in its applicability to other than the exact circumstances for which it was measured. Models which incorporate such parameters are necessarily limited in their ability to predict mass transfer behavior for alternative conditions, and are essentially restricted to data analysis and event simulations.

Quasi-steady-state mass transfer models based upon assumptions similar to those discussed above have been applied to describe the transfer of solute between two different phases, the transport of solute between fluid regions through relatively small pores, and solute transfer into intraparticle or aggregate regions. These models generally relate the flux through and out of one of the types of domains discussed to the rate of change in the solute concentration of the adjacent phase, whether that phase is a fluid or a solid (Weber et al., 1991).

Application of this modeling approach to description of solute mass transfer across a hypothesized Type I immobile "boundary" layer of fluid immediately adjacent to the external surfaces of a sorbent yields the so-called "external film model". This particular mass transfer step has been taken as rate determining in many modeling descriptions of solute uptake from bulk solution by liquid and solid sorbents.

CONTAMINANT FATE AND TRANSPORT
THE ADR EQUATION

The sorption rate and equilibrium models presented above, whether mechanistic or phenomenological, have been developed on what may be termed a local or microscopic scale; that is, by describing processes at a molecular or particle level. Their ultimate utility for characterizing and predicting the behavior and eventual fate of contaminants in subsurface systems depends on: *i*) the relative importance of sorption processes in the context of other reaction and transport processes operative in subsurface environments; and, *ii*) our ability to determine the level of complexity required to describe accurately the impact of microscopic processes on overall fate and transport at the macroscopic scale. It is not within the scope of this lecture to examine macroscopic transport models in detail, but there is value in considering different levels of model sophistication required to capture the effects of sorption processes and reflect them in predictions or estimations of solute transport under field-scale conditions. To do this, several examples are selected to demonstrate that adequate macroscale characterization of solute behavior in typical subsurface environments requires thoughtful consideration and choice of appropriate microscale sorption models.

Macroscopic models for transport in subsurface systems incorporate advection, dispersion processes, and transformation or reaction processes. Such models are generally structured on principles of mass conservation applied on a "volume averaged" or otherwise statistically averaged basis. Generated on a differential scale, the continuity relationship yields the following advection-dispersion-reaction (ADR) equation:

$$\frac{\partial C}{\partial t} = - \mathbf{v} \cdot \text{grad } C + \text{div } (\mathbf{D_h} \cdot \text{grad } C) + \left(\frac{\partial C}{\partial t} \right)_r + S(C) \qquad (11)$$

The term $\mathbf{D_h}$ in Equation 11 is a second-rank hydrodynamic dispersion tensor, $\mathbf{v}$ is a pore-velocity vector, C is the solution-phase concentration of solute, and S(C) is a fluid-phase solute source term. The right hand term subscripted with an r in Equation 11 denotes the time-rate of change in concentration associated with a microscale reaction process, such as adsorption, which may in turn be represented by reaction rate models or by models describing microscale mass transfer processes. The significance of any particular microscale reaction on macroscopic solute transport can be estimated by: i) conducting controlled investigations to determine rate and equilibrium parameters for that reaction; and, ii) incorporating these parameters into the reaction term in the ADR equation and performing field-scale sensitivity analyses to determine the relative impact of the reaction vis-à-vis the advection and dispersion processes.

COMMON ASSUMPTIONS AND APPROXIMATIONS

In laboratory and controlled field scale investigations of transport and transformation processes the experimental design is commonly structured in a manner that will allow close approximation of system behavior with a one-dimensional form of Equation 11. There are certain non-controlled field applications in which use of a one-dimensional form of the ADR equation may also be justified, although this simplification is generally inadequate for describing field-scale transport. Nonetheless, the one-dimensional simplification does afford a convenient means for evaluating the relative effects of reaction and macro-transport processes in various contamination scenarios. Consider, for example, the finite control volume, V, represented schematically in Figure 3 for a system of fluid flow through a porous matrix comprised by a stationary sorbent phase. The flux of any dissolved fluid phase component, i, entering or leaving the control volume includes advective and dispersive components. Within the control volume, the component or solute of interest may undergo reaction (transformation) and/or sorption (phase transfer/exchange) with the sorbent phase. If the sorption reaction term is described by the following Type IV domain model (Weber et al., 1991):

$$\left(\frac{\partial C}{\partial t}\right)_r = \frac{(1-\varepsilon)}{\varepsilon}\,\rho_s\,\frac{\partial q}{\partial t} \qquad (12)$$

and no other fluid-phase reaction or source terms are considered, the ADR relationship given in Equation 11 simplifies in one-dimensional (z) form to an advection-dispersion-sorption model:

$$\left(\frac{\partial C_i}{\partial t}\right)_z = -v_z\frac{\partial C_i}{\partial z} + D_h\frac{\partial^2 C_i}{\partial z^2} - \frac{\rho_s(1-\varepsilon)}{\varepsilon}\left(\frac{\partial q_i}{\partial t}\right)_z \qquad (13)$$

The hydrodynamic dispersion term, D_h, in Equation 13 is now a simple numerical coefficient, v_z is the component of fluid-phase pore velocity in the z direction, and q_i is the volume averaged sorbed-phase solute concentration of component i expressed as a mass ratio.

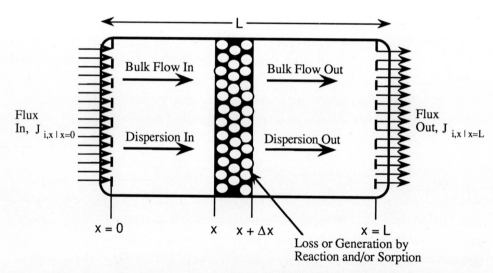

Figure 3.   Control volume for one-dimensional transport through porous media (after Weber, et al., 1991).

The sorption term, $\partial q_i/\partial t$, can assume a variety of forms comprising different rate and equilibrium components. The most simplistic approach to representation of sorption phenomena in a contaminant transport model is to assume that the time scales associated with the microscopic processes of diffusion and sorption are very much smaller than those associated with the macroscopic processes of fluid transport. This effectively assumes equilibrium locally; that is, that the time rate of change of the sorbed phase concentration, q_i, at any point z is instantaneously reflected in the time rate of change of the solution phase concentration, C_i, at that point. This yields the so called local equilibrium model (LEM). A further simplification is to assume that the relationship between q_i and C_i involves a direct proportionality of the type typically associated with simple partitioning or absorption processes. Expressing this proportionality in terms of the distribution coefficient, K_D, defined by Equation 1 gives the following relationship:

$$\left(\frac{\partial q_i}{\partial t}\right)_z = K_{D,i}\left(\frac{\partial C_i}{\partial t}\right)_z \tag{14}$$

The macroscopic transport model which results upon rearrangement of Equation 13 and substitution of the relationship for $(\partial q/\partial t)_z$ given in Equation 14 is termed the linear local equilibrium model (LLEM):

$$\left(1 + \frac{\rho_s(1-\varepsilon)}{\varepsilon}K_D\right)\left(\frac{\partial C}{\partial t}\right)_z = R_{f,i}\left(\frac{\partial C_i}{\partial t}\right)_z = D_h\frac{\partial^2 C}{\partial z^2} - v\frac{\partial C}{\partial z} \tag{15}$$

4.3   ANALYSES OF FIELD-SCALE SENSITIVITY

Equation 15 provides a reasonable means for first-cut assessment of the potential impact of sorption processes under field-scale conditions. For example, Figure 4 presents an LLEM field-scale simulation of the concentration profiles for subsurface transport of a moderately hydrophobic solute through a moderately low organic content soil under representative conditions of fluid flow and hydrodynamic dispersion. The fluid velocity and dispersivity are assumed constant thoughout the domain to further simplify the evaluation. These profiles represent concentration patterns at a point 10-meters downgradient of a pulse input of contaminant as a function of time after addition of that input. Parameter values utilized in this 10-meter simulation are tabulated in Table 1. Concentration-time profiles simulated by neglecting sorption and/or dispersion with all other conditions the same are also presented in Figure 4. Comparison of these several profiles demonstrates that both sorption and dispersion must be accounted for in the one-dimensional ADR equation to adequately describe contaminant distribution in systems where these processes are operative at the levels represented in this simulation, which are reasonably typical of field scale circumstances.

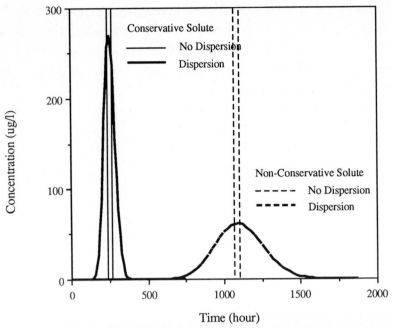

Figure 4. Ten-meter field scale simulations for a moderately hydrophobic
contaminant using simplified forms of the ADS transport model
(after Weber et al., 1991).

Table 1		
Field Scale Case Model Input Parameters		
Parameter	Value	Units
Initital Concentration	1000	μg/l
Velocity	1	meter/day
Dispersion Coefficient	0.1	meter^2/day
Time of Solute Input	1	day
Soil Density	2.67	gr/cm^3
Void Fraction	0.4	

Although the LLEM version of the ADR equation has been widely employed for
describing solute retardation by sorption in subsurface systems (Faust and Mercer, 1980;
McCarty et al., 1981; Pinder, 1984), it has become increasingly apparent that this model

frequently fails to provide adequate representation of the effects of sorption processes on contaminant transport. Inclusion of more sophisticated non-linear equilibrium models often provides better representation of sorption phenomena. Consider, for example, the potential error associated with macroscopic transport model predictions employing linear partitioning models in a system for which the actual equilibrium data are better characterized by a non-linear isotherm model. For this particular example we will assume that local equilibrium conditions prevail; this to examine singularly the effects which accrue to the choice of the equilibrium model itself. The solute-soil system selected for analysis is comprised by tetrachloroethylene (TTCE) and a fairly typical aquifer material from the Michigan area, Wagner soil, for which completely mixed batch reactor (CMBR) equilibrium data collected in our laboratories are presented in Figure 5. TTCE, a slightly polar chlorinated solvent of high volatility, has a moderate degree of hydrophobicity (log K_{ow} = 2.8) and the Wagner soil an organic carbon content of 1.2%. Figure 5 shows the "best fits" to the equilibrium data afforded by linear regression with a simple partitioning model and by non-linear regression with the Freundlich

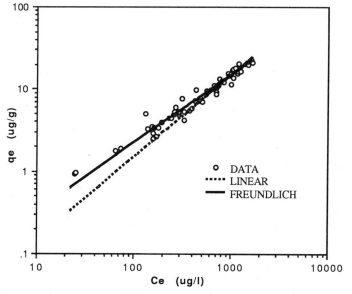

Figure 5. Sorption isotherms for TTCE and Wagner soil (after Weber et al., 1991).

model. Clearly the Freundlich model provides a better overall representation of the data, although portions of that data are reasonably well fit by the linear model. Projections of TTCE transport in a 1-meter field scale simulation made using two different forms of the ADR transport model incorporating these two alternative isotherm models calibrated with the data given in Table 1 are presented in Figure 6. It is apparent from comparison of these simulations that the use of a linear relationship to represent the equilibrium sorption behavior of the TTCE with respect to the Wagner soil results in a substantially different projection for contaminant transport than does use of the Freundlich isotherm model. The most significant

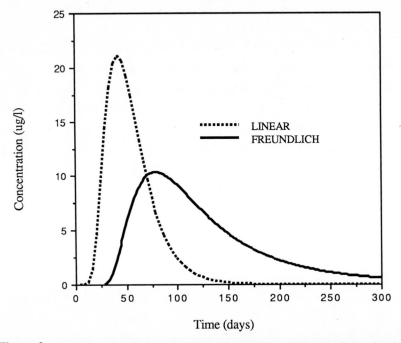

Figure 6. One-meter field scale simulations for TTCE and Wagner Soil using linear and non-linear forms of the ADR transport model (after Weber et al., 1991).

difference between the two projections is the slower rate of travel of the center of mass when the Freundlich model is employed. This is reflective of the model's ability to account for the higher sorption capacities observed at lower solution phase concentrations (Figure 5). The increased retention at low concentrations and the decreased slope of the Freundlich model fit leads to greater asymmetry or "tailing" of the solute pulse. It is important to note that these effects are significant even though the data are not grossly ill-fit by the linear model. Indeed, the linear model would seem quite adequate had the experimental data set been limited to a narrower range, say to only those data above about 150 ug/l. The importance of both measuring and accurately representing equilibrium sorption data over the entire range of interest in any given situation should be abundantly clear.

The second consideration to be made here relates to the validity of assuming that sorption time scales are not important with respect to transport time scales. A number of investigations have shown that field scale contaminant transport may be rate controlled. In such cases, overall contaminant dispersion is due to a combination of macroscopic and microscopic effects. The most simplistic approach for incorporating rate phenomena into model descriptions of contaminant transport is to assume that macroscopic and microscopic effects on front spreading are additive, and that an effective or apparent dispersion coeffficient,

$D_{h,a}$, can be incorporated into Equation 15 to take account of these additive effects. The relative contributions of the reaction rate and hydrodynamic dispersion mechanisms are dependent on flow velocity and microporous particle or aggregate radius. The sensitivity of this apparent dispersion coefficient to particle or aggregate size and the related effective internal diffusion coefficient, D_e, for a Type IV domain is shown in Figure 7, which was generated from a mathematical relationship between internal diffusion and hydrodynamic dispersion developed by Parker and Valocchi (1986). This relationship demonstrates that dispersion due to mass transfer will dominate for systems comprised of large microporous particles or aggregates. In a similar manner, it can be shown that mass transfer dominates at higher flow velocities. Sensitivity analyses performed by a number of investigators have shown that use of an apparent dispersion coefficient reasonably reproduces contaminant breakthrough profiles for large Peclet numbers. In contrast, this approach fails to describe accurately the asymmetry due to mass transfer effects for systems dominated by hydrodynamic dispersion.

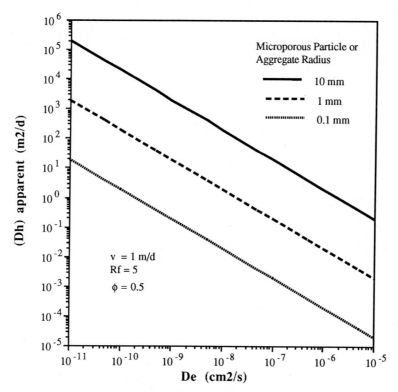

Figure 7. Relationship between effective intraparticle molecular diffusion (De) and hydrodynamic dispersion (Dh) to give equivalent effects on initial spreading (after Weber et al., 1991).

The effect of rates on concentration profile asymmetry is apparent for chemically controlled rate phenomena as well as for mass transfer controlled sorption. These effects are illustrated in Figure 8. Equilibrium conditions for this particular example were described using a linear isotherm model, and a range of first-order rate constants were tested. These models, calibrated for the conditions described in Table 1 yield the one-meter field-scale simulations presented in Figure 8. Comparison of the simulations clearly demonstrates the potential error that might be associated with the use of a local equilbrium model for a system in which sorption rates are not truly negligible at the time scale of associated transport processes. It is readily apparent that rate limitations for the sorption process result in earlier arrival of the contaminant front at the one-meter downgradient point, as well as increased tailing of the front to increase the amount of total "dispersion" of the contaminant plume compared to the LLEM model, which accounts only for macroscopic or hydrodynamic dispersion .

The conditions employed in developing the simulations for Figure 8 represent reasonable values for assessing the anticipated effects of rate-controlled sorption under typical background flow conditions. Such nonequilibrium effects may play an even more important

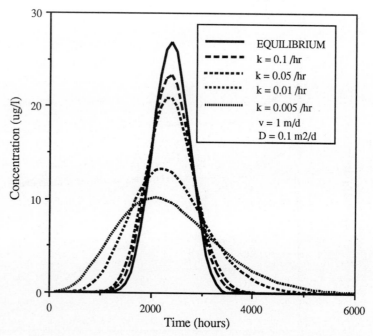

Figure 8.   One-meter field scale simulations showing the effects of first-order
reaction rates under background flow conditions (after Weber et al., 1991).

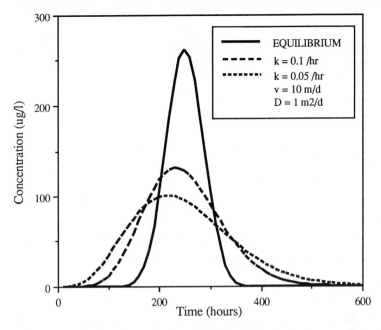

Figure 9.   One-meter field-scale simulations showing effects of first-order reactions rates in remediation flow situations (after Weber et al., 1991).

role in flow situations associated with typical "pump and treat" remediation strategies. Consider for example, the impact of rate-limited sorption on the volume of water to be treated in a remediation scheme in which the fluid velocity is increased by pumping to 10 m/d. As shown in Figure 9, the effect of nonequilibrium conditions at a particular rate constant are even more significant in this scenario, and the potential error associated with erroneously employing a LLEM even greater. The rate-controlled sorption process in this case translates into more than a 100% increase in the volume of contaminated water which must be treated.

CLOSURE

Sorption processes in subsurface systems are complex, often involving non-linear phase relationships and rate-limited conditions. It has been demonstrated that these processes impact reactive solute behavior under typical field scale conditions, and must therefore be considered in attempts to model or otherwise predict contaminant fate and transport in the subsurface. It is further apparent from the examples we have considered that the thoughtful selection of appropriate microscopic equilibrium and rate models, models which adequately describe the inherently complex and system specific dynamics of sorption processes, is an imperative for accurate fate and transport modeling.

REFERENCES

Barrow NJ, Bowden J (1987) A comparison of models for describing the adsorption of anions on a variable charge mineral surface. J of Coll Int Sci 119:236-250.

Carter CW, Suffet IH (1983) Interactions between dissolved humic and fulvic acids and pollutants in aquatic environments. In: Fate of Chemicals in the Environment, Am Chem Soc 215-219.

Chin YP, Weber WJJr (1989) Estimating the effects of dispersed organic polymers on the sorption of contaminants by natural solids I: A predictive thermodynamic humic substance-organic solute interaction model. Env Sci Tech 23:978-984.

Chin YP, Weber WJJr, Eadie BJ (1990) Estimating the effects of dispersed organic polymers on the sorption of contaminants by natural solids II: Sorption in the presence of humic and other natural macromolecules. In press, Env Sci Tech

Chiou CT, Malcolm RL, Brinton TI, Kile DE (1986) Water solubility enhancement of some organic pollutants and pesticides by dissolved humic and fulvic acids. Env Sci Tech 21:1231-1234

Chiou CT, Porter PE., Schmedding DW (1983) Partitioning equilibria of nonionic compounds between soil organic matter and water. Env Sci Tech 17:227-231.

Eriksson E (1952) Cation exchange equilibria on clay minerals. Soil Sci 74:103-11

Faust CR, Mercer JW (1980) Ground-water modeling: recent developments. Gr Wat 18:569-577

Garbarini DR, Lion LW (1986) Influence of the nature of soil organics on the sorption of toluene and trichloroethylene. Env Sci Tech 20:1263-1269

Hayes KF, Leckie JO (1987) Modeling ionic strength effects on cation adsorption at hydrous oxide/solution interfaces. J Coll Inter Sci 115:564-572

James RO, Parks GA (1982) Characterization of aqueous colloids by their electrical double-layer and intrinsic surface chemical properties. Surf Coll Sci 12:119-217

Karichkoff SW (1984) Organic pollutant sorption in aquatic systems. J Hydr Engr 110:707-735

McCarty PL, Reinhard M, Rittmann BE (1981) Trace organics in ground water. Env Sci Tech 15:40-51

Pinder GF (1984) Groundwater contaminant transport modeling. Env Sci Tech 18:108A-114A

Schwarzenbach RP, Westall J (1981)Transport of nonpolar organic compounds from surface water to groundwater, laboratory sorption studies Environ Sci Tech 15:1360-1367

Sposito G (1984) The chemistry of soils. Oxford University Press, New York

Stumm W, Morgan, JJ (1981) Aquatic chemistry. John Wiley and Sons, New York

Parker JC and Valocchi AJ (1986) Constraints on the validity of equilibrium and first-order kinetic transport models in structured soils. Wat Res Res 22:399-407

Weber WJJr, van Vliet BM (1981a) Synthetic adsorbents and activated carbons for water treatment: overview and experimental comparisons. J Am Wat Wks Assoc 73, 8, 420-426

Weber WJJr, van Vliet BM (1981b) Synthetic adsorbents and activated carbons for water treatment: statistical analyses and interpretations. J Am Wat Wks Assoc 73, 8, 426-431

Weber WJJr, McGinley PM, Katz LE (1991) Sorption phenomena in subsurface systems: concepts, models and effects on contaminant fate and transport. Wat Res 25:499-528

Weber WJJr, McGinley PM, Katz LE (1992) A distributed reactivity model for sorption by soils and sediments: I. conceptual basis and equilibrium assessments. Env Sci Tech in press

Theoretical and Practical Aspects of Soil Chemical Behaviour of Contaminants in Soil

S.E.A.T.M. van der Zee and J.C.M. de Wit
Department of Soil Science and Plant Nutrition
Agricultural University Wageningen
P.O. Box 8005, 6700 EC Wageningen, The Netherlands

Natural processes, agricultural practice as well as environmental contamination continuously perturb the chemical composition of soil. This pertubation may be intended (liming, fertilization) or inadvertent (pollution). The effect of the pertubation depends on the buffering capacity of the soil system to counter changes. The buffering capacity depends on the chemical reactivity and is compound-specific. This implies that it differs for each compound as the involved reactions and reaction rates differ. Hence, the quantification of the buffering capacity requires understanding of the speciation processes (complexation, adsorption/desorption, precipitation/dissolution, degradation) for the compounds of interest. By these processes, the inflow of relatively large concentrations of solutes is countered: the reactions in soil lead to a decrease in concentration of some solutes, but may result in a simultaneous increase in the concentration of other solutes. In many cases the concentration in solution is the key variable with regard to effects because it usually reflects the biological availability of the compound (for plants) and the leaching hazard to groundwater. During the past decades a massive research effort has been devoted to speciation processes in soils. Since a review is hardly possible in one chapter, we consider only a few aspects in a little more detail, e.g., metal ion binding, heterogeneity at different scales and relations between chemistry and transport.

NATO ASI Series, Vol. G 32
Migration and Fate of Pollutants in Soils and Subsoils
Edited by D. Petruzzelli and F. G. Helfferich
© Springer-Verlag Berlin Heidelberg 1993

General Background of Metal Ion Binding

Metal ions bind mainly to clay minerals, metal oxides and soil organic matter (Sposito, 1984). The binding properties of these soil constituents differ significantly and are influenced by environmental conditions such as ionic strength and pH. The plate sites of clay minerals are characterized by a constant negative charge due to isomorphic substitution. Because the overall soil system is uncharged, the negative charge is screened by the mobile ions in solution. Near a (negatively) charged interface a so-called diffuse double layer will develop (Sposito, 1984). In it the concentrations of the cations are larger whereas the anion concentrations are smaller than in the bulk of solution. The accumulation of the cations and the exclusion of the anions depends on the charge of the clay surface and the valency of the ions. The specific interaction between ions and surface groups is unimportant for most clays. The non-specific adsorption by clay minerals can be described rather well with diffuse double layer models (Bolt et al., 1991). In practice, often semi-empirical ion exchange relations are used such as for instance the Gaines-Thomas equation. In these equations the total exchange capacity and (empirical) selectivity coefficients have to be specified.

Soil organic matter, oxides and edges of clay minerals, which behave like oxides, have a pH-dependent charge. In general, non-specific binding to those constituents is of minor importance compared to specific adsorption. In specific adsorption, ions form surface complexes with the functional groups of the constituents. The specific binding depends on (1) the types of functional groups or chemical heterogeneity, (2) the type of ion, (3) the electrostatic effects and (4) environmental conditions and composition of the soil solution (Bolt et al., 1991). The formation of a surface complex can be given by:

$$aS^s + bM^m = S_a M_b^{as + bm} \qquad (1)$$

where S is a surface group and M a dissolved compound. M is not necessarily a free ion such as Cd^{2+}, it can also be a certain complex in solution like $CdCl^+$. In Eq. (1), s and m are the valencies of S and M, including sign, and a and b are stoichiometric coefficients. When applying the mass action law to Eq. (1) and taking into account the electrostatic effects the following affinity constant K can be defined:

$$K = \frac{\{S_a M_b\}}{\{S\}^a (M_s)^b} \qquad (2)$$

where the braces {} refer to site fractions and M_s is the metal ion activity near the functional groups at the location of binding. The relationship between M_s and its activity in the bulk, (M^m), is given by Boltzmann's distribution law: $M_s = (M^m) \, exp(-mF\psi_s/RT)$ where ψ_s is the surface potential. The exponential term is known as the Boltzmann factor. Except for electrostatic effects, the formulation for surface complexation looks very similar to the formulation for the complexation in solution. With tabulated affinity constants and appropriate expressions for activity coefficients to account for the ionic strength effects, the chemical speciation in solution can be calculated with chemical equilibrium models (Westall et al., 1976). Due to chemical heterogeneity, the a priori unknown stoichiometry of the binding equations and variable charge effects, the calculation of the surface speciation is more complex and involves additional assumptions. In the next sections these complications are addressed.

Chemical Heterogeneity

We illustrate the effect of heterogeneity with calculations for proton binding by a carboxylic type of functional group COO^-

$$S_i O^- + H^+ = S_i O\,H \quad ; \quad K_i \qquad (3)$$

We define the protonated fraction $\theta_{i,H}$ of the sites of type i:

$$\theta_{i,H} = \frac{\{S_i O\,H\}}{\{S_i O^-\} + \{S_i O\,H\}} \qquad (4)$$

With the help of the affinity constant, K_i, (4) can be rearranged to a Langmuir-type isotherm equation:

$$\theta_i = K_i H_s /(1 + K_i H_s) \qquad (5)$$

where the proton activity at the surface, H_S, incorporates the Boltzmann factor and the valency is +1. If only sites of type i are present, the surface is chemically homogeneous, and Eq. (5) gives the total proton binding, $\theta_{T,H}$. When there are different types of sites present we call the surface chemically heterogeneous. Then Eq. (5) gives the local adsorption, and holds for type i only. For heterogeneous surfaces with a discrete number of sites types the binding is given by a weighted summation over i ($\theta_{T,H} = \Sigma f_i \theta_{i,H}$) where f_i is the fraction of the sites of type i and the total number of sites at the surface. The log K-distribution for a discrete heterogeneous surface is characterized by a set of (infinitely) narrow peaks. For a surface with a continuous log K distribution the total binding follows from integrating according to:

$$\theta_{T,H} = \int \theta_{i,H} f(\log K) dlog K \qquad (6)$$

In Eq. (6), $\theta_{i,H}$ is the local isotherm and $f(\log K)$ is the affinity distribution function. Only for a few distributions the solution of the integral equation results in an analytical expression for $\theta_{T,H}$ (Van Riemsdijk et al., 1991). A well know example is the Langmuir-Freundlich equation:

$$\theta_{T,H} = \frac{(\tilde{K} H_s)^m}{1 + (\tilde{K} H_s)^m} \qquad (7)$$

which is derived for a semi-Gaussian symmetrical distribution function (Van Riemsdijk et al., 1991). Log $\tilde{K}$ determines the position of the distribution on the log K axis, and the parameter m ($0<m \leq 1$) its width. In the case $m=1$ the surface is homogeneous and the Langmuir-Freundlich equation is identical to the Langmuir equation. If $(\tilde{K} H_s)^m \ll 1$ and binding is described by the numerator, we obtain the Freundlich equation. Note that the Langmuir-Freundlich equation provides a theoretical foundation of the widely used empirical Freundlich equation. An advantage of the Freundlich relation is its simplicity. A disadvantage is that it does not describe binding for a range of environmental conditions and the competition between different ions.

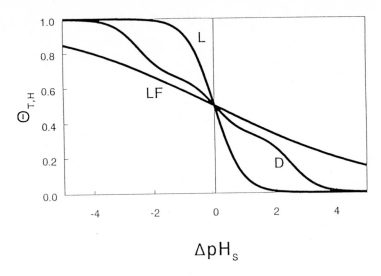

ΔpH_s

Fig. 1.   Overall protonation $\theta_{T,H}$ as function of ΔpH_s, $\Delta pH_s = \log K_{H,\,median} - pH_s$, where $\log K_{H\text{-}median}$ is median value of affinity distribution. L: homogeneous surface (Langmuir) $\log K_{H,\,median} = \log K_H = pH$ at $\Delta pH_s = 0$. D: discrete heterogeneous with three types of functional groups, $f_1 = f_2 = f_3$ and $\log K_1 = pH$ at $pH_s - 2.5$; $\log K_{H,\,median} = \log K_2 = pH$ at $pH_s = 0$; $\log K_3 = pH$ at $pH_s = 2,5$. Langmuir-Freundlich: continuous heterogeneous sunface according to Eq. (9) with $\log K_{H,\,median} = \log \tilde{K}_H = pH$ at $\Delta pH_s = 0$.

In Figure 1, $\theta_{T,H}$ as a function of ΔpH_S is calculated for a homogeneous surface (curve L), for a discrete heterogeneous surface (curve D) and for a continuous heterogeneous surface (curve LF). ΔpH_s is defined as the pH_S minus the log K value for the median value of the distribution.

The chemical heterogeneity influences the shape and the slope of the proton binding curve. The larger the degree of heterogeneity, the smaller is the (absolute) slope of the binding curve. Because the chosen distributions were symmetrical, the curves of Figure 1 have a common intersection point at $\Delta pH_s = 0$ or $\theta_{T,H} = \frac{1}{2}$. All three curves have also inflection points at $\Delta pH_S=0$, which correspond to a peak in the affinity distribution. The discrete heterogeneous surface, characterized by three peaks, shows two more inflection points at $\Delta pH_s=-2.5$ and $\Delta pH_s=2.5$.

The observation that the degree of heterogeneity and the shape and slope of the binding curve are related is used in techniques which obtain affinity distributions from binding curves. The most simple procedure is the condensation approximation, in which the affinity distribution is directly related to the first derivative of the binding curve (e.g., Van Riemsdijk et al., 1991).

Electrostatic Effects

So far we have not taken the electrostatic effects explicitly into account. Binding is calculated as a function of pH_S, a quantity which is not experimentally accessible. The Boltzmann equation shows that ψ_s must be known to calculate ion binding as a function of the measured ion activity in solution. Because ψ_s cannot be measured directly, we have to rely on theoretical expressions (Sposito, 1984). In the theoretical expressions ψ_s follows from the surface charge, the composition of the bulk solution and the ionic strength. Westall (1976) has illustrated how to incorporate the electrostatic calculations in "classical" chemical equilibrium calculations.

In Figure 2 the protonation of a homogeneous surface as a function of ΔpH is calculated for three ionic strength values. For reference, $\theta(\Delta pH_s)$ of Figure 1 is given (curve L). The curves in Figure 2 clearly show ionic strength effects on proton binding. At low salt levels the electric field is not efficiently screened by countercharge. The electrostatic effects are larger, and this results in weaker proton binding at low than at high ionic strength. The electric effects have also influenced the shape of the curves. The steepness of the curves have become considerably smaller. The smaller slope suggests heterogeneity, but actually reflects the electrostatic effects. Because both chemical heterogeneity and electrostatic effects may influence the shape of the curves similarly, it is often not possible to distinguish objectively between both effects. This holds especially if the variation in environmental conditions for which binding is studied is small.

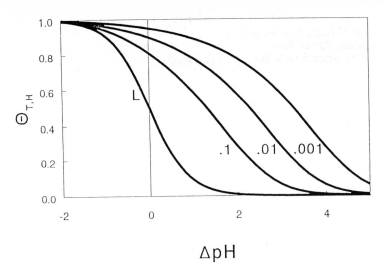

Figure 2. Overall protonation $\theta_{T,H}$ for homogeneous surface as function of ΔpH for three ionic strength levels. (I=0,1, I=0.01 M and I=0,001 M). Electrostatics taken into account. $\Delta pH = logK_H - pH$; $logK_H = pH$ at $pH_s = 0$. Electrostatics: flat plate site density $N_s = 1$ sites/nm^2. Curve L: homogeneous curve without electrostatics. Curve corresponds to fig. 1, curve L.

Competition

Ion binding in natural systems is a multicomponent process (Bolt et al., 1991). In aqueous systems protons (or hydroxyl ions, OH$^-$) are by definition present. Because protons highly determine the state of the functional groups and the charge of the surfaces, at least a proton effect on the metal ion binding is to be expected. Additionally, a cocktail of different metal ions is present at most polluted sites and the soil system often contains specifically bound multivalent ions such as Ca^{2+}, Mg^{2+} and Al^{3+}. Competitive or multi-component binding can be calculated if a set of different binding equations (Eq. 1) are taken into account. The effect of competition is illustrated for the binding of a divalent metal ion:

$$SO^- + M^{2+} = SOM^+ \tag{8}$$

in combination with proton binding according to Eq. (3). In Figure 3 the metal binding is calculated for a three pH values. In Figure 3a the electric effects are not taken into account. At low pH the surface is protonated and there is a strong competition between the proton and the metal ion. The competition results in a pH effect. At high pH the surface becomes fully deprotonated and the pH dependency of the metal ion binding vanishes.

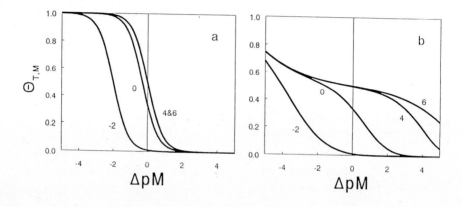

Fig. 3.   The metal binding $\theta_{T,M}$ for a homogeneous surface as a function of ΔpM at several ΔpH values. $\Delta pM = \log K_{M, median} - pM$. Curve a, no electrostatics, Curve b, with electrostatics, flat plate site density $N_s = 1$ sites/nm². In both graphs $\log K_H = \log KM$

Irrespective of the pH, the curves in Figure 3a have the same shape, only their position on the ΔpM axis is shifted as a function of the pH. The shift of the position implies that at low pH the surface apparently has a lower affinity or $\log K$ that at higher pH. In Figure 3b the electrostatic effects on ion binding are taken into account besides competition effects, yielding different curves than those Figure 3a. The pH effect of the binding is more pronounced and the shape of the curve has become pH-dependent. Even at high pH and high metal ion concentrations the surface is not saturated with metal ions. The metal ions add two positive charges to the surface and thus strongly increase the surface charge upon binding. The increased charge hinders further metal binding.

Specific Binding to Oxides

Most of the important metal oxides are amphoteric. The amphoteric behaviour is often described by a two-step protonation of the oxide surface sites:

$$SO^- + H^+ = SO\,H \quad ; \quad K_1 \tag{9a}$$

$$SOH + H^+ = SO\,H_2^+ \quad ; \quad K_2 \tag{9b}$$

An alternative to the $2\,pK$ model (Stumm, 1987) is the $1\,pK$ model (Van Riemsdijk et al., 1991), where protonation is described by a single step:

$$SO\,H^{1/2-} + H^+ = SO\,H_2^{1/2+} \quad ; \quad K_H \tag{10}$$

For both models the binding of metal ions is given by association equation analogous to Eqs. (2) and (8), and the binding constants are fitted. The effects of chemical heterogeneity of the oxide surface are in general not taken into account explicitly. Recently, Hiemstra et al. (1989) proposed the MUSIC model. In this model, the existence of different functional groups and different crystal planes is explicitly taken into account. Furthermore the protonation constants ($\log K_i$) of the functional groups and their fraction (f_i) are predicted on the basis of the structure of the crystal planes. The electrostatic interactions of binding to oxides are described with a flat plate double layer model. In addition to the diffuse double layer, the existence of a Stern layer can be taken into account (Bolt et al., 1991) that corrects for finite ion size effects on the structure of the double layer close to the surface. The number of functional groups of oxide surfaces is about 4-10 sites.nm^{-2} and somewhat larger than was used in the calculation in Figures 2 and 3. Because the electric effects on the binding to oxides are rather large, oxides are neither fully protonated nor fully dissociated under natural conditions, and the observed maximum metal ion binding is in general much smaller than the total number of surface sites available.

Specific Binding to Organic Matter

Soil organic matter is a polydisperse mixture of complex organic polyelectrolytes (Aiken et al., 1985, Hayes et al 1989). Organic matter is non-amphoteric; owing to the dissociation of acidic groups it has a pH-dependent negative charge. The variety in the different types of functional groups is large, and the chemical heterogeneity is important. The number of functional groups per unit surface area of the organic matter is about ten times lower than for oxides. Compared to oxides, the electric effects of ion binding to organic matter are weaker. Despite some differences, the concept of modelling ion binding to oxides is well established. For binding to organic matter this is not the case and the variety in the available models is much larger. In many models the chemical heterogeneity of the organic matter is modelled by choosing an arbitrary set of different types of functional groups. The binding constants for the different groups are taken equal to the constants for the corresponding free ligand in solution. The binding is described by adjustment of the contribution or fraction of the site types. An alternative to the arbitrarily chosen site classes is the use of equations for a continuous heterogeneity. Although these equations are mathematically somewhat more complex, the number of adjustable parameters necessary to describe the binding is in general smaller than in the case of the discrete sites. With the help of the so-called mastercurve procedure, it was established that the electric effects of proton binding to organic matter, especially to humic acids and fulvic acids, can be described reasonably well with either a cylindrical or a spherical double layer model (De Wit et al., 1990, 1991).

In soil systems the organic matter is strongly associated with the mineral soil constituents; it partly coats the clay minerals and the oxide surface. The interactions between the soil constituents influence the binding properties (Bolt et al., 1991). The binding to a soil systems is therefore not simply the addition of the binding for the soil constituents separately. This makes the application of models calibrated for the "purified" organic matter, oxide or clay minerals to binding in soil systems not always straightforward.

Moreover, part of the organic matter is suspended in the soil solution. Although binding in general reduces mobility, binding to this colloidal organic matter facilitates ion-mobility. The fraction suspended is in equilibrium with the solid fraction. Depending on the environmental conditions the suspended matter will coagulate or the solid organic matter will dissolve. The dynamics of the organic matter fraction is not yet well understood, but will become an important research topic in the near future.

Sorption by soil

The complexity of reactions in real soil is due to the many different soil compounds, that behave differently and interact with each other, as well as the multicomponent nature of solute and soil interactions. Hence, we are seldomly able to comprehend all details for real soil. Detailed understanding is commonly limited to model systems that are more or less similar to natural soils (consisting of e.g. clay/hydroxide suspensions or purified organic matter in equilibrium with a well defined electrolyte). Measurements for real soil may be interpreted with models developed for well defined systems but usually this inverse

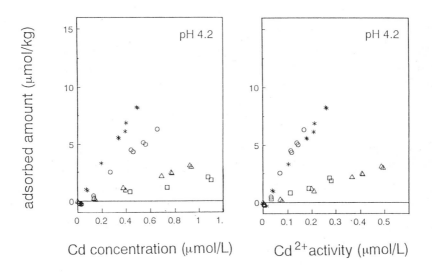

Figure 4: Adsorbed Cd as function of Cd-concentration (4a) and Cd-activity (4b), at pH 4.2 (from Boekhold et al., 1992). Background solutions $CaCl_2$ (square), $CaNO_3$ (triangle), NaCl (circle), $NaNO_3$ (star).

modeling yields several plausible descriptions. For these reasons, the chemistry for real soils is often described empirically without nuances such as electrostatics and microscopic heterogeneity.

An example is sorption and complexation of cadmium by sandy soil, in which the main sorbent is organic matter. To relate the adsorbed Cd (q) with the Cd-activity (a) in solution, Boekhold et al. (1992) used the Freundlich equation ($n=0.75$).

$$q = ka^n \tag{11a}$$

$$k = \bar{k}.oc.(\text{H}^+)^{-0.5}.(\text{Ca}^{2+})^{0.41} \tag{11b}$$

with $\bar{k}$ and n are parameters, oc is organic carbon (% by mass), and bracketed terms are activities in solution. Because Eq. (11) was obtained by fitting, interpretation has to be done with care. For example, since the pH affects almost all soil chemical processes, one cannot conclude for instance that the H^+-term arises from $\text{H}^+/\text{Cd}^{2+}$-competition alone. Boekhold et al., (1992) measured Cd-sorption in various background electrolytes (cations Na^+, Ca^{2+}; anions NO_3^-, Cl^-) of the same ionic strength. The isotherms (Figure 4a) expressed as a function of total Cd-concentration were very dependent on the chloride concentration due to Cd/Cl-complexation. Complexation lead to an increase of dissolved Cd, of which only Cd^{2+} adsorbs. When the isotherms are given as a function of Cd-activity in solution then complexation is accounted for and the curves merge (Figure 4b). The model (11) was able to describe Cd-adsorption well for acidic conditions. In combination with a transport algorithm it predicted breakthrough excellently. For a field soil (cropped with barley) the values of oc, pH, and dissolved as well as total Cd were measured along a transect. For pH and dissolved Cd the background electrolyte was fixed (0.01M CaCl$_2$). The value of $\bar{k}$ was assessed for two samples of the transect. The patterns of all variables (as well as their spectra) differed profoundly. However, with the local data and (11), total Cd could be predicted well (Figure 5). This is of interest because Dutch soil quality standards for Cd expressed in total content apparently need at least a model of the complexity of (11) to assess effect (i.e., bio-availability: dissolved Cd). Although CdCl-complexes do not sorb (they only regulate the dissolved Cd-fractions) other complexes may sorb. For elements such as Fe, Al, Pu, and U we may have to account for hydrolysis and polymerization with the hydroxyl anion.

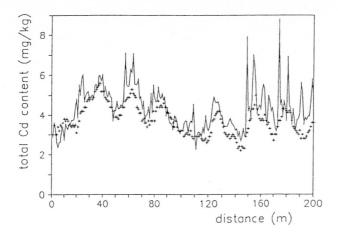

Figure 5: Predicted Cd_T (points) and measured Cd_T (line) using data of *oc*, *p*H, dissolved
Cd and (11) (from Boekhold et al., 1991).

The formed polymers have a tendency to adsorb onto the practically constant charge clay
surface. This process is dependent on *p*H and solution composition and ionic strength.
Furthermore it is poorly reversible and yields a variable charge surface. Simplifying, we
may visualize that the clay minerals are partly coated by an oxide-type of polymer. A
consequence is that heavy metals may sorb more specifically and less reversibly onto the

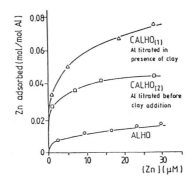

Figure 6: Zinc sorption for clay/aluminum hydroxide (CAlHO) and an aluminum
hydroxide (AlHO) system (from Keizer and Bruggenwert, 1991). Titration with
base to *p*H 6.5.

clay/polymer complex than onto the clean clay surface. This specific binding depends more strongly on the pH (Keizer and Bruggenwert, 1991) and leads to non-additive behaviour. As an example Figure 6 shows the effect of preparation of the model soil on Zn-sorption.

In view of the limited space we give little attention to precipitation/dissolution phenomena such as considered by Meeussen et al. (1992). Nevertheless such phenomena are important, e.g. with regard to soil acidification, mineral weathering and sodic soil reclamation.

A model that may be applicable to describe a fast and irreversible reaction (e.g., precipitation) in a particle is the unreacted shrinking core model (Crank, 1957, Wen, 1968). In this model reactant solute diffuses through a converted shell to the interface where the reaction occurs. As a result of the reaction the unconverted core, where the reactant solute concentration is zero, shrinks as a function of time provided the chemical equilibrium concentration (c_e) is exceeded. Solving the unreacted shrinking core model gives for the fractional conversion (M/M_∞)

$$M/M_\infty = f_1(I); \quad I = \gamma \int_0^t (c(t) - c_e)dt \tag{12}$$

where f_i is a function, γ is a parameter, and $c(t)$ is the exterior concentration (larger than c_e). For $c(t) = c_0$ and constant, I becomes $\gamma(c_0 - c_e)t$. The function f_1 may be either linear or non-linear with respect to concentration, depending on e.g. particle shape. In most cases f_i has to be measured as an explicit expression for $f_1(I)$ can only be obtained for particular geometries and reactions. At least for natural soil samples measurements are usually needed because geometry and size as well as mineral purity of particles may vary significantly in one sample. As will be shown in an example, the possibility of non-linearity of f_1 and the simplicity of (12) make the unreacted shrinking core model a suitable alternative for the spherical diffusion model with internal linear adsorption which is currently popular in soil science. In the latter model no sharp interface exists and concentrations are non-zero throughout the particle. The amount that has diffused into the particle divided by the maximum amount when the entire particle has the final constant concentration (c_0) is

$$M/M_\infty = f_2(t) \tag{13}$$

where f_2 is a function of time. Because for linear adsorption, M_∞ is linear with respect to concentration, Eq. (13) is unable to describe the data presented in Figure 7. In this Figure we show the conversion of chalcopyrite

$$CuFeS_2 + Cu^0 + 2H^+ \; \underset{\rightarrow}{\overset{\leftarrow}{}} \; Cu_2S + Fe^{2+} + H_2S \tag{14}$$

The conversion measured by Hiskey and Wadsworth (1974) gives a fan of curves for different proton concentrations. However, as a function of $I = ct$ where c is the proton concentration these curves merge in agreement with (12). We observe that at a particular time (Figure 7a), conversion is clearly non-linear in c. Similarly good results were obtained for phosphate sorption by soil and feldspar weathering (Van der Zee, 1991).

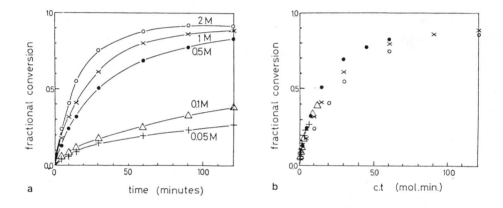

Figure 7: Chalcopyrite conversion as a function of time (a) and exposure I (b); with H_2SO_4 concentrations given in Fig. 10a. Data of Wadsworth et al., (1974).

The importance of heterogeneity has been emphasized a number of times. Among others it complicates the assessment of bio-availability. Bio-availability is often assessed by selective extraction (with water or $CaCl_2$-solution) of a "mixing sample". In such a sample, equal weights of samples taken in the field are mixed. The original samples may be expected to differ with regard to nutrient or contaminant content as well as retention capacity. This variability causes a bias in the assessment of availability. A mixing sample yields systematically a lower availability parameter e.g. for phosphate if all samples are analyzed independently and subsequently averaged. An illustration is provided in Table

1 for water-extracted phosphate (P_w). Qualitatively, the biass can be explained from a mass balance for the mixing sample. For instance, when sample I (low P-content) and II (large P-content) have the same P-Langmuir isotherm, their combination in one mixing sample extracted with water will lead to P-redistribution. Some of the P desorbing from sample II will be adsorbed by sample I. The latter fraction is absent when samples I and II are analyzed separately, hence the lower effectively desorbed (available) P for the mixing sample. Which of the two quantities (mixed; non-mixed average) is more appropriate remains a question. After all, plants effectively scour a soil volume for P that is larger than commonly used quantities in extraction analyses. A mixing sample representing a particular volume that differs for various crops and vegetations may therefore be appropriate.

Table 1: P_w-values for mixing samples and mean value for the same N samples: confidence levels 95% (*) and 99% (**), valid for the difference between the two values.

N	P_w-mix	mean P_w
3	0.90	1.17
3	0.3*	0.48
5	0.34**	0.51
8	0.37*	0.43
12	0.66*	0.88
20	0.44**	0.78
33	0.45**	0.74

Models that were developed to describe chemical interactions may be fitted on available data sets. Their applicability can be verified by independent prediction of transport behaviour and comparison of the results with measurements. Chemical interactions may often be accounted for through a retardation factor $R(c) = 1 + \rho q'(c)/\theta$. The term q' accounts for all chemical interactions. Only for a linear relationship $q(c)$ will $q'(c)$ and R be constants. If $R(c)$ is a non-linear function of c, two situations may occur. The front may be self-sharpening and converge to a traveling wave (or constant pattern front in the terminology of Helfferich, this volume), or it may be non-sharpening (DeVault, 1943). In the first case, $R(c)$ is larger for the downstream part of the front than for the upstream part. In the second case, the upstream part of the front experiences the largest retardation. As follows from the discussion by Helfferich (this volume), the sharpening behaviour of fronts has already received much interest in various research disciplines. Understanding of the phenomena is of interest from a fundamental as well as practical point of view. For instance, the review by Bolt (1982) illustrates the importance of the sharpening behaviour for soil salinization and sodication and the reclamation of saline soils.

In case of ion exchange, a front is self-sharpening when the entering ion is preferred by the adsorbing surface. Assuming exchange of Ca^{2+} by Zn^{2+} we obtain the Zn-fronts of Figure 8. For linear adsorption ($K=1$) a front is calculated with a steepness intermediate between those of self-sharpening and non-sharpening front (K > 1 and K < 1, respectively).

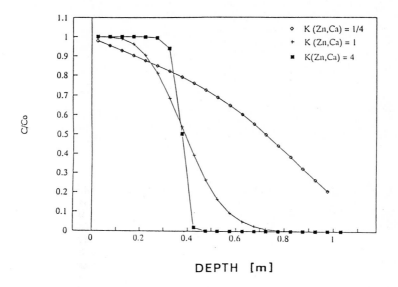

Figure 8:   Fronts of dissolved zinc for three Zn/Ca exchange coefficient values calculated between 0 and 1 m depth.

At the field scale, deterministic predictions of transport are generally in error because they do not account well for heterogeneity (Figure 4 and Valocchi, this volume). Stochastic analyses are more appropriate (Ganoulis, this volume). Examples were given by Van der Zee and Van Riemsdijk (1987), and Destouni and Cvetkovic (1991) who assumed a field to consist of non-interacting parallel vertical streamtubes. Despite field scale heterogeneity a detailed chemical analyses at smaller scales is not useless. Rather, "microscopic" models show what effects should be considered in averaging to a larger scale and how this should be done. Nevertheless, at a particular scale the complexicity of the chemical model should be related to our knowledge of the system and its variation. And this holds not only for chemical modeling.

References

Aiken, G R, McKnight D M, Wershaw R L and MacCarthy P, (Eds.) (1985) Humic Substances in Soil, Sediment, and Water, Wiley, New York

Boekhold A E, Temminghoff E J M and van der Zee S E A T M, (1992), Influence of electrolyte composition and pH on cadmium sorption by an acid sandy soil, J. Soil Sci. (submitted)

Boekhold A E, van der Zee S E A T M and de Haan F A M, (1991), Spatial patterns of cadmium contents related to soil heterogeneity, Water Air Soil Poll., 57-58, 479-488

Bolt G H, De Boodt M F, Hayes M H B and McBride M B, (eds.), 1991, Interactions at the Soil Colloid - Soil Solution Interface, NATO ASI Series Series E: Applied Sciences-Vol. 190. Kluwer Academic Publishers, Dordrecht

Bolt G H, (1982), Movement of solutes in soil: principles of adsorption/exchange chromatography, in Soil Chemistry B, Physico-Chemical Models, Elsevier, New York, 285-384, 1982

Crank J, (1957), Trans Faraday Soc. 53, 1083

DeVault D, (1943), J. Am. Chem. Soc., 65, 532

De Wit J C M, Van Riemsdijk W H, Nederlof M M, Kinniburgh D G and Koopal L K, (1990) Analytica Chim. Acta 232:189-207

De Wit J C M, Van Riemsdijk W H and Koopal L K, (1991) Finnish Humus News 3:139-144

Destouni G, and Cvetkovic V, (1991), Field scale mass arrival of sorptive solute into the ground water, Water Resour. Res., 27, 1315-1325

Hayes M H B, MacCarthy P, Malcolm R L and Swift R S, (Eds.) (1989) Humic Substances II: In Search of Structure, Wiley, New York

Hiemstra T, De Wit J C M and Van Riemsdijk W H, (1989), Multisite proton adsorption modeling at the solid/solution interface of (hydr)oxides: a new approach II, J. Colloid Interface Sci. 133:105-117

Hiskey B J, and Wadsworth M E, (1974), Galvanic conversion of chalcopyrite, in: F.F. Aplan et al. (eds.), Solution Mining Symposium, A.I.M.E., New York, 422-445, 1974

Keizer P, and Bruggenwert M G M, (1991), Adsorption of heavy metals by clay-aluminum hydroxide complexes, in: G H Bolt et al. (eds.), Interactions at the soil colloid-soil solution interface, 177-203, Kluwer Ac. Publ.

Meeussen J C L, Keizer M G and De Haan F A M, (1992) Chemical stability and decomposition rate of iron cyanide complexes in soil solutions, Environ Sci and Technol 26, 511-516

Reiniger P, and Bolt G H, (1972), Theory of chromatography and its application to cation exchange in soils, Neth. J. Agric. Sci., 20, 301-313

Stumm W (Ed) (1987) Aquatic Surface Chemistry, Wiley, New York

Van der Zee S E A T M, (1991), Reaction kinetics and transport in soil: compatibility and differences between some simple models, Transp. Porous Media, 6, 703-737

Van der Zee S E A T M, and Van Riemsdijk W H, (1987), Transport of reactive solute in spatially variable soil systems, Water Resour. Res., 23, 2059-2069

Van Riemsdijk W H, Bolt G H and Koopal L K, (1991) in Bolt G H, et al. (eds.), Interactions at the Soil Colloid - Soil Solution Interface, NATO ASI Series Series E: Applied Sciences- Vol. 190. Kluwer Academic Publishers, Dordrecht, Ch.3

Wen C Y, (1968), Non-catalytic heterogeneous solid fluid reaction models, Industr. Eng. Chemistry 60, 34-54, 1968

Westall J C, (1976) MICROQL II, Technical Report, Swiss Federal Institute of Technology, EAWAG, Duebendorf, Switzerland

Organic Pollutant Migration in Soils as Affected by Soil Organic Matter. Molecular and Mechanistic Aspects

N. Senesi
Istituto di Chimica Agraria
Università di Bari
Via Amendola, 165/A
70126 Bari
Italy

The soil can receive a substantial quantity of organic pollutants that are either applied on purpose or incidentally deposited from a variety of sources. Pesticides are applied directly by aerial or surface treatment. Other pollutants can be deposited either by wet or dry deposition from the vapor phase, or in forms sorbed to atmospheric particulates (i.e. fly ash) originating from waste chemicals and solid waste incineration. Pollutant chemicals can also reach the soil through municipal and industrial wastes, landfill effluents, and composts of various nature and source applied to the soil as fertilizers or amendments.

Once on soil surface, the parent chemicals and their degradation products may be subjected to various fates. They can enter the soil or be transported to the aquatic environment by runoff into surface waters and by erosion of contaminated soils into streams. When in the soil, the pollutant can percolate through the top layer, then penetrate through the unsaturated zone, and eventually reach groundwater in its original form or as a breakdown product. The rate and amount of pollutant penetration through the soil is controlled by several factors including the physical and chemical properties of the pollutant, amount of water arriving to the field soil, thickness and nature of the soil, and type and extent of interactions the pollutant experiences with various components it encounters in each phase in the soil system. These

NATO ASI Series, Vol. G 32
Migration and Fate of Pollutants in Soils and Subsoils
Edited by D. Petruzzelli and F. G. Helfferich
© Springer-Verlag Berlin Heidelberg 1993

components are substantially inorganic, organic and biologic materials that interact differently with organic pollutants depending on their intrinsic properties.

Several studies have suggested that organic pollutants show a greater affinity for organic surfaces than for mineral surfaces. The content and nature of soil organic matter play a major role in the performance of organic pollutants in soil. Groundwaters underlying soils with low organics content are highly vulnerable to contamination by organic pollutants (Perry et al., 1989).

The most ubiquitous and widespread natural non-living organic materials in soils, as well as in all terrestrial and aquatic environments, are humic substances. A number of physical, chemical, biochemical and photochemical properties (see next section) qualify these as privileged natural compounds in the interaction with organic chemicals. Humic substances may affect the fate of organic pollutants in soil in several ways which include adsorption and partitioning, solubilization, degradation by hydrolysis, and photo-decomposition. All these processes are of fundamental importance in determining the fate of the pollutant and its subsequent transfer to groundwater, i.e., the rate and amount of pollutant penetration through the unsaturated zone. These processes also have important implications in biodegradation and detoxication, bioavailability and phytotoxicity, bioaccumulation, and residue persistence and monitoring of organic pollutants in soil.

The main purpose of this review is to summarize and discuss molecular and mechanistic aspects of the most important modes of interaction between humic substances and some classes of organic pollutants, with emphasis on pesticides and their degradation products.

Soil Humic Substances

Humic substances consist of a physically and chemically heterogeneous mixture of relatively high-molecular-weight, yellow to black organic compounds of mixed aliphatic and aromatic nature, formed by secondary synthesis reactions (humification) during the decay process and transformation of biomolecules that originate from dead organisms and microbial activity (Stevenson, 1982). These materials are exclusive of undecayed plant and animal tissues, their partial decomposition products, and the soil biomass. Approximately 60-70% of the total soil-organic carbon occurs in humic substances. Estimated levels of soil organic carbon on the earth surface occurring as humic substances are 30×10^{14} Kg (Stevenson, 1982).

On the basis of their solubility in water at various pH, humic substances are divided into two main fractions, that are: (i) humic acids, the portion that is soluble in dilute alkaline solution and is precipitated upon acidification to pH 2; and (ii) fulvic acid, the portion that is soluble at any pH value, even below 2 (Stevenson, 1982).

Humic and fulvic acids cannot be regarded as single chemical entities and described by unique, chemically defined molecular structures. Both are operationally defined by a model structure constructed on the basis of available compositional, structural, functional, and behavioral data and containing the same basic structural units and the same types of reactive functional groups that are common to all the single, indefinitely variable and unknown molecules (Stevenson, 1982). In Figure 1, "typical" model molecules proposed for soil humic and fulvic acids are presented (Stevenson, 1982; Langford et al., 1983). The macromolecular structure consists of aromatic, phenolic, quinonic and heterocyclic "building blocks" that are randomly condensed or linked by aliphatic, oxygen, nitrogen, or sulphur bridges. The macromolecule bears aliphatic, glucidic, aminoacidic, and lipidic surface chains as well as chemically

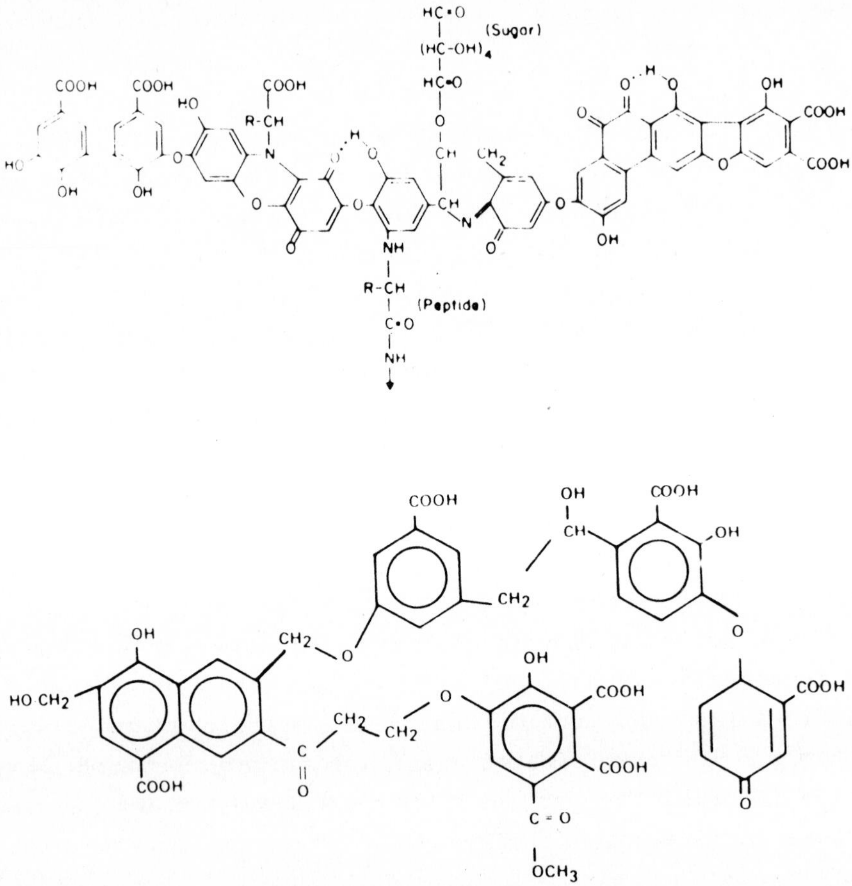

Figure 1. "Typical" model molecules proposed for soil humic acid (top, from Stevenson, 1982) and fulvic acid (bottom, from Langford et al., 1983).

active functional groups of various nature (mainly carboxylic, and also phenolic and alcoholic hydroxyls, carbonyls, etc.) which render the humic polymer acidic. Hydrophilic as well as hydrophobic sites are present. Structure and composition of humic acids are more complex than those of soil fulvic acids, that generally feature a lower aromatic and higher aliphatic character, more oxygen-containing functional groups (preferentially carboxylic), lower molecular weight, and higher water solubility.

Humic materials are polydisperse and exhibit poly-electrolytic behaviour in aqueous solution (Schnitzer, 1978; Hayes and Swift, 1978; Stevenson, 1982). Surface activity is an important property of humic substances that promotes interactions especially with hydrophobic organic pollutants. The amphophilic character and surface activity increase with increasing pH as -COOH and -OH groups form more hydrophilic sites (Schnitzer, 1978). High pH and high concentration of humic substances depress the surface tension of water. This increases soil wettability and promotes interaction of humic substances with both hydrophobic and hydrophilic organic pollutants in solution. Humic and fulvic acids with their relatively open, flexible, sponge-like structure may trap and even fix organic pollutants that can fit into the voids and holes (Schnitzer, 1978).

Humic substances also contain relatively high amounts of stable free radicals, probably semiquinones, which can bind certain organic pollutants (Senesi, 1990). The increase in free radical content of humic and fulvic acids in aqueous media with increasing pH or by visible-light irradiation enhances markedly the chemical and biochemical reactivity (Senesi, 1990).

Organic Pollutants in Soil

Organic pollutants in soil include parent compounds and their biological, chemical and photochemical degradation products that feature a wide variety of physical, chemical and biological properties and belong to widely differing chemical classes. Despite of this, they can be classified into two groups, depending on whether their principal interactions with humic substances involve specific chemistry or unspecific physical forces. Members of the former group are hydrophilic, ionic or ionizable, basic or acidic compounds; among those of the latter group are hydrophobic, nonionic, mainly nonpolar compounds.

Important cationic pesticides used in agriculture are bipyridilium herbicides, diquat and paraquat, that are generally applied in the form of dibromide and dichloride salts of high solubility in water and low volatility. Others are the fungicide phenacridane chloride, the germicide thyamine, the plant-growth regulator phosphon, and chlordimeform. The widely applied herbicides amitrole and s-triazines are weak-base, relatively volatile compounds of low solubility in water that can protonate at pH < 5.0 in soil. Weak-acid, widely used herbicides and insecticides with carboxyl or phenolic functional groups capable of ionizing in aqueous media include highly water-soluble chloro-aliphatic acids, alkali salts of dinitrophenols dinoseb and dalapon, pentachlorophenol (PCP), and phenoxyalkanoic acids such as 2,4-D and 2,4,5-T, halogenated benzoic acids chloramben, dicamba and TBA, and the pyridine derivative picloram. Other ionic or ionizable organic pollutants found in soil systems are several surfactants and detergents such as linear alkyl sulphonates.

Nonionic pesticides vary widely in their composition, physical properties, and chemical reactivity. Basic information on physical and chemical properties of these pesticides are provided elsewhere (Senesi and Chen, 1989). Polynuclear aromatic hydrocarbons (PAHs) and phtalic acid diesters (PAEs) are widespread contaminants, with neutral, nonpolar hydrophobic character practically immiscible with water, that can reach the soil system through various environmental pathways. Other important hydrophobic pollutants that can reach the soil are several organic solvents and alkanes.

Adsorption Mechanisms

Adsorption represents probably the most important mode of interaction of organic pollutants with soil organic matter. The effect of adsorption on pollutant migration in soil depends on whether the adsorption occurs on insoluble, immobile organic fractions such as humic acids, or on dissolved or suspended, mobile fractions such as fulvic acids. Soil organic matter in the form of humic acids or fulvic acids can, therefore, either "attenuate" or "facilitate" transport.

Organic compounds can be sorbed by soil OM through physical-chemical binding by specific mechanisms and forces with varying degrees of strengths. These include ionic, hydrogen and covalent bonding, charge-transfer or electron-donor acceptor mechanisms, dipole-dipole and Van der Waals forces, ligand exchange, and cation and water bridging. Adsorption of nonpolar (hydrophobic) organic compounds can also be described in terms of non-specific, hydrophobic or partitioning processes between soil water and the soil organic phase.

The type and extent of adsorption will depend on the amount and properties of both the chemical and the humic molecule. The various chemical properties of the pollutant molecule, such as number and type of functional groups, acidic or basic character, polarity and polarizability, ionic nature and charge distribution, water solubility, shape and configuration, will often result in several possible adsorption mechanisms that may operate in combination.

For any given chemical, a sequence of different mechanisms may be responsible for adsorption onto humic substances. The organic molecule may be sorbed initially by sites that provide the strongest binding, followed then by progressively weaker sites as the stronger adsorption sites become filled. Once adsorbed, the chemical may be subject to other processes that can affect retention. For instance, some chemicals may further react to become covalently and irreversibly bound, while others may become only physically trapped into the humic matrix. Adsorption processes may thus vary from complete reversibility to total irreversibility, that is, the adsorbed chemical may be easily desorbed, desorbed with various degrees of difficulty, or not at all.

Ionic bonding (cation exchange)

Ionic bonding involves the electrostatic attraction of cationic organic functional groups such as amines and heterocyclic nitrogen atoms (proton acceptors) with carboxylic or phenolic groups (proton donors) at the surface of the organic matter.

Infrared (IR) (Burns et al., 1973; Khan, 1974; Maqueda et al., 1983) and potentiometric-titration (Narine and Gouy, 1982) data support ionic binding as the dominant mechanism for adsorption of diquat, paraquat, and chlordimeform by humic substances. Divalent cationic bipyridilium herbicides can react with two negatively charged sites on HS, e.g., two -COO⁻ groups or a -COO⁻ plus a phenolate ion. However, not all negative sites on HS seem to be positionally available to bind large organic cations, probably because of steric hindrance. Phosphon and phenacridane chloride are also reported to be adsorbed onto soil organic matter through ionic bonding (Weber, 1972).

IR and differential thermal analysis show that s-triazines and amitrole may protonate at the amino-group or heterocyclic nitrogen at a pH close to the pK_a of the compound, and form ionic bonds with carboxylate and phenolate groups of humic substances (Sullivan and Felbeck, 1968; Turski and Steinbrich, 1971; Senesi and Testini, 1980, 1982; Senesi et al., 1986b, 1987). The substituent in the 2-position and the alkyl groups at the 4- and 6- amino groups of the s-triazines affect the basicity and steric hindrance of the latter, and thus their reactivity with humic substances (Turski and Steibrich, 1971; Senesi et al., 1987).

Hydrogen bonding

Multiple oxygen- and nitrogen-containing sites are available on soil organic surfaces that can be used to form hydrogen bonds with several organic pollutants containing suitable complementary groups. A strong competition with water molecules is, however, expected for such sites.

Heat-of-formation, IR and DTA data provide evidence for H-bonding between C=O groups of humic acids and secondary amino-groups of s-triazines (Sullivan and Feldbeck, 1968; Li and Feldbeck, 1972; Turski and Steinbrich, 1971; Senesi and Testini, 1982). Acidic pesticides such as chlorophenoxyalkanoic acids and esters, asulam, and dicamba can be adsorbed by H-bonding onto humic substances at pH values below their pK_a through -COOH, -COOR and similar groups (Khan, 1973; Senesi et al., 1986a). H-bonding plays an important role in the adsorption onto HS of several nonionic polar pesticides including substituted ureas and phenylcarbamates, alachlor, metolachlor, cycloate, malathion, and glyphosate (Senesi and Testini, 1983; Senesi et al., 1986b; Miano et al., 1992).

Charge-transfer (electron donor-acceptor)

Humic macromolecules contain both electron-deficient structures such as quinones, and electron-rich moieties such as diphenols. This can account for their electron-donating and -accepting properties. Electron donor-acceptor mechanisms and formation of charge-transfer complexes will thus be possible with organic pollutants possessing electron donor or electron acceptor features.

Strong spectroscopic evidence exists for the formation of charge-transfer complexes between several humic acids and s-triazines (eq 1), substituted ureas, and amitrole. The shift to lower wave number observed in the IR for the out-of-plane deformation vibration of the heterocyclic donor triazine is ascribed to the decreased electron density in the ring resulting from the formation of a complex with electron-deficient structures of humic acid (Muller-Wegener, 1987; Senesi, 1981; Senesi and Testini, 1982; Senesi et al., 1987). The increase in free-radical concentration measured by electron spin resonance (ESR) in the interaction products of humic acid with s-triazine, substituted ureas, and amitrole is ascribed to the formation of semiquinone free radical intermediates in the single-electron donor-acceptor transfers occurring between the electron-rich amino or heterocyclic nitrogens of the herbicide and the electron-deficient, quinone-like structures in the humic acid (Senesi, 1981; Senesi and Testini, 1982, 1983; Senesi et al., 1986b, 1987). Charge-transfer humic acid-s-triazine systems may also originate under the effect of light, which may induce an unpairing of electrons involved in the interaction and produce the observed increase in the ESR signal (photo-induced transfer) (Senesi and Chen, 1989).

$$\text{(1)}$$

| (s-triazine; electron-donor, | (humic quinone; electron-acceptor) | (radical cation and anion; charge-transfer complex) |

A number of structural and chemical properties of both the humic acid and the organic pollutant affect to some extent the efficiency of formation of the electron donor-acceptor system. The electron-acceptor tendency of the humic molecule seems to be related directly to the quinone content and inversely to the total acidity and the content of carboxyls and phenolic hydroxyls (Senesi and Testini, 1982; Senesi et al., 1987). The presence of activating electron donors, such as a methoxyl group on the 2-position of the ring and isopropyl substituents on the amino groups, renders prometone the most efficent electron donor among s-triazines (Senesi and Testini, 1982; Senesi et al., 1987). The absence of chlorine atoms on the phenyl ring of fenuron qualifies this urea as an electron donor stronger than ureas containing one or more deactivating electron-withdrawing chlorine atoms on the ring (Senesi et al., 1987).

IR frequency shifts of the CH out-of-plane bending vibration of the electron acceptors paraquat, diquat and chloridimeform, measured in their products of interaction with humic substances, are ascribed to the formation of charge-transfer complexes (Burns et al., 1973; Khan, 1974). UV-difference spectroscopy provides evidence of the formation of a charge-transfer complex between electron donor moieties of humic substances and the electron acceptor chloranyl (eq 2)

(Melcer et al. 1989). Other important organic pollutants such as DDT, dioxins, and polychlorobiphenyls (PCBs) possess electron-accepting properties and may interact with humic substances by charge-transfer mechanisms similar to those previously described. Such a mechanism has been postulated to be responsible for the photolysis of DDT in the presence of humic substances (Miller and Narang, 1970).

Covalent binding

The presence of several types of reactive functional groups in humic substances offers multiple possibilities for their chemical reaction with suitable functional groups of the pollutant molecule. These reactions are often mediated by chemical, photochemical or enzymatic catalysts and result in the formation of covalent bonds with incorporation of the pollutant or, more likely, its intermediates and products of degradation, into the HS macromolecule.

Anilines, catechols, and phenols released by degradation of several classes of pesticides are subject to formation of covalent bonds with soil organic matter, with or without the intervention of microbial enzymes. Chloroanilines can be chemically bound to soil organics by two possible mechanisms including oxidative-coupling reactions to phenols, and nucleophilic addition to carbonyl and quinone groups (eqs 3, 4, 5). These reactions may lead to the formation of hydrolyzable (e.g., anil and anilinoquinone) or nonhydrolyzable (e.g., heterocyclic rings or ethers) bound forms (Hsu and Bartha, 1976; Parris, 1980). Nucleophilic addition is suggested for the binding to HS of benzidine, -naphthylamine and para-toluidine (Graveel et al., 1985).

A marked quenching of free-radical concentration was measured by ESR spectroscopy in the interaction products of soil HA with water-dissolved chlorophenoxyalkanoic acids and

$$(3)$$

$$(4)$$

$$(5)$$

esters. This suggests the occurrence of cross-coupling reactions between indigenous humic free radicals and phenoxy or aryloxy radical intermediates generated photochemically or by chemical or enzymatic catalysis in the partial oxidative degradation of chlorophenoxyalkanoic compounds (eq 6) (Senesi

$$(6)$$

Humic Hydroquinone
(Electron Donor)

Semiquinone Radicals
(Electron Donor-Acceptor System)

Chloranil
(Electron Acceptor)

et al., 1984; 1986a). The coupling efficiency results indirectly related with both the carboxyl content and the carboxyl to phenolic hydroxyl ratio of the humic acids and with the number of chlorine atoms on the phenoxy ring of the herbicides (Senesi et al., 1986a).

Ligand exchange (cation bridge)

Adsorption by ligand exchange mechanisms involves the displacement of hydration water or other weak ligands partially holding a polyvalent metal ion associated to soil organic matter by a suitable functional group of the pollutant. This group may act as ligand to the metal, with the formation of an inner-sphere complex. S-triazines and anionic pesticides such as picloram are likely to bind humic substances by this mechanism (Nearpass, 1976).

Dipole-Dipole and van der Waals Forces

Although limited experimental evidence is available on the occurrence of these forces in the interaction between soil humic substances and organic pollutants, they are considered to be involved, possibly in addition to stronger binding forces, in all adsorbent-adsorbate interactions (Senesi and Chen, 1989). These forces will assume particular importance, however, in the adsorption of nonionic and nonpolar pollutants. Since Van der Waals forces are additive, their contribution increases with the size of the pollutant molecule and with its capacity to adapt to the surface of the humic molecule.

Hydrophobic adsorption and partitioning

Hydrophobic adsorption on the surface or trapping within interior pores of the humic macromolecular sieve has been proposed as an important nonspecific mechanism for retention of nonionic, nonpolar, organic pollutants that interact weakly with water. Hydrophobic active sites of humic substances include aliphatic side chains or lipid portions and aromatic lignin-derived moieties with high carbon content and bearing a small number of polar groups. Water molecules do not compete effectively with hydrophobic molecules for these sites.

Hydrophobic adsorption by soil organic matter is suggested to be important for DDT and other organochlorine insecticides, PAEs, PAHs, PCBs, and many other organic pollutants, while it is considered a possible, additional mechanism for some s-triazines and phenyureas (Senesi and Chen, 1989).

Hydrophobic retention need not to be regarded as an active adsorption mechanism and is often considered as a "partitioning" between water and a non-specific organic phase. In contrast to adsorption, the term partitioning describes a process in which the adsorbate permeates, i.e., dissolves, into the network of the organic phase. Thus, partitioning is distinguished from adsorption by the homogeneous, aspecific distribution of the sorbed material throughout the entire volume of the organic phase (Chiou et al., 1983). Partitioning is modeled as an equilibrium process, similar to that between two immiscible solvents, say, between water and an organic solvent such as 1-octanol. Humic substances both in the solid and dissolved phase are thus considered as a non aqueous solvent into which the hydrophobic pollutant can partition from water.

Thermodynamic arguments have been proposed, but may be too simplified to distinguish between partitioning and adsorption and may lead to erroneous conclusions (Mingelgrin and Gerstl, 1983). The wide variation observed by several authors in the strength of interactions (expressed as partition coefficients) between hydrophobic pollutants and humic substances of different source and nature is attributed to the variation in the chemical and structural composition of the latter (Davis and Delfino, 1992). This confirms that strong and "specific" chemical interactions are involved and should be accounted for in any predictive modeling of hydrophobic pollutant-humic substance association (Chin and Weber, 1989).

Solubilization Effects

Interactions of certain organic pollutants with dissolved organic matter, mainly fulvic acids, can significantly modify their solubility and thus affect their mobility and migration.

The physical and chemical properties of the pollutant (the solute) and the humic substance can markedly affect its solubility. At a given concentration of dissolved humic substances, relatively water-insoluble, nonionic organic pollutants, e.g., DDT, PCBs, PAHs, n-alkanes, PAEs, are most easily solubilized (Fig. 2) (Wershaw et al., 1969; Boehm and Quinn, 1973; Chiou et al., 1986; Carter and Suffet, 1982; Landrum et al., 1984; Shinozuka et al., 1987). The increase in solubility may be the result of direct pollutant adsorption or partitioning, or of an overall increase in solvency (Kile and Chiou, 1989).

The effect of solubility enhancement decreases with increasing intrinsic water solubility of the pollutant (Kile and Chiou, 1989). This trend has been observed for relatively water-insoluble higher alkanes in comparison with more water-soluble aromatic compounds (Boehm and Quinn, 1973) and for DDT versus lindane (Caron et al., 1985).

The observed pollutant-solubility enhancement has been described as possibly resulting from a partition-like interaction between solute and dissolved organic matter of high molecular weight. The latter is regarded as a "microscopic organic phase", similar to a micelle (Chiou et al., 1986). The effectiveness of this interaction appears to be largely controlled by a number of properties of the humic molecule such

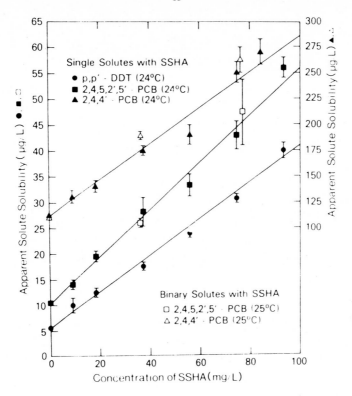

Figure 2. Dependence of apparent water solubilities of DDT and some PCBs on concentration of soil humic acid at pH 6.5 (from Chiou et al., 1986).

as size, polarity, configuration, conformation, composition, total acidity, and surface activity (Shinozuka et al., 1987; Chiou et al., 1986; Carter and Suffet, 1982). The fact that humic acids in soil enhance the water solubility of DDT and PCBs more effectively than do fulvic acids and river-water humic acids is attributed to higher molecular weight, carbon content, and aromaticity as well as large nonpolar volume and lower oxygen and polar-group content and hydrophilicity (Kile and Chiou, 1989). Low molecular weight organic matter or a highly polar organic pollutant would not exhibit a strong effect.

The magnitude of pollutant solubility enhancement is also dependent on the concentration of dissolved humic substances (Fig. 2), solution pH and temperature. The water solubilities of compounds such as DDT, some PCBs and higher alkanes are sensitive to low levels of dissolved humic substances; those of highly water-soluble pollutants are less so. The solubility enhancement is decreased if the pH is raised for DDT and PCBs (Carter and Suffet, 1982; Chiou et al., 1986). At alkaline pH, the hydration and concomitant repulsion of ionized acidic groups of the humic molecule would lead to more expanded conformations and thus decrease the extent of nonpolar regions available for the interaction with the hydrophobic pollutant. A temperature decrease from 25 to 9 °C increases twofold the binding constant of DDT to dissolved humic substances, thus increasing its apparent water solubility (Carter and Suffet, 1982).

All these results suggest that the solubility effect exerted by dissolved humic substances on hydrophobic organic pollutants cannot be completely and satisfactory explained on the basis of a simple partitioning model of interaction. Specific binding interactions involved in such a complex process should be duly accounted for in any general model by introduction of suitable, additional parameters and mechanisms that can better account for these effects.

Hydrolysis catalysis

Hydrolysis is an important degradative reaction that leads to products with solubility, volatility, reactivity, and other physical and chemical properties different from those of the parent pollutant molecule, thus affecting its migration and fate in soil.

Humic substances, particularly in the dissolved phase, are able to exert a catalytic or inhibitory effect in the abiotic hydrolysis of a number of herbicides. In the presence of fulvic acids or humic acids, the acid hydrolysis rates of n-alkyl esters of 2,4-D and chloro-s-triazines are enhanced (Struif et al., 1975; Khan, 1978), whereas that of alkaline hydrolysis of the n-octyl ester of 2,4-D (2,4-DOE) is lowered (Perdue and Wolfe, 1982). Sorption on humic substances has little or no effect on the rates of pH-independent (i.e., non acid-base catalyzed) hydrolysis of some organophosphorothioate esters such as chlorpyrifos and diazinon, halogenated alkenes, aziridine derivatives, and others (Macalady et al., 1989).

The catalytic effects of humic and fulvic acids on the dechlorohydroxylation of the chloro-s-triazines simazine, atrazine, and propazine are attributed to specific interaction through H-bonding between the surface carboxylic groups of the humic molecule and the side-chain nitrogens of the triazine (eq 7). This would enhance the electron-withdrawing effect from the electron-deficient carbon bearing the chlorine atom, thus reducing the activation energy barrier for hydrolytic cleavage of the C-Cl bond and promoting its replacement by the weak nucleophile water (eq 7) (Li and Felbeck, 1972; Khan, 1978). No

evidence of catalytic action was observed for weak-acid functional groups of fulvic acid, such as phenolic OH (Gamble and Khan, 1985). The retardation of base-catalyzed hydrolysis of 2,4-DOE by solution-phase humic substances is envisioned in terms of the changes in solution chemistry in the vicinity of the negatively charged (at neutral and alkaline pH values) surfaces of humic molecules (Perdue and Wolfe, 1982).

A tentative general mechanism has been proposed for the overall effects of humic substances on hydrolysis kinetics of hydrophobic organic pollutants. The model is based on an analogy with micellar catalysis, with only a minor contribution from general acid-base catalysis (Perdue, 1983). The model has been successfully tested with experimental data obtained for base-catalyzed hydrolysis of 2,4-DOE (Perdue and Wolfe, 1982) and acid-catalyzed hydrolysis of atrazine (Li and Feldbeck, 1972). The effects in the two cases are attributed to electrostatic stabilization of the transition state for the acid catalysis, in which the substrate becomes more positively charged, and to the destabilization of the transition state for base-catalyzed hydrolysis, in which the substrate becomes more negatively charged (Perdue, 1983). On the basis of this model, one would expect the effect on base-catalyzed hydrolysis to be small for parathion, which associates only weakly with humic substances, but large for compounds such as DDT that associates strongly with them (Perdue, 1983).

Inhibition by humic acids of hydrolytic enzymes in soils is indicated as an additional mechanism that may operate in certain reactions (Malini de Almeida et al., 1980; Mulvaney and Bremner, 1978).

In conclusion, a general model to predict the effects of humic substances on the hydrolysis rates of organic pollutants seems not yet to be available, the proposed ones appearing over-simplified or incomplete.

Photosensitization

Sunlight-induced photochemical transformations are important pathways in the abiotic degradation of organic pollutants in the top layer of soil and waters. Photoreactions may modify the physical and chemical properties of pollutants and significantly affect their fate and migration into the bulk soil (Miller et al., 1989). Light-induced transformations can be distinguished in direct photolysis processes, initiated by direct absorption of light by the chemical, and indirect or sensitized photolysis, involving light absorption by natural "photosensitizers" or producers of photoreactants.

Humic substances can strongly absorb sunlight and may behave, therefore, as initiators of photoreactions, some of which involve reactive secondary products that have important implications in accelerating, increasing or even determining light-induced transformation of non-absorbing, photochemically-stable organic pollutants. For example, photolysis of atrazine in water is more extensive, although initially retarded, in the presence of 0.01 % dissolved fulvic acid (eq 8) (Khan and Schnitzer, 1978).

$$(8)$$

Humic substances principally act as sensitizers or precursors for the production of highly reactive, short-lived species (so-called "photoreactants") such as the solvated electron, e^-_{aq}, singlet oxygen, $^1O_2$, superoxide anion, $O_2^{-\cdot}$, peroxy radicals, $RO_2\cdot$, hydrogen peroxide, H_2O_2, and redox-active species including excited-state of humic molecule and humic organic radicals. However, humic substances can also function as a scavenger of other phototransients such as hydroxy radicals, $HO\cdot$ (Cooper et al., 1989; Hoigné et al., 1989).

The formation of e^-_{aq} and humic-derived cation radicals in irradiated solutions of humic substances is suggested to result from photoejection of an electron from the excited-states $^1HS^*$ or $^3HS^*$, according to:

$$^1HS \xrightarrow{\text{light}} {}^1HS^* \longrightarrow {}^3HS^* \tag{9}$$

$$^1HS^* \text{ or } {}^3HS^* \longleftrightarrow [HS^{+\cdot} + e^-] \underset{-H_2O}{\overset{+H_2O}{\rightleftharpoons}} HS^{+\cdot} + e^-_{aq} \tag{10}$$

The solvated electron is a powerful reductant that may react rapidly with electronegative xenobiotics such as chlorinated organics, e.g., dioxins (Cooper et al., 1989; Hoigné et al., 1989). The dominant reaction of solvated electrons are, however, with O_2, to form $O_2^{-\cdot}$ [superoxide ions (reaction 11)] which, in turn, may disproportionate and react with water to form hydrogen peroxide (reaction 12). Excited (triplet-state) humic substances [$^3HS^*$ (reaction 9)] can transfer energy directly to ground-state oxygen ($^3O_2$) to form singlet oxygen ($^1O_2$), which, in turn, may generate humic-acid cation radicals and superoxide ion (reaction 13). The latter two species can also be formed directly by reaction of $^3HS^*$ with $^3O_2$. Photochemically-derived humic-substance cation radicals may generate organoperoxy cation radicals by reaction with $^3O_2$ (14).

$$e^-_{aq} + {}^3O_2 \longrightarrow O^-_2\cdot \tag{11}$$

$$2\ O^-_2\cdot \longrightarrow O_2 + O_2^{2-} \xrightarrow{2\ H_2O} O_2 + H_2O_2 + 2OH^- \tag{12}$$

$${}^3HS^* + {}^3O_2 \longrightarrow {}^1HS^* + {}^1O_2 \longrightarrow HS^{+}\cdot + O^-_2\cdot \tag{13}$$

$$HS^{+}\cdot + {}^3O_2 \longrightarrow HSO_2^{+}\cdot \tag{14}$$

Singlet oxygen and superoxide radicals are efficient, but selective photoreactants for the transformation of various organic pollutants. For example, humic substances are able to catalyze the photoxidation to the corresponding sulfoxide of the sulfide group of the thioether insecticides disulfoton, fenthion, methiocarb, and butocarboxim at the soil surface, but not of the sulfur group in methylthio-triazenes, thiocarbamates, or dithiolane insecticides (Miller et al., 1989). Furthermore, humic-substance derived peroxy radicals are important photooxidants for alkyphenols (Hoigné et al., 1989).

The photoexcited humic parent macromolecule can also undergo direct photochemical reactions with some organic pollutants, thanks to the presence of conjugated structures such as keto and quinone groups. The three major possible pathways are energy transfer (or photosensitization), charge-transfer, and photoincorporation (Cooper et al., 1989). The first pathway [see reaction (15) below] is a type of indirect photolysis with humic substances (HS) in a photo-excited triplet state acting as sensitizer. They transfer energy to previously bound acceptor molecules and so catalyze the degradation of some organic pollutants (OP) that have low excited-state energies, but cannot absorb sunlight themselves (Cooper et al., 1989). DDT is such a compound.

$${}^3HS\text{-donor}^* + OP\text{-acceptor} \rightleftharpoons (HS\text{---}OP)^* \rightleftharpoons HS + OP^*$$

$$\downarrow$$

$$OP\text{-photoproducts} \tag{15}$$

The process is reversible, that is, humic substance can also act as acceptor of energy from the excited pollutant, thus quenching the photodegradation of the latter. Charge-transfer, i.e., photoinduced electron transfer reactions, may occur between irradiated humic substances systems and PAHs, s-triazines and paraquat (Cooper et al., 1989; Senesi and Chen, 1989). Finally, direct photoincorporation of organic pollutants such as polychlorobenzenes into the humic macromolecule may occur through radical combination or cycloaddition (Cooper et al., 1989).

Conclusions

Organic pollutants may interact with humic substances in the solid and dissolved phases of soil in several ways. The interactions affect to a greater or lesser degree the behavior and fate of the pollutants, in particular, their migration into subsoil and groundwater. Adsorption processes by soil organic matter directly or indirectly control all the other processes in soil that affect organic chemicals by determining how much of the organic chemicals is solubilized and moves into the aqueous and gaseous phase, degrades, or is consumed by organisms. Humic substances can also catalyze some chemical degradation reactions such as hydrolysis, and affect photodegradation of organic pollutants. These effects lead to the formation of reaction intermediates and products having physical and chemical properties different from the parent compound and behaving differently with respect to migration in soil.

Rarely is a unique process involved in the interaction. More often, several processes with different mechanisms occur side by side, one or a few of which may dominate for a given pollutant under given conditions. The type and extent of interaction may change with time and may ultimately result in irreversible modification of the chemical and biological properties of the pollutant. This might include complete immobilization of the latter. Alternatively, the pollutant can attach itself reversibly to mobile organic soil fractions and migrate with these, constituting a potential time-delayed source of contamination that might affect distant ground. Any simplistic representation of the interaction, such as depicting it as a partitioning, may therefore not be indicative of the phenomena actually involved in the interaction of most organic pollutants with soil organic matter.

A better knowledge of the chemical nature and reactivity of humic substances, which are the major natural compounds interacting with organic pollutants in soil, and a better understanding of the mechanisms of their interactions with organic pollutants can be expected to help in the development of a quantitative thermodynamic description and kinetic modeling of pollutant migration in soils. Advanced techniques that are available and show promise for the molecular and mechanistic investigation of these interactions include Fourier-transform infrared, nuclear magnetic resonance, electron spin resonance and fluorescence spectroscopies, and differential scanning calorimetry.

References

Boehm PD, Quinn J (1973) Solubilization of hydrocarbons by the dissolved organic matter in seawater. Geochim Cosmochim Acta 37:2459-2477

Burns IG, Hayes MHB, Stacey M (1973) Spectroscopic studies on the mechanisms of adsorption of paraquat by humic acid and model compounds. Pestic Sci 4:201-209

Caron G, Suffet IH, Belton T (1985) Effect of dissolved organic carbon on the environmental distribution of non polar organic compounds. Chemosphere 14:993-1000

Carter CW, Suffet IH (1982) Binding of DDT to dissolved humic materials. Environ Sci Technol 16:735-740

Chin Y-P, Weber JW Jr (1989) Estimating the effect of dispersed organic polymers on the sorption of contaminants by natural solids. 1. A predictive thermodynamic humic substance-organic solute interaction model. Environ Sci Technol 23:978-984

Chiou CT, Porter PE, Schmedding DW (1983) Partition equilibria of nonionic organic compounds between soil organic matter and water. Environ Sci Technol 17:227-231

Chiou CT, Malcolm RL, Brinton TI, Kile DE (1986) Water solubility enhancement of some organic pollutants and pesticides by dissolved humic and fulvic acids. Environ Sci Technol 20:502-508

Cooper WJ, Zika RG, Petasne RG, Fischer AM (1989) Sunlight-induced photochemistry of humic substances in natural waters: major reactive species. In: Suffet IH, MacCarthy P (eds) Aquatic Humic Substances. Influence on Fate and Treatment of Pollutants. Advances in Chemistry Series 219, Am Chem Soc, Washington, pp 333-362

Davis WM, Delfino JJ (1992) Interaction among hydrophobic organic pollutants and humic substances. Am Chem Soc-Div Environ Chem, S. Francisco, 5-10 April 1992, pp 7-10

Graveel JG, Sommers LE, Nelson DW (1985) Sites of benzidine, -naphthylamine and p-toluidine retention in soils. Environ Toxicol Chem 4:607-613

Gamble DS, Khan SU (1985) Atrazine hydrolysis in soil: catalysis by the acidic functional groups of fulvic acid. Can J Soil Sci 65:435-443

Hayes MHB, Swift RS (1978) The chemistry of soil organic colloids. In: Greenland DJ, Hayes MHB (eds) The Chemistry of Soil Constituents. Wiley, New York, pp 179-320

Hoigné J, Faust BC, Haag WR, Scully FE Jr, Zepp RG (1989) Aquatic humic substances as sources and sinks of photochemically produced transient reactants. In: Suffet IH, MacCarthy P (eds) Aquatic Humic Substances. Influence on Fate and Treatment of Pollutants. Advances in Chemistry Series 219, Am Chem Soc, Washington, pp 363-381

Hsu TS, Bartha R (1976) Hydrolysable and nonhydrolysable 2,4-dichloroaniline-humus complexes and their respective rates of biodegradation. J Agric Food Chem 24:118-122

Khan SU (1973) Interaction of humic acid with chlorinated phenoxyacetic and benzoic acids. Environ Lett 4:141-148

Khan SU (1974) Adsorption of bipyridilium herbicides by humic acids. J Environ Qual 3:202-206

Khan SU (1978) Kinetics of hydrolysis of atrazine in aqueous fulvic acid solution. Pest Sci 9:39-43

Khan SU, Schnitzer M (1978) UV irradiation of atrazine in aqueous fulvic acid solutions. J Environ Sci Health 3:299-310

Kile DE, Chiou CT (1989) Water-solubility enhancement of nonionic organic contaminants. In: Suffet IH, MacCarthy P (eds) Aquatic Humic Substances. Influence on Fate and Treatment of Pollutants. Advances in Chemistry Series 219, Am Chem Soc, Washington, pp 131-157

Landrum PF, Nihart SR, Eadie BJ, Gardner WS (1984) Reverse-phase separation method for determining pollutant binding to Aldrich humic acid and dissolved organic carbon of natural waters. Environ Sci Technol 18:187-192

Langford CH, Gamble DS, Underdown AW, Lee S (1983) Interaction of metal ions with a well characterized fulvic acid. In: Christman RF, Gjessing ET (eds) Aquatic and Terrestrial Humic Materials, Ann Arbor Science Publ, Ann Arbor-MI, pp 219-237

Li GC, Felbeck GT Jr (1972) A study of the mechanism of atrazine adsorption by humic acid from muck soil. Soil Sci 113:430-433

Macalady DL, Tratnyek PG, Wolfe NL (1989) Influences of natural organic matter on the abiotic hydrolysis of organic contaminants in aqueous systems. In: Suffet IH, MacCarthy P (eds) Aquatic Humic Substances. Influence on Fate and Treatment of Pollutants. Advances in Chemistry Series 219, Am Chem Soc, Washington, pp 323-332

Malini de Almeida R, Pospisil F, Vockova K, Kutacek M (1980) Effect of humic acid on the inhibition of pea choline esterase and choline acyltransferase with malathion. Biol Plant 22:167-175

Maqueda C, Perez Rodriguez JL, Martin F, Hermosin MC (1983) A study of the interaction between chlordimeform and humic acid from a typic chromoxevert soil. Soil Sci 136:75-81

Melcer ME, Zalewski MS, Hassett JP, Brisk MA (1989) Charge-transfer interaction between dissolved humic materials and chloranil. In: Suffet IH, MacCarthy P (eds) Aquatic Humic Substances. Influence on Fate and Treatment of Pollutants. Advances in Chemistry Series 219, Am Chem Soc, Washington, pp 173-183

Miano TM, Piccolo A, Celano G, Senesi N (1992) Infrared and fluorescence spectroscopy of glyphosate-humic acid complexes. Sci Total Environ (in press)

Miller LL, Narang RS (1970) Induced photolysis of DDT Science 169:368-370

Miller GC, Hebert VR, Miller WW (1989) Effect of sunlight on organic contaminants at the atmosphere-soil interface. In: Sawhney BL, Brown K (eds) Reactions and Movement of Organic Chemicals in Soils, SSSA Special Publication 22, Soil Sci Soc Am Inc, Madison-WI, pp 99-110

Mingelgrin U, Gerstl Z (1983) Reevaluation of partitioning as a mechanism of nonionic chemicals adsorption in soils. J Environ Qual 12:1-11

Müller-Wegener U (1987) Electron donor acceptor complexes between organic nitrogen heterocycles and humic acid. Sci Total Environ 62:297-304

Mulvaney FL, Bremner JM (1978) Use of p-benzoquinone for retardation of urea hydrolysis in soils. Soil Biol Biochem 10:297-302

Narine DR, Guy RD (1982) Binding of diquat and paraquat to humic acid in aquatic environments. Soil Sci 133:356-363.

Nearpass DC (1976) Adsorption of picloram by humic acids and humin. Soil Sci 121:272-277

Parris GE (1980) Covalent binding of aromatic amines to humates. 1. Reactions with carbonyls and quinones. Environ Sci Technol 14:1099-1106

Perdue EM (1983) Association of organic pollutants with humic substances: partitioning equilibria and hydrolysis kinetics. In: Christman RF, Gjessing ET (eds) Aquatic and Terrestrial Humic Materials. Ann Arbor Sci Publ, Ann Arbor-MI, pp 441-460

Perdue EM, Wolfe NL (1982) Modification of pollutant hydrolysis kinetics in the presence of humic substances. Environ Sci Technol 16:847-852

Perry AS, Muszkat L, Perry RY (1989) Pollution hazards from toxic organic chemicals. In: Gerstl Z, Chen Y, Mingelgrin U, Yaron B (eds) Toxic Organic Chemicals in Porous Media. Springer-Verlag, Berlin, pp 16-33

74

Schnitzer M (1978) Humic substances chemistry and reactions.
In: Schnitzer M, Khan SU (eds) Soil Organic Matter,
Elsevier, Amsterdam, pp 1-64
Senesi N (1981) Free radicals in electron donor-acceptor
reactions between a soil humic acid and photosynthesis
inhibitor herbicides. Z Pflanzen Bodenkd 144:580-586
Senesi N (1990) Application of electron spin resonance (ESR)
spectroscopy in soil chemistry. Advances in Soil Science
14:77-130.
Senesi N, Chen Y (1989) Interactions of Toxic Organic Chemicals
with Humic Substances. In: Gerstl Z, Chen Y, Mingelgrin U,
Yaron B (eds) Toxic Organic Chemicals in Porous Media.
Springer-Verlag, Berlin, pp 37-90
Senesi N, Testini C (1980) Adsorption of some nitrogenated
herbicides by soil humic acids. Soil Sci 10:314-320
Senesi N, Testini C (1982) Physico-chemical investigations of
interaction mechanisms between s-triazine herbicides and
soil humic acids. Geoderma 28:129-146
Senesi N, Testini C (1983) Spectroscopic investigations of
electron donor-acceptor processes involving organic free
radicals in the adsorption of substituted urea herbicides by
humic acids. Pestic Sci 14:79-89
Senesi N, Miano TM, Testini C (1986a) Role of humic substances
in the environmental chemistry of chlorinated
phenoxyalkanoic acids and esters. In: Pawlowski L, Alaerts
G, Lacy WJ (eds) Chemistry for Protection of the Environment
1985, Studies in Environmental Science 29, Elsevier,
Amsterdam 1986, pp 183-196
Senesi N, Padovano G, Loffredo E, Testini C (1986b) Interaction
of amitrole, alachlor and cycloate with humic acids. In:
Proc 2° Int Conf "Environmental Contamination", Amsterdam
1986, CEP Cons, Edimburgh, pp 169-171
Senesi N, Testini C, Miano TM (1987) Interaction mechanisms
between humic acids of different origin and nature and
electron donor herbicides: a comparative IR and ESR study.
Org Geochem 11:25-30
Shinozouka N, Lee C, Hayano S (1987) Solubilizing action of
humic acid from marine sediment. Sci Total Environ 62:311-
314
Stevenson FJ (1982) Humus Chemistry: Genesis, Composition,
Reactions. Wiley, New York, p 443
Struif B, Weil L, Quentin KE (1975) The behavior of herbicides
phenoxy acetic acids and their esters in waters. Wom Wasser
45:53-73
Sullivan JD, Felbeck GT (1968) A study of the interaction of s-
triazine herbicides with humic acids from three different
soils. Soil Sci 106:42-50
Turski R, Steinbrich A (1971) Studies on the possibilities of
binding herbicides of triazine derivatives by humic acids.
Polish J Soil Sci 4:120-124
Weber JB (1972) Interaction of organic pesticides with
particulate matter in aquatic and soil systems. Adv Chem
Series 111:55-120
Wershaw RL, Burcar PJ, Goldberg MC (1969) Interaction of
pesticides with natural organic matter. Environ Sci Technol
3:271-273

Solid-Phase Characteristics and Ion Exchange Phenomena in Natural Permeable Media

D. Petruzzelli, A. Lopez.
Istituto di Ricerca sulle Acque
Consiglio Nazionale delle Ricerche
5, Via De Blasio
70123 Bari
Italy.

The distribution of ionic species between liquid- and solid- phases in natural permeable media is known to be governed primarily by ion exchange [Bolt and Bruggenwert 1978, Bolt 1982, Sposito 1981].

Although theoretical and experimental methods of equilibria determinations are well established for pure systems (e.g., synthetic reactive polymers), this is not the case for natural porous media where the extreme complication of the systems is the main point of uncertainty.   In this context authors still adopt unverified assumptions concerning, for example, the structure of the exchanging counterions and the binding sites (e.g., in terms of hydration energy of charges, charge density and polarizability of ions), these parameters strongly influence the free energy of exchange.   Of particular importance are those factors related to the geometry of the binding sites which have to accommodate properly the exchanging counterions, as well as factors related to the dimensions and the volume of the hydrated counterions.   This is true for clays and zeolites where the rigidity of the crystalline lattice is a discriminating condition for ionic interaction, thus rendering the accompanying electrostatic interaction particularly selective and specific.   In some cases this results in a kind of sieving effect due to the need for proper correspondence between spatial orientation of the binding sites and coordinative bond sites of exchanging counterions.

The above mentioned and several other questions will be discussed in this chapter.

Physico-chemical properties at mineral-water interface.

Geochemical behaviour of solutes at liquid-solid interfaces in natural permeable media is so complex that understanding of the overall phenomena relies essentially on models which include hydrological, physico-chemical, chemical and biochemical aspects among others.

NATO ASI Series, Vol. G 32
Migration and Fate of Pollutants in Soils and Subsoils
Edited by D. Petruzzelli and F. G. Helfferich
© Springer-Verlag Berlin Heidelberg 1993

Hydrological models usually include physico-chemical, chemical and biochemical aspects as submodels, such submodels may entail simple evaluation of the partitioning properties of chemicals among phases, or evaluation of a first-order degradation of the substrates. In the real world, however, systems are so complex (see Figure 1) that they should properly be described in terms of a combination of elementary phenomena such as ion exchange, adsorption, precipitation, dissolution, chemical and biochemical transformations, with each contribution assuming its own importance which differs from case to case.

The reactivity at the solid-liquid interface is determined by chemical and physico-chemical properties which distinguish this region from the rest of the bulk solution. As reported by Davis and Hayes [1986] these properties can be summarized:

a) Water is more structured in the interfacial region;

b) Ions and water molecules are less mobile because of electrostatic constraints;

c) The dielectric constant of the water is depressed;

d) Electric potentials may develop at the surface, leading to the formation of a diffuse loud of ions at decreasing concentration from the surface of the solid-phase.

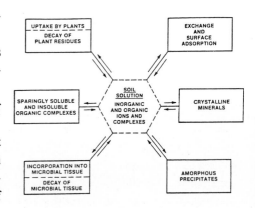

Behaviour of ionic species will also vary as a function of changed environmental conditions.

Fig.1 : Schematic diagram of possible pathways of pollutants in soils and subsoils. (from Page et.al. 1984)

Soil constituents and origin of charges.

Ion exchanging materials in soils include "inorganic", "organic", and "organo-mineral" components, the last being a combination of the first two [Talibudeen 1981].

Inorganic constituents.

Inorganic soil constituents are produced from rocks by "weathering" (rain and wind action). Soil minerals are mixed with a large number of extraneous materials in their natural environment. A simplified listing of the inorganic components commonly found in soils is reported in Table 1, where the classification, beyond oxides and alumosilicates, is essentially based on the anionic composition [Bolt, 1978].

Table 1. Inorganic constituents of soils. (from Bolt and Bruggenwert, 1978)

Oxides	
Al oxides-hydroxides	Gibbsite, boehmite, diaspore.
Fe oxides-hydroxides	Goethite, limonite, hematite.
Si oxides	Quartz, tridymite.
Alumosilicates	
Nesosilicates	Augite, hornblende.
Inosilicates	Talc biotite, muscovite,
Phyllosilicates	Clays: Illite, kaolite, montmorillonite, vermiculite.
Tectosilicates	Zeolite, anortite, orthoclase,
Carbonates	Calcite, dolomite.
Sulphates	Gypsum.
Halides	Halite, sylvine, carnallite.
Sulphides	Pyrite.
Phosphates	Apatite, vivianite.
Nitrates	Soda niter, niter.

Natural **oxides** (of Si, Al, Fe, Ti) occurring in soils are the stable endpoint of the weathering phenomena of the natural rocks and basically costitute the soil skeleton.

From the thermodynamic point of view the surface species (atoms, molecules,ions) of natural solids generally suffer from an imbalance of chemical forces. This imbalance is expressed in terms of surface energy that is usually minimized either by reduction of the surface area (by agglomeration of the particles) or by sorption of molecules and ions from the adjacent liquid phase (adsorption, ion exchange).

Oxide particles are present in the soil mixture in a state of extreme subdivision. In this context, since ion exchange is a "surface" phenomenon, the most important property of Pyrite. these materials is their exposed surface area which, in turn, depends on the particle size. A limit of 2 μm "equivalent" diameter (assuming spherical particles) corresponds to a surface area of about 1 m^2/g, representing the borderline between coarse and finely divided materials.

The exposed surface area determines the charge densiry and thus the strength of the coulom-big forces, upon which ion exchange equilibria at the solid-liquid interface largely depend. For example, surface charges of hydrated oxides (see below) are associated with the amphoteric behaviour of hydroxyl groups resulting from hydration [Schindler, 1981].

The average exposed surface areas of oxides range between 10 and 100 m^2/g, corresponding to an average charge density of 0.0001 to 0.0005 meq/m^2.

From a crystallographic point of view soil oxides consist of dense packings of metal atoms connected with one another by oxygen atoms in a specific steric arrangements. Figure 2 shows the surface layers of a natural oxide.

The figure clearly shows that the metal ions at the outer surface layer have a smaller coordination number than those in the inner layers (Figure 2a), and thus behave as Lewis acids. As a consequence, in hydrated conditions, the metal ions at the surface of the solid tend to coordinate water molecules (Figure 2b), which, after dissociative chemisorption, leads to the formation of an hydroxylated surface (Figure 2c) [Davidov et.al.,1964]. The number of hydroxyl groups per unit surface area (100 $Å^2$) ranges between 4 and 10 depending on the type of crystal lattice, the cleavage of planes, and pretreatments of the solid-phase [Huang and Stumm, 1973, Huang et al., 1987].

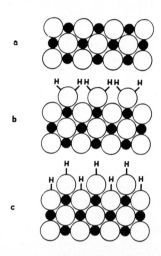

Fig.2: Cross section of the surface of a metal oxide after hydration. ● Metal ion, ○ Oxide atom (from Schindler P.W., 1981).

Upon hydration the hydroxyl groups undergo protonation-deprotonation reactions by specific adsorption of H^+ and OH^-:

$$H_2O + O^- \underset{K_{a2}}{\overset{OH^-}{\rightleftharpoons}} \overset{H}{O} \underset{K_{a1}}{\overset{H^+}{\rightleftharpoons}} \overset{H \quad H}{O^+} \qquad (1)$$

$$\text{cation exchanger} \qquad\qquad\qquad \text{anion exchanger}$$

Accordingly, oxides behave as amphoteric ion exchangers depending on the pH of the solution. They also possess a specific pH of the zero point of charge. At this pH the positive and negative charges resulting from the protonation-deprotonation reactions balance one another.

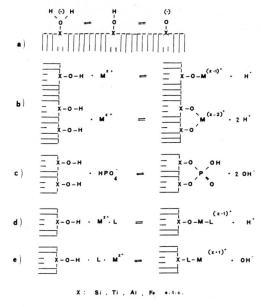

X : Si , Ti , Al , Fe e.t.c.

Fig.3 : Coordination phenomena at
the water-oxide interface
(from Schindler P.W., 1981).

Figure 3 shows a general scheme of the main reactions which can potentially occur at the surface of hydrated oxides. Three types of reaction may be distinguished (i.e., protolytic, complex formation and ligand exchange) [Schindler, 1981].

Clay minerals, together with zeolites, play the most important role in determining ion exchange behaviour in soils and subsoils. Their ion exchange properties stem from either a large exposed surface area or an open crystalline lattice that makes charges in its interior accessible to ions from solution.

Although a large variety of different classes of alumosilicates may contribute to the clay composition of a soil (see Table 1) the most important class in terms of ion exchange properties are the phyllosilicates.

From a crystallographic point of view phyllosilicate clays are essentially made of layers with two main types of sheet- like structures. The 2:1-type clays consist of two layers of tetrahedral SiO_4 sheets sandwiching a closely packed oxygen or hydroxide complex, $[AlO_4(OH)_2]$, in which the coordinating element (Al, also Fe or Mg) is in octahedral coordination. Thus the sheets expose to the liquid only one kind of planar surface, made of -Si-O-Si- units. The 1:1-type clays consist of a tetrahedral SiO_4 sheet sharing O-bonding with an octahedral aluminum complex, $[AlO_2(OH)_4]$. In this case the sheets expose alternatively -Si-O-Si- and Al-OH units.

The combination of different planar sheets of 1:1 and 2:1 clay platelets determines the overall physico-chemical properties of the clay, in particular its ion exchange properties. As an example, the polarity of 1:1 platelets is responsible for the agglomeration, via H-bonding, between the -OH group of one layer with the dehydrated -O- group of the adjacent layer, thus leading to the formation of large crystals. The average specific surface area of kaolinite, a typical representative of this type of clay, is in the order of 10-100 m^2/g, whereas the specific surface area of typical 2:1 clays, with more disperse platelets, is at least ten times larger ($>500-800$ m^2/g).

The source of permanent negative charges, resulting in the cation exchange behaviour of clays stems, from the isomorphic substitution of Mg^{2+} and Al^{3+} or Si^{4+} by equivalent quantities of ions of smaller charge, e.g., Li^+, Mg^{2+} or Fe^{2+}, Al^{3+} respectively. The substitution is possible, provided that the interchanging ions have similar crystalline radii. As a consequence an excess of negative charge on the substituting ion appears. Electroneutrality requires this charge to be compensated by an equivalent amount of cations from the adjacent liquid-phase.

The association of the clay platelets (extension of exposed surface area) and the degree of isomorphic substitution in the crystal lattice (number of charges) are the parameters influencing the surface charge density of the clays and thus the cation exchange properties. For example, a highly polar material such as kaolinite with a 1:1 structure, strong association of platelets, and low degree of substitution of other metals for Si has almost no ion exchange capacity. On the other hand, highly dispersed less polar montmorillonite and vermiculite, which are typical 2:1 clays having a high degree of substitution and large surface area (800 m^2/g), possess a high ion exchange capacity ranging from 50 to 80 meq/100 g.

Table 2 shows some typical variations and range of values of CEC and surface areas in reference to different clay minerals.

Table 2. Cation Exchange Capacities, surface areas and surface charge densities for some clay minerals.

Mineral	Surface area (m^2/g)	Exchange capacity $(meq/100 \text{ g})$	Charge density (meq/m^2)
Allophane	500-700	50-100	$1-2 \times 10^{-3}$
Kaolinite	10-20	2-6	1-6 "
Illite	90-120	20-40	1.5-4 "
Smectite	800	60-120	0.75-1.5 "
Vermiculite	800	120-200	1.5-2 "
Montmorillonite	800	50-80	1.0 "
Hydrated Fe and Al oxides (pH 8) (for comparison)	50	0.5-1	0.1-0.5 "

Zeolites are a class of minerals with remarkable adsorption and ion exchange properties. Their natural occurrence in sedimentary formations is geographically localized. For example they are localized in the U.S., (Nevada, Oregon, Arizona), and Europe (Hungary, Italy, Russia).

Zeolites have the general chemical formula: $(MeO)_{x/2}(AlO_2)_x(SiO_2)_y \cdot zH_2O$ in which Me are exchangeable alkaline-earth or alkali metal ions and z is a nonstoichiometric number of water molecules retained in the crystalline lattice. Six substrates, Mordenite, Erionite, Chabazite, Clinoptilolite, Phillipsite, and Heulandite, assume importance in terms of ion exchange properties. Depending on the SiO_2/Al_2O_3 ratio in the crystal lattice (usually ranging from 2 to 10 in natural materials), zeolites have different physical and physico-chemical properties. Materials with higher Si:Al ratios are mechanically stronger, less hydrophylic and have a lower ion exchange capacity than materials with lower Si:Al ratio [Breck, 1977].

In other words, zeolites are three-dimensional alkaline or alkaline earth alumosilicates with a regular, rigid network of channels intersecting the lattice (Figure 4a) [Peterson 1965, Barrer 1963]. As in clays, the isomorphic substitution of Si^{4+} by Al^{3+} ions determines the excess negative charge responsible for the cation exchange properties. As opposed to clays, zeolites have a more rigid crystalline structure with "cavities" connected by "windows" of strictly controlled apertures in the range of 3 to 10 Å, through which water and hydrated ions freely diffuse (Figure 4b). The ion exchange selectivities of these materials is caused by the exclusion of hydrated ions that are too large to pass through the windows (sieving effect).

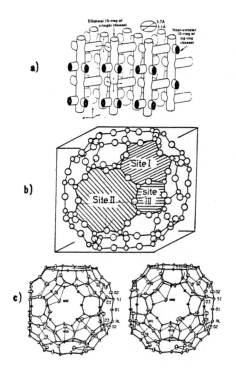

Open crystal lattices with cavities and windows arise because of the tendency of aluminate and silicate tetrahedra to share oxygen atoms to form four-, five-, six-, eight-, or twelve-membered rings (even up to sixteen-membered in some synthetic zeolites) [Smith 1980, Peterson 1965] (Figure 4c).

Fig.4 : Crystalline structure of zeolites with channels and windows of controlled size.
(from Smith J.V., 1980)

Generally speaking, the properties on which the sorption behaviour of zeolites mainly depends can be summarized as follows [Breck 1977]:

a) The basic framework structure in terms of internal pore volume determining the size exclusion of ions.

b) The exchange selectivity of mobile cations depending on the location of charged functionalities in the structure and on the charge density of counterions (i.e., net charge and hydrated volume).

c) The uniform internal pore structure leading to many different situations especially when the size of the hydrated molecular species approaches that of the windows (sieving effect).

The organic and organo-mineral fraction of soils.

The **organic fraction** of soils is related to the biochemical and chemical degradation of the organic substrates resulting from vegetal biomass. The term "humification" refers to the complex of those transformations of the biomass in soil humus which results in materials having a certain stability to further (bio)degradation.

In this context soil organic matter is classified into substrates in which the parent complex structure is still recognizable (humus and related humic substances), and substrates which have completely decayed structures. Referring to the former, because their macromolecular coiled structure, they offer a limited surface area to the liquid phase and consequently are less effective as adsorbents than they could potentially be if the entangled polymer chain were relaxed. The conformation of these polymers, however, is quite flexible so that the coils can be easily stretched by interaction with mineral fractions (clays, zeolites, oxides), this enlarges the exposed surface area and improves accessibility of ionic species to the functional groups.

Dissociation of acidic functional groups (carboxylic, carboxy-phenolic, and phenolic hydroxyls) is responsible for the cation exchange properties of these materials. Dissociation depends on solution pH, the pKa value of the functional groups, and from the synergistic effects of the nearby functionalities. A fairly continuous range of pKa values of dissociation is present due to the complex molecular structure of the compounds and to the large number of functional groups distributed on the polymer chain. Therefore it is quite difficult to define a pH range in which the structure can or cannot be considered completely dissociated.

Another important feature of these substrates is the easy formation of metal-complex structures, because the above mentioned functional groups are strong ligands for most transition metals. Of paramount importance is the spatial geometry of the functional groups on the polymeric matrix; this geometry has to fit the coordination geometry of the transition metal.

In the **organo-mineral complex** the organic fraction plays a key role in determining ion exchange properties. This is because the functional groups associated with the inorganic components are essentially engaged in binding the organic substrates to the mineral surface. The organic surface layer constitute an intricate entangled structure through which the ions migrate and interact with the charged sites. The layer is subject to sensible volumetric variations (swelling-shrinking) of the matrix, which influences the general kinetic and thermodynamic properties of these materials.

The selectivity sequence toward ionic species, as well as overall exchange capacity of the components (organic and inorganic) taken individually can be considerably modified [Talibudeen 1981, Bunzl 1991], depending on the synergistic effects of the functional groups.

Carboxylic and phenolic groups mainly contribute to the exchange properties of these systems. In the common pH range (3.5-8) encountered in soils both groups are dissociated. Depending on pKa values of the phenolic hydroxyl groups, these groups can be active at pH higher than 7.0, whereas carboxylyc groups are substantially dissociated at more acidic pH values.

The pH dependence of charges.

Cation exchange.

With the only exception of the negative permanent charges due to the isomorphic substitution in the crystal lattice of clays and zeolites, which can reasonably be considered fully dissociated in the entire pH range, all the other charges in natural porous materials are associated with weak (mainly acidic) functional groups whose dissociation is strictly dependent on pH. Permanent charges on the alumosilicate structure allow the ion exchange properties to be explicated in almost the entire pH range, with the only exception being the highly acidic environments (pH < 2.0) where hydrolytic reactions tend to break-down the crystal lattice [Barrer 1978]. These drastic conditions, however, are not common in the natural systems under consideration.

pH dependent negative charges are associated to the carboxyl (pKa ≈ 3-6) and hydroxyl (pKa ≈ 9) groups of soil organic matter, as well as with the mineral fraction after surface hydration of the oxides. Equilibrium pK values for oxides in reference to protolitic reactions (1) range from 3 to 9 depending on the type of oxide in consideration. Lower pK ranges occur in aluminum oxide (gibbsite) and ferric oxide (goethite) than in silica oxides [Schindler 1981, Davis 1978]. pH is generally adopted as the master variable controlling the extent of the adsorption phenomena on oxides [Huang, 1987]. Accordingly, cation adsorption begins at a specific pH value and reaches a maximum over a pH range.

This range of pH is termed the "retention (or adsorption) edge" and is the result of concurrent phenomena such as hydrolisys and precipitation of the interacting metal ions, as well as
to the protonation-deprotonation properties of the oxide. [Kinninburgh, 1981, Huang,
1987).

Variation of the density of charges on the solid phase, even higher than that potentially
associated with the complete dissociation of the weakest groups (pKa ≈ 12), is detected on
some mineral components when dispersed in solutions at pH > 12-13. This is the effect
associated to the increased dispersion of the soil minerals in the liquid-phase thus exposing
larger surface areas for potential interaction with ionic species [Greenland 1978].

Anion exchange.

Anions are predominantly excluded by Donnan potential from natural exchangers. Calculations based on thermodynamic considerations confirm that, under prevailing conditions encountered in natural soils, the anion exchange capacity is only 1-5% of the cation exchange
capacity with peaks of up to 15% for saline soils [Bolt, 1978].

Soil particles, however, although predominantly negatively charged, may also carry
positive charges. The proton accepting capacity of hydrated hydroxyls at the surface of
oxides increases considerably as the solution pH drops below the isoelectric point of zero
charge. The main source of positive charges for anion exchange is thus essentially associated with the protonation of hydroxyl groups of hydrated oxides, principally those of alu-
minium and iron. Other sources of anionic functionalities are associated with the edges of
clay minerals particles, and to the protonation of amino groups present in the organic fraction of the soils.

Retention of anions is very sensitive to pH and ionic strength; the selectivity of the positively charged sites toward anions is much higher than the selectivity toward cations.
Accordingly, experimental evidence shows that in natural oxides silicate and phosphate ions
are, for example, strongly preferred over sulphate nitrate and chloride ions, in that order
[Talibudeen, 1981]. In temperate regions where silicate and phosphate ions are commonly
present in soils, it is unlikely that sulphates and chlorides would be retained except in
presence of high contents of Al and Fe oxides and low pH values, a situation which is more
common in tropical than in temperate regions.

Cation exchange capacity.

Ion exchange phenomena in natural permeable media can be schematically represented by:

$$\text{Solid-phase - A} + B^{n+} \rightleftharpoons \text{Solid-phase - B} + A^{n+} \tag{2}$$

where A and B are the exchanging counterions. The "reaction" is reversible and, unlike other "sorption" phenomena, must be a stoichiometric process, because of the preservation of the electroneutrality condition of the system (i.e., an equivalent amount of charges have to transfer at the liquid-solid interface).

The total amount of exchangeble cations per unit weight of solid-phase is, by definition, the Cation Exchange Capacity, Q, of the exchanger and is usually expressed in terms of milliequivalents per 100 g of dry solid-phase (meq/100g). In other words it represents the overall quantity of charges available (as exchanging sites) on the exchanger material. It is related to the surface density of charges, Γ, through the specific surface area, S,:

$$Q = S \times \Gamma \tag{3}$$

For montmorillonite, kaolinite, and illite, respectively, typical values for Γ are 1×10^{-3}, 2×10^{-3} and 3×10^{-3} meq/m^2, and for S are in excess of 700, 15 and 100 m^2/g. Accordingly, average figures for the cation exchange capacity for clay materials range in the order of tens of meq/100 g (see Tab.II). For comparison, average figures for humic materials range in the order of several hundreds of meq/100 g.

As previously defined the cation exchange capacity corresponds to the overall quantity of exchangeable cations in the solid-phase. As the solid phase contains also soluble salts, the cations released by dissolution from the solid phase must be subtracted from the total amount to have a reliable figure for the exchangeable cations. Accordingly, it is a good practice to consider the equivalent quantity of cations associated to anions in the surface-phase as those resulting from dissolution phenomena. Thus the cation exchange capacity of the solid-phase can be determined as:

Q = (Equivalents of cations in the liquid-phase) - (Equivalents of anions in the liquid-phase)

In the "surface phase" the diffuse cloud of ions adjacent to the solid phase (see below) includes cations and anions for an equivalent net charge compensating the negative charge of the solid-phase. Accordingly, the excess of cations in the volume of surface solution bounded between the solid surface and double layer thickness constitutes the Cation Exchange Capacity. The deficit of anions resulting from Donnan esclusion in the same volume of solution is termed "negative adsorption" of anions.

Thermodynamics of ion exchange in natural permeable media.

Ion exchange equilibria have been extensively studied over the last several decades. Early theoretical formulations described equilibrium in terms of the mass action law, and other approaches considered ion exchange as an adsorption phenomena, whereas strictly thermo-dynamic approaches incorporated the effects of physico-chemical properties of the system on the state of equilibrium [Helfferich, 1962].

Prediction of ion exchange phenomena is essentially based on the Gibbs-Donnan model which assumes the equality of the electrochemical potentials of the species in the solid- and in the contacting liquid-phase [Gregor, 1948, 1951]. Also, empirical correlations are used by some authors [Hogfeldt, 1984; Boyd and Bunzl, 1967]. More recently, surface excess theories, traditionally adopted by soil chemists, have been applied to ion exchange equilibria on reactive polymers [Marton and Inczedy, 1988; Horst and Hoell, 1990].

The Electric Double Layer Model and general thermodynamic considerations

The interaction of a solution with a natural porous material leads to the equilibration of the respective thermodynamic potentials. Thermodynamic potentials are related to structural rearrangements of the interacting substrates on each side of the plane of contact. In other words, the boundaries of the phases in contact are both modified in their structure and composition, thus it is reasonable to assume a new phase, "the surface phase" with its own peculiarities.

Referring to the ionic composition at the surface, ions dissociated from the solid tend to remain in the vicinity of the interfaces and cause an electrostatic interaction which pre-vents free diffusion toward the bulk solution. The overall interactive energy among the charges (electric potential) and the resulting electric field will depend on the relative dis-tance of the ions from the surface of the solid. The electric field will attract ions from the bulk solution which will accumulate at the liquid-solid interface.

The distribution of the ions in the surface phase is interpreted in terms of enthalpy, entropy and overall free energy of the system.

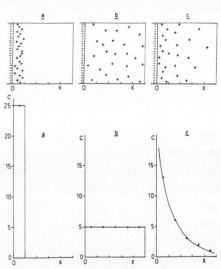

Fig.5 : Distribution of ions in the Diffuse Double Layer.
a: condition of minimum energy,
b: condition of minimum entropy,
c: condition of minimum free energy.

As shown in Figure 5 distribution of ions tends spontaneously toward the condition of minimum energy (condition a) on one side, and toward the condition of maximum entropy (condition b) on the other side. Equilibrium corresponds to the minimum free energy and will be reached when cations diffusively distribute with decreasing concentration away from the solid surface, thus establishing a "Diffuse Double Layer" (DDL) of ions adiacent to the surface of the solid-phase. The diffuse double layer corresponds to the zone near the solid surface in which the concentration of ions is different from that of the bulk solution.

The distribution of ions in the diffuse double layer is statistically described by the Boltzmann equation:

$$C/C° = \exp(- E/kT) \tag{4}$$

where E is the difference in electric potential induced by the same ionic distribution, kT is the average kinetic energy of ions, k the Boltzmann constant, espressed as the ratio of the Universal gas constant and the Avogadro number, C and C° are respectively the concentrations of the ion of interest in the double layer and in the bulk solution.

The width of the diffuse double layer, or "Gouy layer", is variable and depends on conditions in the liquid-phase, in particular the ionic strength, surface charge of the solid, charge densities of counterions, and dielectric constant of the medium. The thickness of the Gouy layer varies from a minimum of 10 Å in concentrated Ca^{2+} solutions ($\approx 1M$), to hundreds of Angstroms (500 Å) as in the case of monovalent ions in $< 10^{-3}$ M solutions. An empirical equation given by Schofield [1947] accounts for most of the mentioned parameters in calculating the average width of the diffuse ionic atmosphere:

$$x = 1/(z\beta)^{0.5} (q/C^{0.5} - 4/(z\beta)^{0.5}\Gamma \tag{5}$$

where x is the width of the DDL, q is a factor depending on the cation/anion charge ratio, z is the cation charge, Γ is the surface charge of the solid-phase, C is the liquid-phase concentration, and ß is a constant accounting for the temperature and dielectric constant of the medium.

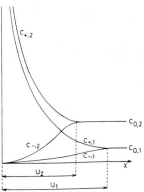

Figure 6 reports typical concentration profiles in the Gouy layer for cations and anions at two different bulk concentrations.

Fig.6 : Concentration profiles of ions in the Diffuse Double Layer.
(from Bolt, 1978).

An inner two dimensional layer of molecular thickness, the so-called Stern layer, also accomodates cations or anions at specific binding sites; such ions are held at molecular distance from the surface with specific strong electrostatic bonds. The presence of these ions strictly bound to the surface influences the effective electrical potential in the diffuse double layer and consequently varies the overall ion exchange capacity (either negatively, by subtracting functional groups to the cation interaction, or positively, by binding ligands carrying negative charges).

The retention of the ionic species in the Stern layer depends on the ion-functional group electrostatic energy, the relative concentration of ions in the bulk solution, and the dehydration energies of the counterions. These parameters are typical of the classical ion exchange theory as described by Gregor [1948], Eisenman [1961], Marinsky [1961]. In this context it is worth mentioning here that a model traditionally applied to the description of ion exchange phenomena on mineral surfaces, i.e., the surface complex formation model [Schindler, 1981], has been recently applied to the description of equilibria on ion exchange synthetic polymers [Horst et al., 1990].

In general terms, ion exchange phenomena in natural porous media are governed by electrostatic interactions of counterions modified by geometrical factors related to the hydration and polarizability of hydrated ions which, in turn, influence the distance of closest approach of the charges at the exchanging site. Accordingly, three contributing terms can reasonably be assumed as the most important in determining the free energy of exchange; these are related to the coulombic (coul) interaction, to the hydration (hydr), and to the polarizability of ions (pol).

$$G_{ex} = G_{coul} + G_{hydr} + G_{pol} \qquad (6)$$

Good references for a comprehensive mathematical description of ion exchange physicochemical models, with particular emphasis upon the thermodynamics of the mineral water interface, are the books by Bolt [1982] and Sposito [1981].

Highly selective retention of cations in natural porous media.

It is well known from ion exchange equilibrium theory [Eisenman, 1962] that a stronger electrostatic interaction (ion-functional group) is manifested toward less hydrated ions. The fitting of the hydrated ions into the structural cavities of the exchanger is another important factor, as well. This latter contribution is much more pronounced in determining the selectivity of ion exchange in natural systems.

Hydration energy plays an important role in determining ion exchange selectivity in natural exchangers. The lighter ions, having larger and more polarizable hydration shells, are thus unable to penetrate the structural windows and cavities which open through the crystalline lattice. Accordingly among alkaline ions the selectivity is generally manifested in the order:

$$Cs^+ > Rb^+ > K^+ > Na^+ > Li^+ \tag{7}$$

A second factor determining ion exchange selectivity in soil minerals is directly related to the isomorphic substitution in alumosilicate derivatives. The isomorphic substitution between lighter ions (i.e., Mg-Al or Li-Mg with respect to Al-Si) causes the residual negative charge to be located on a smaller ion thus allowing for closer proximity of the charges to the surface layer. As a consequence a stronger electrostatic interaction, i.e., greater affinity, will be implemented between counterions and charged site.

A third factor for cation selectivity is the occurrence of faults, cleavages and cracks in the crystalline lattice. These structural faults open additional possibilities for local isomorphic substitutions, and, at the same time, create room for specific fitting of hydrated ions. The spatial location of the acidic functional groups is another factor affecting selectivity. As already mentioned, the position of the functional groups on the solid matrix has to match with the coordination geometry of the exchanging ions. Moreover, the extent of dissociation phenomena of functional groups is related to the environment in which they are inserted in the solid matrix. Single functionalities influence each other and their respective dissociation depends also on the relative position on their solid matrix. This latter consideration, of course, applies primarily to organic substrates.

In general terms, since the solid matrix is a rigid crystalline network not all the potentially exchanging sites are available for interaction with counterions. Free migration of ions is strongly limited by steric hindrances (e.g., size exclusion in zeolites). Accordingly, factors determining the volume of hydrated ions (hydration-dehydration energy) and the presence of specific interacting sites of particular geometry favoring the selective fitting of ions play a major role in determining general ion exchange selectivity.

Specific retention of cations in clay lattices.

Very high selectivities toward specific ions leading to almost irreversible reactions are often found in in clay minerals. The high affinity is, once again, attributed to geometrical factors. With reference to 2:1 packings, the layered alumosilicate structure has specific niches open for ions in between the layers. The fitting of the hydrated ions into these cavities will strongly depend on the dehydration energy (the larger the ion the lower the energy) and from its polarizability. Accordingly, among alkaline metals, for example, the "softer" Cs^+, Rb^+ and K^+ are strongly preferred over Na^+ and Li^+, the latter ions having a stronger hydration energy and a lower polarizability. The manifestation of this phenomenon is much more pronounced in phyllosilicates (illites, vermiculites) where the layered structure hosts very selectively potassium ions in the hexagonal spacings between the layers. Potassium ion is retained almost irreversibly in a closed structure between the packings. Only in particularly drastic conditions (high ionic strength and liquid phase K concentration) does a form of lattice relaxation allows these ions to be at least partially released.

Selective ion exchange reactions in zeolites.

Chemically zeolites have the same composition as clays, but they differ essentially in their physical structure (see Figure 4). As already mentioned, the crystal lattice of zeolites is made of a rigid network of tetrahedral aluminate and silicate anions linked by sharing oxygen atoms. Exchanging cations fill the cavities to achieve electroneutrality; water fully permeated the structure.

The silicate/aluminate ratio in the crystallographic cell yields different cavities of rigidly controlled dimensions open in structure (see above). The cavities behave differently in terms of site specificity toward ions and the related free energy of exchange. The stability of the ion-functional group complex is dependent on the dimensions of the accomodating cavities. It has been demonstrated that in reference to most zeolites 6-membered ring complex are more stable than 8-membered and 4-membered rings respectively [Barrer 1978, Maes and Cremers, 1986].

The structural water determines the hydration phenomena of exchanging ions; accordingly, the smaller ions with a denser surface charge will hold their hydration shell more tenaciously than the larger ions, thus hindering the electrostatic interaction. The larger ions will have a greater affinity of exchange onto zeolites, provided they fit into the channels of the porous structure.

Finally, the distance between permanent charges is another factor influencing the specificity of exchange. The longer the distance between charges the stronger the affinity for monovalent cations with respect to divalent cations.

Conclusions

The knowledge of the physico-chemical properties of the solid-liquid interface in natural permeable media is still advancing. The main problems reside essentially in the complexity of the systems and in the characterization of the phases constituting the composite media.

Due to the synergistic effects of the different contributing phenomena at the liquid-solid interface, however, the understanding of the overall phenomenon by the observation of the integrated behaviour of the real world is definitely a formidable task. Accordingly, an accurate characterization of the solid-phase, together with experiments on simplified systems (may be purified clays, oxides or humic fractions) may be a preliminary step. Reconstruction of the natural porous media by mixing balanced fraction of the pure minerals and comparing their behaviour to real samples would be a subsequent step toward approaching the real world.

References.

Barrer RM, Bartholomew RF, Rees LCV (1963) Ion exchange in porous crystals. II. The relationship between self and exchange diffusion coefficients. J Phys Chem Solids 24:309-317

Barrer RM (1978) Zeolites and Clay Minerals as Sorbents and Molecular Sieves. Academic Press, London.

Barrer RM (1980) Zeolites exchangers some equilibrium and kinetic aspects. Proc.5th Int.Conf. Zeolites, Naples 2-6,1980.

Bolt GH, Bruggenwert MGM (1978) Soil Chemistry Vol.5A Basic elements. Elsevier Sci Pub Co Amsterdam.

Bolt GH (1982) Soil Chemistry Vol.5B Physico-chemical models. Elsevier Sci Pub Co Amsterdam.

Bowden JW (1973) Generalized model for anions and cations adsorption at oxide surfaces, Nature 245:81-83.

Boyd GE, Bunzl K (1967) The Donnan equilibrium in cross-linked polystyrene cation and anion exchangers, J Am Chem Soc 89:1176-1180.

Breck DW (1977) Advances in Adsorption. Chem Eng Proc Oct 1977:43-64.

Bunzl K, Schimmack W (1991) Rates of Soil Chemical Processes S.S.S.A. Special Pub n°27, Soil Science Soc. of America, Madison, WI. Chap 5

Davis JA, James RO, Leckie JO (1978) Surface ionization and complexation at the oxide-water interface computation of electric double layer properties in simple electrolytes J.Colloid Interface Sci 63:480-499.

Davis JA, Hayes KF (1986) Geochemical properties at mineral surfaces. ACS Symp Ser 323: Chap 1 Davydov V, Kieselev AV, Zhuravlev LT (1964) Surface and bulk hydroxyl groups of silica by infrared spectra and D_2O exchange. Trans Farad Soc 60:2254.

Dugger DL (1964) The Exchange of twenty metal ions with the weakly acidic silano group of silica gel. J Phys Chem 68:757-760.

Eisenman G (1961) On the elementary atomic origin of equilibrium ionic specificity. Symp.Membrane Transport and Metabolism, Acad.Press NY, p.163-179.

Greenland DJ, Mott CJB (1978) Surfaces of soil particles, in Greenland DJ, Hayes MBH (eds) The Chemistry of Soil Constituents. J.Wiley, New York, NY.

Gregor HP (1948) A general thermodynamic theory of ion exchange processes. J Amer Chem Soc 70:1293-1305.

Gregor HP (1951) Gibbs Donnan equilibria in ion exchange resin systems. J Am Chem Soc, 70:1293-1305.

Hogfeldt E (1984) A useful method for summarizing data in ion exchange I. Some illustrative examples. React Polym 2:19-30.

Huang CP, Stumm W (1973) Specific adsorption of cations on hydrous Al_2O_3, J Colloid Interface Sci 43:409.

Huang CP, Hsieh YS, Carapcioglu MO, Bowers AR, Elliot HA (1987) Chemical interactions between heavy metal ions and hydrous solids. in Patterson JW, Passino R. (eds). Metals Speciation Separation and Recovery. Lewis Pub Inc, Chelsea, MI, USA.

Helfferich F (1962) Ion Exchange McGraw-Hill NY,NY. Chap. 5 (out of print, available from University Microfilms, Ann Arbor, MI,USA

Horst J, Hoell W, Eberle SH (1990) Application of the surface complex formation model to exchange equilibria on ion exchange resins. Part I. Weak acid resins. React Polym. 13:209-231.

James RO, Healy TW (1972) Adsorption of hydrolyzable metal ions at the oxide water interface. J Colloid & Interf Science 40:53-64.

Kinninburgh DG, Jackson ML (1981) Cation adsorption by Hydrous metal oxides and clay, in Anderson MA, Rubin J (eds) Adsorption of inorganics at solid-liquid interfaces Ann Arbor Sci Pub Co. Ch.3.

Maes A, Cremers A (1986) Highly selective ion exchange in clay minerals and zeolites, in Davis JA, Hayes KF (eds) Geochemical Processes at Mineral Surfaces. ACS Symp Ser n°323. Chap.13

Marinsky JA (1961) Prediction of ion exchange selectivities, J Phys Chem 71:1572-1578.

Marton A, Inczedy J (1988) Application of the concentrated electrolyte solution model in the evaluation of ion exchange equilibria, in Streat M ed. Ion Exchange for Industry, Ellis Horwood Chichester, p.326-329.

Morel FMM (1981) Adsorption models. A mathematical analisis in the framework of general equilibrium calculations, in Anderson MA, Rubin AJ (eds). Adsorption of inorganics at the solid-water interface, Ann Arbor Sci Pub MI USA.

Page AL (1984) Fate of wastewater consituents in soils and groundwater: Trace elements, in Pettigrove S, Asano T (eds) Irrigation with municipal reclaimed wastewaters. A guidance manual. Lewis Pub Co., MI, USA

Peterson DL, Helfferich F, Blytas GC (1965) Sorption and ion exchange in sedimentary zeolites, J Phys Chem Solids, 26:835-848.

Schindler PW (1981) Surface complexes at oxide-water interfaces, in Anderson MA, Rubin JA (eds) Adsorption of Inorganics at Solid-Liquid Interfaces. Ann Arbor Sci Pub Inc Ann Arbor, MI,USA

Schofield RK (1947) Calculation of surface areas from measurement of negative adsorption, Nature,60:408-10.

Smith JV (1980) Review of new crystal structures and mineralogy of zeolites and related materials. Proc.5th Int.Conf.on Zeolites, Naples 2-6, 1980.

Sposito G (1981) The Thermodynamics of Soil Solutions, Oxford Univ.Press, Oxford. Stumm W, Huang CP (1973) Adsorption of cations on hydrous Al_2O_3, J Colloid and Interf Sci, 43:409-420.

Talibudeen O (1981) Cation Exchange in soils, in Greenland D.J. Hayes MBH (eds) The Chemistry of soil Processes J.Wiley, NY,NY.

Interactions of Toxic Organics with Subsoils Components

A. Lopez, D. Petruzzelli
Istituto di Ricerca Sulle Acque
Consiglio Nazionale delle Ricerche
5, Via De Blasio
70123 Bari
Italy

Technological progress has certainly made life more pleasant but it has also increased the danger of environmental pollution. Groundwater is among the most seriously threatened environmental resources. Although groundwater contamination has occurred for centuries, population demands and agricultural activities, as well as increased industrialization, have certainly exacerbated the problem. Much of the concern over groundwater quality is due to the occurrence of anthropogenic organic compounds in these water resources; this concern is reflected by the U.S. priority pollutants list in which organic compounds are the majority $(114/129 \approx 88\%)$ [Callahan et al. 1979].

For such contaminants the route of entry into the subsurface may include agricultural practices, accidental spills, and improper surface and subsurface disposal of liquid or solid chemical wastes. The behaviour of organic chemicals in soils, subsoils and aquifers materials is governed by a variety of bio-physico-chemical processes among which *sorption* is one of the most relevant [Weber et al. 1991]. Actually, sorption is the result of different types of forces, attractive as well as repulsive, among each component of the ternary system, contaminant/liquid-phase/solid-phase; these forces together contribute to cause different pollutant concentrations at the liquid-solid interface and in the bulk solution. Because of soil heterogeneity it is difficult to distinguish among different types of specific interactions. The term sorption is then used to describe any accumulation of dissolved contaminant by the solid phase, regardless of the specific forces and/or mechanisms causing such accumulation.

The ability to predict the nature and extent of sorption phenomena is essential for solving serious groundwater pollution problems. Several attempts to find reliable relationships between sorption extent and physico-chemical properties of each component of the ternary system (solute-solvent-sorbent) have been made. However, except for highly hydrophobic pollutants and high organic content soils [Karichoff et al. 1979, Means et al. 1980, Chiou et al. 1979, Schwarzenbach et al. 1981, Briggs 1981], such attempts have only been partially successful owing the variety of conditions in natural environments.

NATO ASI Series, Vol. G 32
Migration and Fate of Pollutants in Soils and Subsoils
Edited by D. Petruzzelli and F. G. Helfferich
© Springer-Verlag Berlin Heidelberg 1993

In the subsurface environment, where the organic carbon content is usually very low, the factors that play the main role in determining sorption extent are the physico-chemical properties of each component of the above ternary system. Moreover, interactions between organic contaminants and inorganic solid phases are often the result of specific forces as chemical bonds, ion-exchange, hydrogen bonds, etc.. Accordingly, to evaluate their nature and magnitude, a case by case accurate characterization of the ternary system is necessary (e.g., pH, ionic strength, temperature, sorbent and solute physico-chemical properties).

In this chapter the main physico-chemical properties of each component of the solute-sorbent-solvent system to be taken into account during such characterization will be reviewed, and references to in-depth studies in each area will be provided.

Physico-chemical properties of organic contaminants (solute).

The global production of anthropogenic compounds is estimated to be about 300×10^6 tons/year. As for organics, their environmental fate will mainly depend upon their chemical structure which, in turn, will determine their physico-chemical properties. From the environmental pollution standpoint, the most important categories of organic contaminants can be restricted to the following classes: mineral oil products (complex mixtures of predominantly aliphatic and aromatic hydrocarbons), combustion products of different materials (polyciclic aromatic hydrocarbons, dioxins), chlorinated compounds (solvents, pesticides, polychlorinated biphenyls, polychlorinated phenols, trihalomethanes), detergents (surfactants, bleaching agents, optical brighteners), and plasticizers (phtalates). Typical chemical structures of some of these classes are shown below:

(PCB)   (PHTALATES)

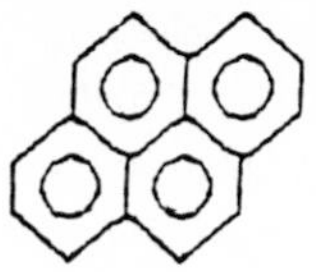

PYRENE (PAH)

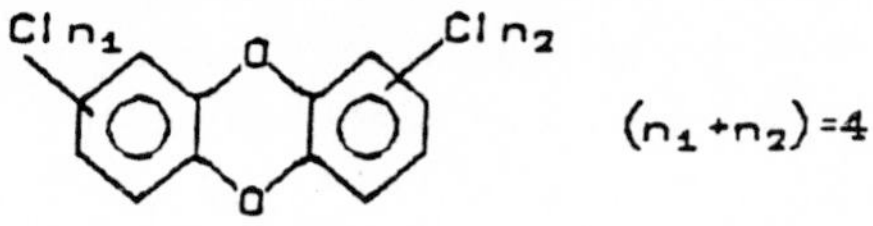
Cl n1
Cl n2
(n1+n2)=4

TETRACHLORODIBENZODIOXINS (TCDD)

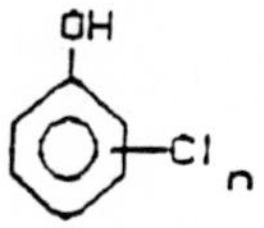
OH
Cl n

(PCP)

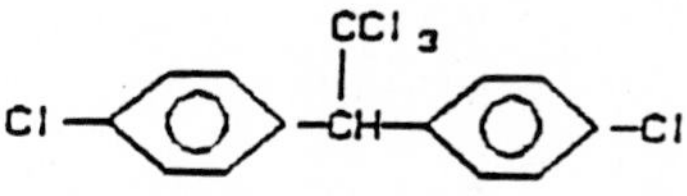
CCl3
Cl
CH
Cl

DDT

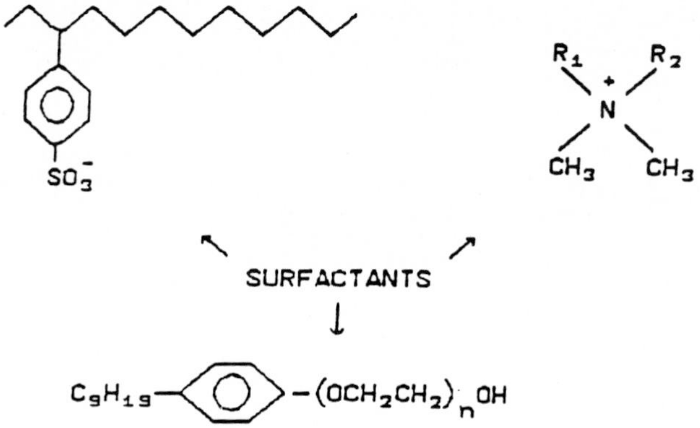
SO3-
R1
R2
N+
CH3
CH3
SURFACTANTS
C9H19
(OCH2CH2)nOH

As for their physico-chemical properties, the following set can be considered a real "fingerprint" :

- Melting point;
- Boiling point;
- Acidity constant;
- Vapour pressure;
- Water solubility;
- Henry's law constant;
- Octanol/Water partition coefficient.

A detailed treatment of the relative importance of each of the above parameters in determining the environmental distribution of a contaminant, together with their physico-chemical definitions and numerical values are reported in specialized texts [Lyman et al. 1982, Weast et al. 1990, Howard 1989]. In the above list, the last four parameters are representative of equilibria between two phases and provide an idea about the compound's compatibility with one phase with respect to the other.

In addition to the above properties, there are some others (dipole moment, polarizability and Hammett constant) related to the behaviour of electrons inside the molecule that are particularly significant in the case of low energy interactions (van der Waals, dipole-dipole, hydrogen bonding).

The **dipole moment** (μ) is a measurement of the displacement of the centres of gravity of positive and negative charges of a molecule. It is the product of a charge (q) times a distance (r) and is important because it provides a means for comparing the electrical dissymetry of a molecule. It is caused by the different electronegativities of each atom present in the molecule. Many molecules having atoms with different electronegativities, however, are not polar ($\mu=0$) because of their steric symmetry (e.g. CCl_4). Of course, the presence of a permanent dipole in a molecule will favour interactions (ion-dipole, dipole-dipole, dipole-induced dipole) with sites which are electrically charged or characterized by their own permanent or induced dipole moment. The energy associated with such interactions is inversely proportional to the distance (r^n) separating the interacting molecules [Israelachvili 1985]:

$$\text{ion-dipole} \quad \alpha \quad (1/r^2)$$
$$\text{dipole-dipole} \quad \alpha \quad (1/r^3)$$
$$\text{dipole-induced dipole} \quad \alpha \quad (1/r^6)$$

Applying an external electrical field to an atom or a molecule will result in an induced electrical dipole moment. The higher the intensity of the applied external electrical field (E) the greater the induced dipole moment (μ): $E = \alpha \cdot \mu$. The constant α is called the **polarizability** of an atom or a molecule and describes the response of its electron cloud to an external electric field [Kortum 1966]. The value of α is obtained, to a good approximation, by multiplying [$3/(4\pi N)$], where N is the Avogadro's number, times the compound's molar refraction R, which is defined as:

$$R = \frac{(n^2 - 2)}{(n^2 + 2)} \cdot (M/d)$$

where: **n** = refractive index; **M** = molecular weight; **d** = compound's density.

The polarizability appears in many formulas for low energy processes involving the valence electrons of a molecule as van der Waals interactions.

The **Hammett constant** (σ) is a dimensionless number that characterizes the electron withdrawing or releasing effect of a substituent in an aromatic molecule, as that sketched on the right, where R is a reaction site in the side-chain of a benzene ring and X is a meta or para substituent [Jones 1979]. The ortho position is excluded because of possible specific steric interactions between the reaction site and substituent. If the acidity constant (Ka) of substituted benzoic acids are compared with that of unsubstituted benzoic acid, it is possible to define σ as:

$$\sigma = \log \ \frac{Ka(X)}{Ka(H)}$$

σ values for selected substituents are given in the following list:

Substituent	σ_m	σ_p
$-CH_3$	-0.07	-0.17
$-F$	+0.34	+0.06
$-Cl$	+0.37	+0.23
$-Br$	+0.39	+0.23
$-I$	+0.35	+0.18
$-OH$	+0.12	-0.37
$-NH_2$	-0.16	-0.66
$-NO_2$	+0.71	+0.78
$-OCH_3$	+0.12	-0.27
$-COOH$	+0.37	+0.45

For each substituent the magnitude of σ will depend on its position (meta or para) in the molecule. Electron withdrawing substituents have positive σ values, electron releasing substituents have negative σ values. Any substituent can withdraw or release electrons by inductive or mesomeric effects. For some substituents these effects are opposite (e.g., -OH) and the advantage of considering σ is that it gives the overall resulting effect. A typical case showing the usefulness of σ is for predicting hydrogen bonding formation between phenolic compounds and mineral surface components [Fenn et al. 1972, Saltzmann et al. 1975]:

(A) $\quad$ X ◄ -C$_6$H$_4$-O-H·····O-(organic matter or minerals)

(B) $X \rightarrow C_6H_4\text{-}O\cdots\cdots H\text{-}O\text{-}$(organic matter or minerals)

   $\backslash$

   H

In case (A) H-bonding formation will be favoured if X is a withdrawing substituent (positive σ). Substituents with negative σ will favour the formation of hydrogen bonding in case (B).

Relevant physico-chemical properties of mineral constituents of subsoils (sorbent).

Natural occurring organic matter, minerals, air and water are the main constituents of soils and subsoils. However, it should be kept in mind that subsoil components are not discrete entities but a complex mixture which behaves as a single heterogeneous surface.

Clays, metal oxides and hydroxides, together with carbonates, sulphates, halides, sulphides, phosphates and nitrates, in crystalline as well as amorphous forms, represent the most common minerals present in subsurface environment. Among these, clays and oxides/hydroxides can be considered the most active sites available for surface interactions [Bolt et al. 1987]; these will be discussed in the following.

Clays.

Clay minerals and their surface properties are described in specialized texts [Petruzzelli's chapter in this book, Dixon et al. 1977, Sposito 1984], accordingly for our purposes we will just summarize their main characteristics:

-   Clays are made up of sheets of silica tetrahedra and alumina octahedra differently layered;

-   There are two basic types of layering: dimorphic, where tetrahedral and octahedral sheets are in a ratio 1:1, and trimorphic, where an octahedral sheet is sandwiched between two tetrahedral sheets in a 2:1 ratio;

-   The thicknesses of single dimorphic and trimorphic minerals are about 7 A° and 10 A°, respectively;

-   Because of isomorphic substitution of Al^{+3} or Fe^{+3} for Si^{+4} in tetraedral layers and of Fe^{+2} or Mg^{+2} for Al^{+3} in octahedral layers, clays are negatively charged; their cationic exchange capacity ranges typically between 2 and 200 meq/100 gr;

- Such negative charges are neutralized by hydrated cations whose exchangeability will increase with increasing hydrated radius and decreasing cation polarizability and cation counter ion charge;

- The volume of liquid phase under the influence of the electrical field caused by the charged surface of a clay is called the *diffuse electric double layer*; in this volume cation concentration is higher than in the bulk solution; the thickness of this layer is inversely correlated to the charge of the exchangeable cations and to the ionic strength of the bulk solution; in environmental conditions the double layer thickness is often less than 10 A°;

- The distance between the silicate layers is called the *basal spacing* and will depend on both the clay hydration state (dehydrated clays have smaller basal spacings than hydrated ones) and the type of cationic species trapped between their crystal lattice (clays with ammonium and potassium ions trapped have smaller basal spacings than clays with different cations); a typical basal spacing is about 10 A°;

- Water and inorganic as well as organic compounds can enter the interlayer spaces causing clay swelling; accordingly, clays can be classified as *expanding, limited-expanding* and *non-expanding*.

Exchangeable ions (Ca^{+2}, Na^+, Mg^{+2}, Al^{+3}, NH_4^+) on clay surfaces play an important role in sorption phenomena. Because of their localized electric charge, such cations polarize and/or dissociate their hydration water thus allowing several surface interactions [Mortland et al. 1970].

In the diffuse double layer region, because of negative charges on the clay surface, H^+ concentration is higher than in the bulk solution thus influencing pH-depending reactions [Bailey et al. 1968].

Sorption of organics by clays can occur in several ways:

- Positively charged organics can replace the naturally occurring exchangeable cations [Philen et al. 1970,1971];

- Anionic organics can be adsorbed at proper pH values when protonation of the amphoteric Si-OH and Al-OH groups occurs [Schofield 1949];

- Neutral as well as electrically charged organics can penetrate between two silicate tetrahedral sheets (interlayer adsorption) in expandable clays [Grim 1968].

In any case, adsorption of organic compounds on clay surfaces modifies clay properties (e.g. hydrophobicity, interlayer access, etc.), particularly the ion exchange capacity [Kown et al. 1969].

In conclusion it can be generalized that clay minerals at naturally occurring pH are negatively charged and possess exchangeable cations at their surfaces. Accordingly, such minerals present ideal sites for electrostatic interactions with positively charged chemical species.

Metal oxides and hydroxides.

Si-oxides (quartz, tridymite), Al-oxides/hydroxides (gibbsite, boehmite, diaspore) and Fe-oxides/hydroxides (goethite, hematite, limonite) are the most common minerals of this class.

The chemistry of such minerals is dominated by their amphoteric properties [Schindler et al. 1976]. In fact, because of their minimal external coordination, the metal ions present on the surface of a dry oxide can be chemically considered as a Lewis acid. Accordingly, in hydrated conditions, such under-coordinated surface metal ions can easily coordinate water molecules. In a second step called *dissociative chemisorption*, a more energetically favoured rearrangement of the hydrogen atoms of the coordinated water molecules occurs leading to a surface surrounded by hydroxyl groups (-OH). The amphoteric behaviour of such metal oxides is associated with the chemical structure of these groups (two lone electron pairs on the oxygen and one ionizable hydrogen):

$$H_3O^+ + Y\text{-}O^- \rightleftharpoons Y\text{-}OH + H_2O \rightleftharpoons Y\text{-}OH_2^+ + OH^-$$

where Y indicates a generic hydroxylated surface. As for every amphoteric compound, a hydroxylated surface will be characterized by its own pH_{zpc} value (zpc is the zero point of charge when $[Y\text{-}O^-]=[Y\text{-}OH_2^+]$). Hydroxylated surfaces become positively charged at pH values lower than pH_{zpc} and negatively charged for pH values lower than pH_{zpc}. For pure minerals pH_{zpc} values for silica, alumina and iron hydroxide are, respectively, 2, 9 and 7 [Stumm et al. 1981]. Accordingly, at neutral pH, the electrical charge of silica, alumina and iron-oxide hydroxylated surfaces will be, respectively, negative, positive and zero.

Of course, permanently charged or ionizable organics will be the class of compounds which will interact more effectively with such charged surfaces. The nature and magnitude of such interactions will mainly depend on the pH of the liquid phase. pH, in fact, will determine the protonation or deprotonation extent of surface hydroxyls, as well as the dissociation or protonation of ionizable organic compounds.

At pH$=7$, since its surface is negatively charged, silica (pH$_{zpc}=2$) will behave as a weak cation-exchanger which will easily interact with positive species as protonated organic bases (e.g., amines) provided that pH$>$pK$_b$. When the protonation is scarce, at pH$\approx$pK$_b$, amines interact with silica via hydrogen bonding:

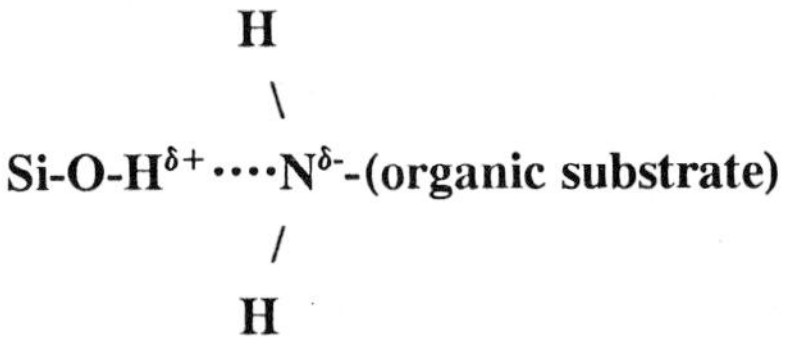

$$
\begin{array}{c}
H \\
\backslash \\
Si\text{-}O\text{-}H^{\delta+}\cdots\cdots N^{\delta-}\text{-(organic substrate)} \\
/ \\
H
\end{array}
$$

If pK$_b$ is much greater than pH, amines will not interact with silica surfaces at all. In fact, if we consider urea (pK$_b=13.8$) at neutral pH, it will be totally in the unprotonated form $_2$HN-CO-NH$_2$. However, because of the following resonance forms:

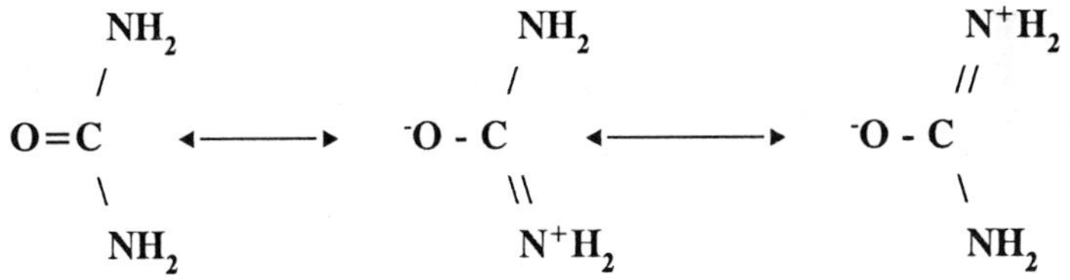

$$
\begin{array}{ccccc}
NH_2 & & NH_2 & & N^+H_2 \\
/ & & / & & // \\
O=C & \longleftrightarrow & ^-O\text{-}C & \longleftrightarrow & ^-O\text{-}C \\
\backslash & & \backslash\backslash & & \backslash \\
NH_2 & & N^+H_2 & & NH_2
\end{array}
$$

the hydrogen atoms will be bonded to a nitrogen atom which, despite its high electronegativity, is positively charged, hindering hydrogen bonding formation.

As for alumina (pH$_{zpc}=9$), its surface at neutral pH will be positively charged. Accordingly, it will easily interact with negative species as dissociated natural organic acids (fulvic and humic substances containing carboxyl as well as phenolic hydroxyl groups). Once adsorbed, however, such natural substances give a negative charge to the alumina surface. Such adsorption has been postulated to occur by the formation of a complex between the dissociated acid groups of natural organic matter and the protonated hydroxyl groups present on the alumina surface according to the following reaction [Davis et al. 1981]:

$$
(AlOH)_x + H_yA^{-n} \Longleftrightarrow [(AlOH_2)_xH_{(y-x)}A]^{-n}
$$

where A generically indicates a natural acid organic substance. Because of excess negative charges present on adsorbed natural organic substrates, the end result of such interaction is that coated alumina behaves as a weak cation exchange resin. Interactions via hydrogen bonding can also occur between alumina surfaces and undissociated acids [Thurman 1985]:

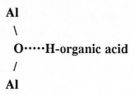

Al
\
O·····H-organic acid
/
Al

Another mechanism called *ligand exchange* has been proposed to explain strong interactions between oxide surfaces and organic anions [Stumm et al. 1981]. In this mechanism a real nucleophic substitution occurs between protonated surface hydroxyl groups (leaving groups) and the carboxyl groups (entering groups) of natural organics. The result is the formation of strong bonds between the oxide surface and the naturally occurring organic matter:

$$
\begin{array}{ccc}
OH_2^+ & & O \\
\backslash \; / & & // \\
Al \; \blacktriangleleft \cdots\cdots \; ^- O\text{-}C \\
/ \; \backslash & & \backslash
\end{array}
$$

(natural organics)

For iron oxide surfaces ($pH_{zpc} = 7$), the most common crystalline form *goethite* behaves as an alumina surface. To explain the strong interactions between natural organics and this oxide, some authors [Tipping et al. 1982] have proposed a mechanism in which a divalent cation (Me), such as magnesium or calcium, is shared between the iron oxide surface and organic matter carboxyl groups as sketched below:

$$
\begin{array}{c}
O \\
// \\
\text{(goethite surface)-Fe-O··Me··O-C} \\
\backslash
\end{array}
$$

(humic organic matter)

Main physico-chemical properties of liquid-phase (solvent).

pH, ionic strength, and temperature are the liquid-phase properties that mainly affect sorption of contaminants to naturally occurring solid phases.

With respect to the influence of **pH** on sorption behaviour of organic contaminants, a distinction has to be made among ionic, ionizable and neutral (with no charge) compounds.

Ionic organics have a molecule with a pH independent permanent charge. If pH-dependent sorption is obtained for such a class of organics, it must be ascribed to the influence of pH on solid phase surface behaviour. In a well-designed investigation (see Fig.1a) it has been demonstrated that, in the pH range 4÷9, the adsorption of differently branched alkylbenzenesulphonates [4-(n'-dodecyl)benzenesulfonate] on alumina decreases as pH and branching increase [Dick et al. 1971].

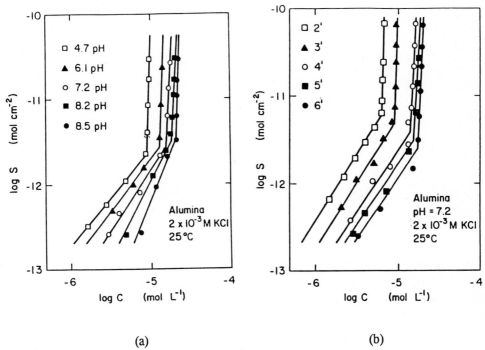

(a)   (b)

Fig. 1 Effect of pH and structure of hydrophobic group on adsorption of anionic surfactants on alumina. [Dick et al. 1971, Westall 1987].

This sorption versus pH trend has been explained by considering that benzene sulfonates are negatively charged species and that the alumina surface is positive at $pH < pH_{zpc} = 9$. The branching effect (see Fig.1b) has been instead attributed to weak hydrophobic interactions whose contribution to the overall sorption extent is lesser than that of electrostatic interactions. This is demonstrated by the fact that sorption occurs with difficulty at $pH \geq pH_{zpc}$.

For *nonpolar organics* no pH influence on sorption extent is expected even though pH might affect sorbent surface charge. In natural environments, for this class of contaminants, sorption phenomena are usually ascribed to the contaminant incompatibility with water (hydrophobicity) [Haymaker et al. 1972]. In aqueous solution, in fact, a molecule of a nonpolar hydrophobic organic compound will force the dipoles of water molecules to assume an ordered disposition around it (*iceberg effect*). In turn, such an entropically unfavourable arrangement of water molecules forces the contaminant molecules to accumulate at the solid-liquid interface [Nemethy et al. 1962]. This mechanism does not imply any specific attraction between the solute and the sorbent surface and explains sorption of nonpolar hydrophobic organics on pure minerals. However, in the case of minerals containing organic matter, the sorption of this class of organics is ascribed to the solute compatibility with the organic matter. For such systems, sorption extent is usually described by a pH-independent partition coefficient Kp ($Kp=[\overline{A}]/[A]$) pH , where A indicates a generic nonpolar organic solute and the overbar indicates a nonaqueous phase. Kp is usually correlated to solid phase organic matter content through the equation: $Kp=f_{oc}\cdot Koc$, where f_{oc} is the fraction of the solid phase organic matter and Koc is a sorption constant characteristic for each compound and independent of the solid phase. Linear free-energy relationships can be used to estimate Koc values from solute properties related to hydrophobicity, such as normal octanol/water partition coefficient (Kow) or water solubility (S). Some of the numerous available relationships are listed below:

logKoc =	0.38·logKow	+ 0.19	[Chiou et al. 1979]
logKoc =	0.52·logKow	+ 0.64	[Briggs et al. 1981]
logKoc =	0.54·logKow	+ 1.377	[Kenaga et al. 1980]
logKoc =	0.72·logKow	+ 0.49	[Schwarzenbach et al. 1981]
logKoc =	0.87·logKow	+ 0.056	[Mingelgrin et al. 1983]
logKoc =	0.94·logKow	− 0.006	[Brown et al. 1981]
logKoc =	1.00·logKow	− 0.21	[Karickhoff et al. 1979]
logKoc =	1.00·logKow	− 0.317	[Means et al. 1980]
logKoc =	1.03·logKow	− 0.18	[Rao et al. 1980]
logKoc =	−0.54·logS	+ 3.72	[Felsot et al. 1979]
logKoc =	−0.54·logS	+ 4.70	[Karickhoff et al. 1979]
logKoc =	−0.56·logS	+ 4.28	[Chiou et al. 1979]
logKoc =	−0.56·logS	+ 4.24	[Mingelgrin et al. 1983]

A critical evaluation of the above listed slopes and intercepts values has been done recently [Gerstl 1989]. In general terms, the slope is related to the nature of the process dominating adsorption, and the intercept to the nature of the organic matter present in the sorbent.

The sorptive behaviour of *ionizable organics* with pure minerals, already described in the previous paragraph, will be dominated by specific interactions which can be roughly predicted knowing the pH of the bulk solution, the solid phase pH_{zpc} and the contaminant pKa or pKb. However, the behaviour of this type of contaminants with minerals containing organic matter is rather different from that of nonpolar organics. For ionizable organics, in fact, instead of a partition coefficient Kp, it is necessary to consider a distribution ratio D which is defined as [Schwarzenbach et al. 1985]:

$$D = ([\overline{AH}] + [\overline{A^-}])/([AH] + [A^-])$$

where AH represents an undissociated organic acid and overbars refer to the nonaqueous phase. For simulating sorption of ionizable organics to naturally occurring organic matter exhaustive investigations have been carried out with substituted phenols in water-octanol systems. Some results, shown in Fig.2, have proved that the distribution ratio varies with pH [Schwarzenbach et al. 1985].

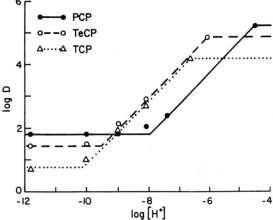

Fig.2
The effect of pH on the distribution ratio of chlorophenols: (PCP: penta-chlorophenol; TeCP: 2,3,4,5-tetra-chlorophenol; TCP: 2,4,5-trichloro-phenol). [Schwarzenbach et al 1985]

At pH values lower than the pKa of AH the predominant species is AH in both phases and the contribution of A^- species to the overall distribution is insignificant. In this pH region $D = [\overline{AH}]/[AH] = Kp$ and D is pH independent. For pH values much greater than pKa the predominant species in both phases is A^- and $D = [\overline{A^-}]/[A^-] = Kp'$ is, once again, pH independent. Of course, Kp' is lower than Kp because of the scarce affinity of the charged species A^- for the nonaqueos phase. In the pH region between the extremes the predominant species in the nonaqueous and aqueous phase are, respectively, AH and A^-. In this pH region D will regularly decrease with pH.

As for **ionic strength**, in general terms it can be said that it affects sorption phenomena by modifying the thickness of the solid phase diffuse double layer and changing the solute activity coefficient [Bolt et al. 1978, Stumm et al. 1981]. The theory states that the diffuse double layer thickness is lower at greater ionic strength. In any case, even though such an effect influences the solid phase physical properties (e.g. hydraulic conductivity), it does not have any influence on sorption mechanisms. As far as the effect of ionic strength on solute activity coefficient is concerned, a distinction has to be made between neutral and charged chemical species.

Sorption of *neutral species* can be influenced by the effect of ionic strength on solute aqueous activity through the so-called *salting-out effect* [Karickhoff 1984]. This effect can vary directly or inversely with ionic strength, depending upon the salt considered. However, in most cases, an increase in sorption occurs with increasing ionic strength [Lee et al. 1990]. This trend can be explained by considering that an increase in the ionic strength will increase the incompatibility of a hydrophobic solute with water so that the solute will be squeezed out of the water phase. Furthermore, the importance of such an effect will depend on the ionic strength range.

If the solute is a *polar or charged* molecule involved in specific electrostatic interactions with sorbent, an increase in ionic strength usually causes a reduction of solute activity because of the shielding effect of the neighbouring ions. In such a case, specific interactions will be proportionally weakened [Bolt et al. 1978]. However, in the case of nonspecific interactions, such as hydrophobic partitioning, an increase in ionic strength often causes a corresponding increase in solute concentration in the nonaqueous phase via a mechanism which is not the salting-out effect. Extensive investigations on the partitioning of ionizable hydrophobic organics (e.g. phenols) between water and n-octanol have clearly demonstrated that, in the pH range in which the compound is dissociated, an ionic strength increase results in a distribution ratio increase (see Fig.3) [Schwarzenbach et al. 1985].

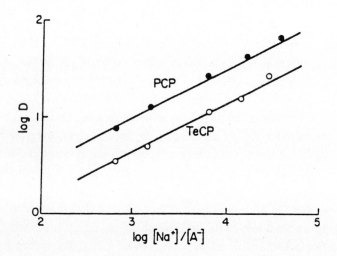

Fig. 3   Octanol/water distribution ratio of pentachlorophenol (PCP) and 2,3,4,5-tetrachlorophenol (TeCP) as a function of the ratio $[Na^+]/[A^-]$. The pH of the aqueous phase is 11.8. [Schwarzenbach et al. 1985]

This result is interpreted in terms of an enhanced partitioning of neutral ion-pairs (organic-anion/associated-cation or organic-cation/associated-anion) into an organic phase with respect to the partitioning of charged organic species. Higher ionic strengths and the presence of divalent cations favour the formation of such ion pairs [Javfert et al. 1990]:

$$A^- + Me^+ \rightleftharpoons A^-Me^+$$

Temperature effects on sorption phenomena will essentially depend on the nature of the interactions involved and on the thermodynamics of the global sorption process [Zettlemoyer et al. 1971]. Adsorption of *neutral hydrophobic organics* on a solid phase is the result of a favourable entropy change with small, if any, enthalpy change. Accordingly, except for a few exceptions [Weber et al. 1983], temperature effects on sorption extent has been reported to be insignificant for such organics [Chiou et al. 1979, Hassett et al.1983]. On the contrary, adsorption of *polar or electrically charged organics* will be affected by temperature because such sorption is often the result of specific interactions characterized by peculiar enthalpy changes [Thomas 1961, Hassett et al. 1981]. A quantitatively rough estimation of the temperature effect can be based on a van't Hoff type equation:

$$\ln [(\overline{[A]}/[A])_2/(\overline{[A]}/[A])_1] = -(\Delta H/R) \cdot (1/T_2 - 1/T_1)$$

where A is a generic organic compound, 1 and 2 refer to different temperatures, overbars refer to the nonaqueous phase, R is the gas constant and ΔH is the enthalpy change of the overall sorption reaction. Often this latter value is replaced by the ΔH value characteristic of the specific solute-sorbent interaction which is thought to be responsible for sorption:

Van der Waals forces	$\approx$	1÷2	Kcal/mol
Hydrophobic bonding	$\approx$	"	"
Hydrogen bonding	$\approx$	0.5÷15	"
Ion exchange	up to	50	"
Chemisorption	up to	30÷190	"

Of course, if the ΔH value is low, insignificant temperature effects will result. In most cases an increase in sorption occurs with decreasing temperature. In a few instances, however, sorption has also been reported to increase with increasing temperature [Horzempa et al. 1983]. Such a trend, the opposite of what would be expected for an exothermic reaction of a sorbed organic compound, can be explained only by taking into account the effect of temperature on various factors (kinetics, solubility, etc.) different from sorption equilibrium.

Conclusions.

In attempting to predict sorption of organic contaminants in subsurface environments the assumption that sorption results only from interactions with naturally occurring organic matter appears questionable.

The content of native organic matter in such deeper-lying materials is usually very low and then, even for highly hydrophobic organic compounds, it is likely that specific interactions with inorganic components of solid matrix can contribute to the overall sorption extent.

In reality, in subsurface environments, sorption of organic pollutants occurs through a combinations of competing interactions with clays and hydrous oxides, as well as with naturally occurring organic substrates.

Particularly for polar compounds, the physico-chemical properties of each component of the solute-sorbent-solvent ternary system will play the main role in determining sorption nature and extent.

In spite of the progress made so far to elucidate how organic chemicals can interact with subsurface components, additional research is necessary for developing a comprehensive approach for predicting the transport and fate of organic contaminants in subsurface environments.

References.

Bailey GW, White JL, Rothberg T (1968) Adsorption of organic herbicides by montmorillonite. Role of pH and chemical character of adsorbate. Soil Sci Soc Am Proc 32:222-234

Bolt GH, Bruggenwert MGM (1978) Soil Chemistry. A. Basic elements. John Wiley & Sons, Inc, New York

Bolt GH, Van Riemsdijk WH (1987) Surface chemical processes in soil. In: Stumm W (ed) Aquatic surface chemistry, John Wiley & Sons, Inc, New York Chichester Toronto Singapore, p 127

Briggs AA (1981) Theoretical and experimental relationships between soil adsorption, octanol/water partition coefficients, water solubilities, bioconcentration factors, and the parachor. J Agric Food Chem 29:1050-1057

Brown DS, Flag EV (1981) Empirical prediction of organic pollutant sorption in natural sediments. J Environ Qual 10:382-387

Callahan MA, Slimak M, Gbel N, May I, Fowler C, Freed R, Jennings P, DuPree R, Whitmore F, Maestri B, Holt B, Gould C (1979) Water related environmental fate of 129 priority pollutants. EPA-44014-79029a,b, NTIS

Chiou CT, Peters LJ, Freed VJ (1979) A physical concept of soil-water equilibria for non ionic organic compounds. Science 206:831-832

Davis GA, Gloor R (1981) Adsorption of dissolved organics in lake water by aluminum oxide. Effect of molecular weight. Environ Sci Techn 15:1223-1229

Dick SG, Fuerstenau DW, Healy TW (1971) Adsorption of alkylbenzene sulfonate surfactants at the alumina-water interface. J Colloid Interface Sci 37:595-602

Dixon JB, Weed SB (eds) (1977) Minerals in soil environments. Soil Science Society of America, Madison, WI

Felsot A, Dahm PA (1979) Sorption of organophosphorous and carbamate insecticides by soil. J Agric Food Chem 27:557-559

Fenn DB, Mortland MM (1972) Interlamelar metal complexes on layer silicates:2.Phenol complexes in smectites. In: Proc. Int. Clay Conf., Madrid, 591-603

Gerstl Z (1989) Predicting the mobility and availability of toxic organic chemicals. In: Gerstl Z, Chen Y, Mingelgrin U, Yaron B (eds) Toxic organic chemicals in porous media. Springer, Berlin Heidelberg New York, p 151

Grim RE (1968) Clay mineralogy. McGraw Hill, New York Hassett JJ, Banwart WL, Wood SG, Means JC (1981) Sorption of α-Naphthol; implications concerning the limits of hydrophobic sorption. Soil Sci Soc Am J 45:38-42

Hassett JJ, Banwart WL, Griffin RA (1983) Correlation of compound properties with sorption characteristics of nonpolar compounds by soils and sediments: concept and limitations. In: Francis CW, Auerback SI (eds) Characterization, treatment and disposal, environment and solid wastes. Butterworth Publishers, p 161

Haymaker JW, Thompson JM (1972) Adsorption. In: Goring CM, Haymaker JM (eds) Organic chemicals in the soil environment, Dekker, New York p 49

Horzempa LM, Di Toro DM (1983) The extent of reversibility of polychlorinated biphenil adsorption. Water Res 17:851-859

Howard PH (ed)(1989) Handbook of environmental fate and exposure data for organic chemicals. Lewis Publishers, Inc, Chelsea, Michigan

Israelachvili JN (1985) Intermolecular and surface forces. Academic Press, London

Javfert CT, Westall JC, Grieder E, Schwarzenbach RP (1990) Distribution of hydrophobic ionogenic organic compounds between octanol and water: organic acids. Enviro Sci Technol 24:1795-1803

Jones RAY (1979) Physical and mechanistic organic chemistry. Cambridge University Press, Cambridge U.K.

Karickhoff S, Brown DS, Scott T (1979) Sorption of hydrophobic pollutants on natural sediments. Wat Res 13:241-248

Karickhoff SW (1984) Organic pollutant sorption in aquatic systems. J Hydraul Eng 110:707-735

Kenaga EE, Goring CAI (1980) Aquatic Toxicology. Eaton JG, Parrish PR, Hendricks (eds), ASTM special publication 770, Philadedelphia

Kortum G (1966) Lehrbuch der elektrochemie. Verlag Chemie, GMBH Weinheim Kown BT, Ewing BB (1969) Effect of the organic adsorption on clay ion exchange properties. Soil Sci 108:321-325

Lee SL, Rao PSC, Nkedi-Kizza P, Delfino JJ (1990) Influence of solvent and sorbent characteristics on distribution of pentachlorophenol in octanol-water and soil-water systems. Environ Sci Technol 24:654-661

Lyman WJ, Reehl WF, Rosenblatt DH (eds) (1982) Handbook of chemical property estimation methods; Environmental behavior of organic compounds. Mc Graw-Hill, New York

Means JC, Wood SG, Hassett JJ, Banwart WL (1980) Sorption of polynuclear aromatic hydrocarbons by sediments and soils. Environ Sci Technol 14:1524-1528

Mingelgrin U, Gerstl Z (1983) Reevaluation of partitioning as a mechanism of nonionic chemical adsorption in soils. J Environ Qual 12:1-11

Mortland MM (1970) Clay-organic complexes and interactions. Adv Agron 22:75-117 Nemethy G, Scheraga HA (1962) Structure of water and hydrophobic bonding in proteins. II: model for the thermodynamics properties of aqueous solutions of hydrocarbons. J Chem Phys 36:3401-3411

Philen OD Jr, Weed SB, Weber JB (1970) Estimation of surface charge density of mica and vermiculite by competitive adsorption of Diquat vs. Paraquat. Soil Sci Soc Am Proc 34:527-531

Philen OD Jr, Weed SB, Weber JB (1971) Surface charge characteristics of layer silicates by competitive adsorption of two organic divalent cations. Clays Clay Miner 19:295-302

Rao PSC, Davison JM (1980) Estimation of pesticide retention and transformation parameters required in nonpoint source pollution models. In: Overcash MR, Davison JM (eds) Environmental impact of nonpoint source pollution. Ann Arbor Science Publischers, Ann Arbor, p 23

Salzman S, Yariv S (1975) Infrared study of the sorption of phenol and p-nitrophenol by montmorillonite. Soil Sci Soc Am J 39:474-479

Schindler PW, Walti E, Furst B (1976) The role of surface hydroxyl groups in the surface chemistry of metal oxides. Chimia 30:107-109

Schofield RK (1949) Effect of electric charges carried by clay particle. J Soil Sci 1:1-8

Schwarzenbach RP, Westall J (1981) Transport of non polar organic compounds from surface water to groundwater. Laboratory sorption studies. Environ Sci Technol 15:1360-1367

Scwarzenbach RP, Westall J (1985) Sorption of hydrophobic trace organic compounds in groundwater system. Wat Sci Tech 17:39-55

Sposito G (1984) The surface chemistry of soils. Oxford University Press, New York Stumm W, Morgan JJ (1981) Aquatic Chemistry. 2nd edn. John Wiley & Sons, Inc, New York London Sidney Toronto

Thomas JM (1961) The existence of endothermic adsorption. J Chem Ed 38:138-139 Thurman EM (1985) Organic geochemistry of natural waters. Nijhoff M,Junk W, Dordrecht Boston Lancaster

Tipping H, Cooke JP (1981) Adsorption of organic matter by alumina. Geochim Cosmoch Acta 46:75-80

Weast RC, Lide DR, Astle MJ, Beyer WH (eds) (1990) Handbook of Chemistry and Physics, 70th edn. CRC Press, Inc, Boca Raton, Florida

Weber WJ Jr, Voice TC, Pirbazari M, Hunt GE, Ulanoff DM (1983) Sorption of hydrophobic compounds by sediments, soils and suspended solids. II. Sorbent evaluation studies. Water Res 17:1443:1452

Weber WJ Jr, McGinley PM, Katz LE (1991) Sorption phenomena in subsurface systems: concept, models and effect on contaminant fate and transport. Wat Res 25:499-528

Westall JC (1987) Adsorption mechanisms in aquatic surface chemistry. In: Stumm W (ed) Aquatic surface chemistry. 1st edn. John Wiley & Sons, Inc, New York London Sidney Toronto, p 3

Zettlemoyer AC, Micale FJ (1971) Solution adsorption thermodynamics for organics on surfaces. In: Faust SD, Hunter JV (eds) Organic compounds in aquatic environments. Marcel Dekker, Inc, New York, p 165

Fate of Persistent Organic Compounds in Soil and Water

M. Mansour, I. Scheunert[1] and F. Korte[2]
GSF - Research Center for Environment and Health
Institut für Ökologische Chemie
Schulstraße 10
D-8050 Freising-Attaching - Germany

The increasing contamination of the environment with organic
chemicals necessitates a scientific appraisal of their
environmental compatibility. In order to detect adverse effects as
early as possible, evaluation parameters must be characterized to
provide a primary indication of any environmental pollution
resulting from the organic chemicals. Whether this is local or
longterm global pollution or what is the consequence to the living
environment from the application of the chemical concerned should
be determined by an appropriate focussing of the data from these
parameters. The parameters include the level of production, the
area of application, the tendency to spread as well as the
knowledge of the transformation of these chemicals under
environmental conditions.

---

[1] GSF - Research Center for Environment and Health
    Institut für Bodenökologie
    Ingolstädter Landstraße 1
    D-8042 Neuherberg - Germany
[2] Technische Universität München
    Institut für Chemie
    D-8050 Freising-Weihenstephan - Germany

NATO ASI Series, Vol. G 32
Migration and Fate of Pollutants in Soils and Subsoils
Edited by D. Petruzzelli and F. G. Helfferich
© Springer-Verlag Berlin Heidelberg 1993

BIOTIC REACTIONS IN SOIL

In soil, xenobiotic chemicals undergo various abiotic and biotic
degradation reactions. In numerous experiments reported, it has
been demonstrated that, in contrast to degradation in aquatic
systems, in soil biotic reactions are the dominant processes in the
transformation and degradation of xenobiotic organic substances.
Therefore, this paragraph deals mainly with biotic processes.
However, up to now, it is still very difficult to distinguish
between abiotic and biotic causes of chemical reactions in soil.
In many cases, the same reaction product partly originates from an
enzymatic and partly from an abiotic process.

Alterations in the chemical structure of a xenobiotic, as
initiated either by biotic or abiotic factors, result in the
formation of new xenobiotic compounds which may be more or less
toxic or ecotoxic than the parent compounds and which represent
unwanted chemical residues in the ecological system. Furthermore,
due to the changes in chemical structure, conversion products may
differ significantly from the parent compounds in physico-chemical
properties and, thus, also in their distribution behaviour in the
environment. Therefore, it is important both to identify their
chemical structure and to determine their quantity including
formation and degradation kinetics.

Contrary to the transformation resulting in persistent organic
products, mineralization of xenobiotic substances in soil is a
desirable process since it is the only pathway for a final
elimination of the xenobiotic from the environment. Mineralization
kinetics govern the time-course of decline of total residues - of
both parent chemicals and all conversion products - in soil.

Formation of Conversion Products

In soil, foreign chemicals undergo numerous biotic reactions, some
of which are listed in Table 1.

Table 1   Biotic reactions of organic xenobiotics in soil
           (Scheunert, 1992)

Type of transformation	Reaction	Examples of xenobiotics
Oxidative processes	C-Hydroxylation	Cypermethrin, carbofuran, aromatic compounds
	ß-Oxidation	Phenoxy alkanoic acids
	Epoxidation	Cyclodienes
	Ketone formation	Carbofuran
	C=C cleavage	Cyclodienes, aromatic compounds
	C-dehydrogenation	Lindane
	N-oxidation	Anilines
	N-demethylation	Phenylureas
	S-oxidation	S-containing pesticides
	Substitution of O for S	Organophosphorous compounds
	Ether cleavage	Phenoxy alkanoic acids
	Oxygenolytic dechlorination	2-Chlorobenzoate, 4-chlorophenylacetate
	Mineralization	Nearly all organic compounds
Reductive processes	C-reduction	Alkenes and alkines
	N-reduction	Nitro compounds
	S-reduction	Sulfoxides, disulfides
	Reductive dehalogenation	Chlorinated aromatics, DDT, cyclodienes, camphechlor, halogenated propanes
Hydrolytic processes	Ester hydrolysis	Carboxylic esters, sulfates
	Carbamate hydrolysis	Carbamate insecticides
	Nitrile hydrolysis	2,6-Dichlorobenzonitrile, cypermethrin
	Epoxide hydrolysis	Dieldrin, intermediates of aromatic C-hydroxylation
	Hydrolytic dehalogenation	Trichloroacetate, halogenated propanes, chloroallylalcohols, chlorinated phenols, 4-chlorobenzoic acid
Synthetic processes	Reaction with natural humic monomers	Anilines and phenols

They may be roughly divided into oxidative, reductive, and
hydrolytic reactions. It should be emphasized, however, that there
exist a number of microbiological reactions which cannot be
classified in this scheme, such as isomerizations, molecular
rearrangements, or the loss or addition of various substituents.
In many cases, the reaction mechanism is not known, and sometimes
conversion products regarded as resulting from hydrolytic attack
are actually produced by mixed function oxidases, as in some cases
of organophosphorous compounds (Brooks, 1972).

In addition to these reactions which are more or less degradative,
synthetic reactions may occur also in soil. For example,
xenobiotic chemicals may participate in natural reactions of humus
formation, taking the place of natural reaction partners. In these
reactions, the participating natural amino compounds may be
replaced by amines of xenobiotic origin, such as chlorinated
anilines which are degradation products of various herbicides in
soil. A lower molecular reaction product between catechol and 4-
chloroaniline has been identified by Adrian et al. (1989) as an
anilinoquinone. The natural phenols may be replaced by xenobiotic
phenols such as chlorinated ones.

When the foreign compound has become part of a natural polymer, it
must be regarded as a soil-bound residue which is no more
extractable by solvents as used, e.g., in pesticide residue
analysis, and thus is no longer accessible to analytical detection
and determination.

Dechlorination of chlorinated xenobiotics - the cleavage of the
foreign C-Cl-bond - is a highly important step in the degradation
of this xenobiotic substance class. Although the C-Cl-bond is not
absolutely foreign in nature but is synthesized by some living
organisms, it is, nevertheless, the main reason for the general
slow degradability of organochlorine compounds and their
persistence in the environment. Dechlorination is achieved by
either oxidative, reductive, or hydrolytic mechanisms, or by
conjugation followed by elimination of the conjugating molecule,
or a part of it.

The multitude of potential reactions in soil may act simultaneously or successively on a xenobiotic chemical, resulting in a large number of conversion products. The quantity of different conversion products demonstrates that upon estimating the consequences of pesticide application to soils, potential conversion reactions should not be neglected.

Total Degradation

Total degradation of foreign organic compounds in soil, resulting in their final elimination from the environment, means their complete mineralization to small inorganic molecules such as CO_2, CO, H_2O, NH_3, H_2S, Cl^-, and other species. In the complex soil community many members may contribute to this degradation process. A final product resulting from the activity of certain species may be degraded further by other species groups, or may be accessible to abiotic attack. Thus, mineralization in soil is a complex process in which various biotic and abiotic factors are involved.

In biotic degradation, besides carbon dioxide, small organic fragments are formed, which join the natural carbon pool. These are released as carbon dioxide only when the natural compounds where they are incorporated are degraded, e.g. during normal respiration or upon decay after the death of the organisms.

The quantification of total degradation of foreign compounds is very important for their ecotoxicological evaluation. The best method to perform it is the quantitative determination of $^{14}CO_2$ evolved from ^{14}C-labelled substances in the laboratory. The time-course of biomineralization is a major topic of current research into biodegradation of man-made organic chemicals. Information on these kinetics is indispensable for an examination of extent and relevance of mineralization.

In biomineralization experiments measuring $^{14}CO_2$, many time curves show apparent lag phases at the beginning followed by an exponential second phase and a third phase where mineralization decreases continuously until a complete stop is reached; complete mineralization is not achieved. As an example, the biomineralization of a surfactant, n-dodecylbenzene sulfonate, in a forest soil at room temperature is shown in Fig. 1 (Scheunert et al., 1990). This sigmoid curve course of $^{14}CO_2$ release from ^{14}C-labelled organic compounds is reported frequently in the literature (Simkins and Alexander, 1984; Scow et al., 1986; Brunner and Focht, 1984). It is a characteristic of biological systems involving growth and/or adaptation in soil. Scow et al. (1986) showed that the lag phase accounts for an adaptation of soil microflora to the foreign compound. The third phase may represent the stage where a substrate level is reached which is too low to sustain the metabolizing microbial fraction. It might also represent the mineralization of formerly assimilated carbon from the xenobiotic either by normal respiration, or by decay after the death of the microorganisms. This evolution of carbon dioxide continues also after the degradation of the original xenobiotic has discontinued.

If relative biomineralization rates are required, in order to compare different chemicals with each other or to compare a new chemical with a well-known reference compound, biomineralization studies are carried out under standardized conditions in the laboratory; measurements of $^{14}CO_2$ then are restricted to a few time intervals. Table 2 gives some data obtained after shaking ^{14}C-labelled chemicals in a soil-water suspension under aerobic or anaerobic conditions at 35°C (Scheunert et al., 1987). $^{14}CO_2$ was trapped by drawing air once a day through the apparatus and absorption of $^{14}CO_2$ in a scintillation liquid containing an organic base. The table lists the sum of $^{14}CO_2$ after 5 days and at the end of the experiment after 14 to 56 days.

The table reveals that all these compounds are degraded, although some of them only at a very low rate. The differences in degradation rates between aerobic and anaerobic conditions are low. For some chlorinated compounds, degradation is somewhat

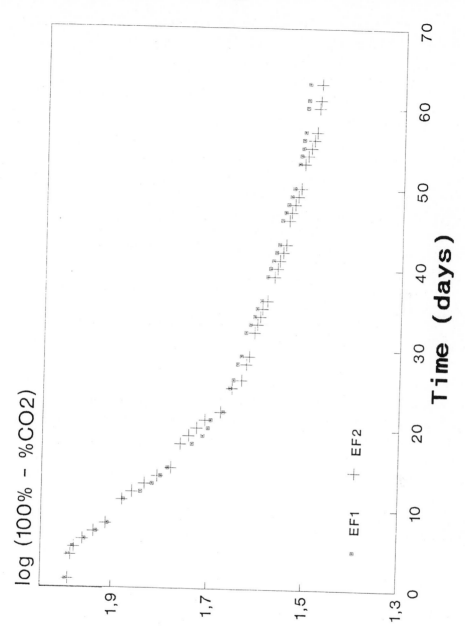

Fig. 1   Time-course of biomineralization of n-dodecylbenzene
         sulfonate in a forest soil (Scheunert et al., 1990)
         (EF = "Ebersberger Forst"; 1 and 2 = replicates)

Table 2   Biodegradation of ^{14}C-labelled organic chemicals in a
soil-water suspension under aerobic and anaerobic conditions (35°C) (Scheunert et al., 1987)

^{14}C-Chemical applied	Experi-mental time (days)	Conditions	$^{14}CO_2$ after 5 days (% of ^{14}C applied)	$^{14}CO_2$ at the end of experiment (% of ^{14}C applied)	^{14}C in water at the end of experiment (% of ^{14}C applied)
Urea	5	aerobic	66.3	66.3	25.3
		anaerobic	70.1	70.1	18.0
Methanol	5	aerobic	53.4	53.4	15.0
		anaerobic	46.3	46.3	13.6
n-Dodecylben-zenesulphonate	42	aerobic	13.8	40.6	7.5
		anaerobic	26.0	51.9	20.2
Di-(2-ethylhe-xyl)phthalate	56	aerobic	5.6	11.6	14.4
		anaerobic	2.9	8.1	14.3
Phenanthrene	14	aerobic	3.0	7.2	0.1
		anaerobic	4.2	6.3	0.4
Lindane	42	aerobic	0.4	1.9	3.1
		anaerobic	0.4	3.0	2.2
DDT	42	aerobic	0.1	0.8	2.4
		anaerobic	0.3	0.7	2.0
Anthracene	14	aerobic	0.1	1.3	1.6
		anaerobic	0.3	1.8	1.8
2,4-D	14	aerobic	0.1	0.5	89.1
		anaerobic	0.2	0.7	83.4
2,6-Dichloro-benzonitrile	42	aerobic	<0.01	0.5	38.9
		anaerobic	<0.01	<0.01	36.8
Hexachloro-benzene	14	aerobic	<0.01	0.4	1.9
		anaerobic	<0.01	0.2	2.1

better under anaerobic than under aerobic conditions. The
reason is the first step of degradation, namely reductive
dechlorination, which is favoured under anaerobic conditions.

It should be mentioned that the levels of $^{14}CO_2$ strongly depend
on soil type, temperature, humidity, pH, soil biomass, and
other soil properties (Scheunert et al., 1990). Therefore,
mineralization rates must not be regarded as absolute values
but, e.g. in Table 2, are valid only for the experimental
conditions used for obtaining these data.

ABIOTIC REACTIONS IN AQUATIC SYSTEMS

Abiotic environmental factors (solar radiation) make an
appreciable contribution to the transformation and degradation
of persistent organic chemicals in aquatic systems (Sundstrom
and Ruzo, 1978; Zafiriou, 1974; Zepp et al., 1975; Mansour et
al., 1986 a).

In this context, abiotic transformation of organic chemicals is
to be investigated with particular consideration given to the
dynamic and catalytic effects due primarily to the state of the
molecule and its interaction with the environment.

The complexity of ecosystems substantially impedes the study of
such reactions. In order to investigate single reactions,
conditions must be created corresponding to those of various
ecosystems. Only in this way a rational interpretation of
experimental results is possible.

Direct and indirect photolytic processes in aquatic systems.

Under the influence of sunlight, which is only relevant in the two uppermost water layers, depending on the degree of turbidity and the depth of the water, photoprocesses occur as shown in Table 3. Two different effects must be considered: on the one hand, direct photolysis of a foreign substance which absorbs UV light in the spectral region of sunlight ($\lambda > 290$ nm) and on the other hand indirect photolysis, in which the light energy is absorbed by other constituents of the water and is either transmitted to the foreign chemical under consideration (sensitization) or leads to the formation of reactive species that enter into a chemical reaction with the foreign chemical (Fig. 2).

As it is known so far from the literature, this essentially involves OH radicals, RO_2 and singlet oxygen (Mansour et al., 1985). The total radiation at the surface of a body of water consists of the direct solar radiation and the diffuse radiation from the sky. The original spectral distribution of sunlight is altered to a greater or smaller extent by its passage through both the atmosphere and the water layers (Fig. 3). The attenuation of a light ray on passage through the aqueous medium is caused by absorption and scatter (Miller and Zepp, 1983).

The amount of scatter depends greatly on the degree of turbidity due to suspended particles, but at a depth of 1 m hardly accounts for more than 10 % of the entire light attenuation (Fig. 4).

A measure of the light attenuation by dissolved substances is the extinction of the water sample. According to the Beer-Lambert-Law, this is given by

Table 3 Natural photoprocesses in the aquatic environment

Substrate	Products	Probable Mechanism and Effects
NO_2^-	$\cdot NO + \cdot OH$	direct photolysis; induces NO air-sea flux.
$CH_3 I$	$\cdot CH_3 + \cdot I$	direct photolysis; air-sea exchanges.
Natural organic chromophores	$\cdot C + HO_2$ $C^+ + \cdot O_2^-$ $\cdot C + O_2 + H_2O_2$	transfer (H, electron, energy) to O_2; initiation of numerous reactions.
Petroleum hydro-carbons (RH, ArH, ArBH$_2$, R$_2$S)	$ROO\cdot$ $R=O, RCO_2^-, ArOH$ $ArCOR, ARCHO,$ $R_2 SO$	oxygen addition; oxidizes or-ganic radicals; direct or sensitized photolysis; changes in oil properties and toxicity.
Pesticides (disulfoton)	disulfoton sulfoxide	singlet oxygen; changes in properties and toxicity.
Herbicides (2,4-D, 2,4,5-T)	oxidation, reduction products	direct photolysis; complex.
Pentachlorophenol	phenols, quinones, acids	direct photolysis; changes in properties and toxicity.
PCB's, DDT	complex	direct or sensitized photo-lysis; dehydrochlorination.
Cu(II)	Cu(I)$^+$	alteration of cycling-toxicity.
Domestic wastes Fe(III)-NTA	Fe(II) + amine + $+ CO_2$	charge transfer to metal; NTA degradation.

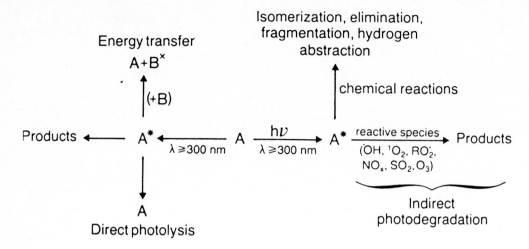

Fig. 2   Reaction pathways in environmental photochemistry

$$\log \frac{Io}{I} = \epsilon \cdot c \cdot d$$

$$I \lambda = Io \ (1 - e^{-2.3 \ \epsilon \cdot c \cdot d})$$

where ϵ is an absorption constant, c is the concentration of the chemical in solution, d is the pathlength of the cuvette containing the solution, Io is the incident light and I is the light intensity after passage through the pathlength d.

The ratio between reflected radiation and that penetrating the water depends on the angle of incident radiation and hence on the season and the time of day.

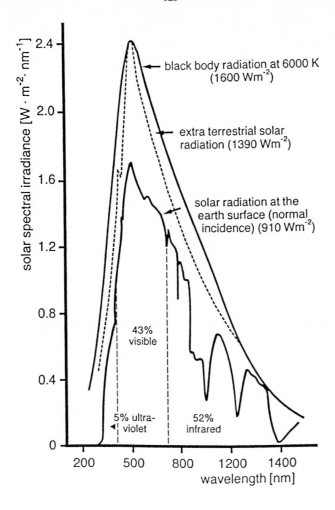

Fig. 3   Solar flux as a function of wavelengths

The radiation losses by reflection are on average 3 % in Central
Europe in summer and 14 % in winter. In flowing water the
reflection losses are somewhat larger because of the moving
surface. Of course, natural water differs from pure water in that
it contains a large number of dissolved and suspended substances.
Directly relevant for photochemistry, however, are those
constituents of water whose UV absorption spectrum overlaps with
the spectrum of sunlight, but only with those substances which
absorb light above roughly 300 nm. The inorganic substances are
mainly chloride, sulfate, and nitrate as well as some transition
metal compounds and complexes (Table 4).

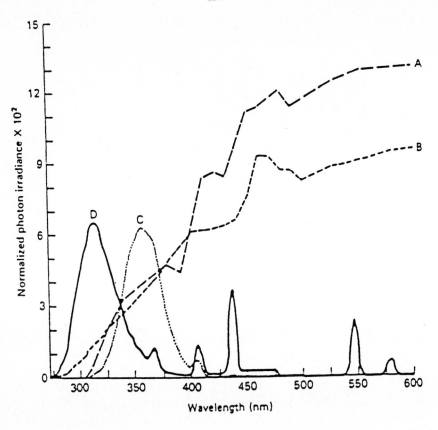

Fig. 4   Emission spectra of several light sources (Miller and Zepp, 1983)

A = fall sunlight, midday lat. 40° N
B = unfiltered xenon lamp
C = black-light
D = sunlamp

Although the latter must be ranked with the substances absorbing sunlight, no appraisal can be made of their significance for photochemistry in bodies of water. Beside the lack of data on the characteristics of the radiation, the very low concentration must also be considered. The concentration range of substances in various waters is appreciable and this is reflected in the range of chemical and in some cases also of physical properties.

Table 4   Properties, diversities and mean values for physico-
chemical characterization of freshwater and sea-water

Characteristics	Freshwater*	Sea-water**
Bromide (Br^-)	$890 \cdot 10^{-3}$	$0.63 - 0.65 \cdot 10^2$
Calcium (Ca^{2+})	$0.6 \cdot 10^2$	$0.4 - 0.48 \cdot 10^3$
Chloride (Cl^-)	$0.12 \cdot 10^2$	$19.35 - 21.385 \cdot 10^3$
Fluoride (F^-)	$150 \cdot 10^{-2}$	$0.009 - 0.12 \cdot 10^2$
Hardness ($CaCO_3$)	$4.57 \cdot 10^2$	$12.47 - 12.87 \cdot 10^3$
Iron (Fe^{2+})	$19 \cdot 10^{-2}$	$0.009 - 1.1 \cdot 10^{-2}$
Sodium (Na^+)	$1.5 \cdot 10^2$	$9.69 - 10.65 \cdot 10^3$
Magnesium (Mg^{2+})	$0.5 \cdot 10^2$	$9.86 - 1.29 \cdot 10^3$
Nitrogen (NO_3^-)	$0.5 \cdot 10^2$	$0.001 \cdot 10^2$
Nitrogen (NO_2^-)	$0.001 \cdot 10^2$	$0 - 0.0045 \cdot 10^{-4}$
Sulfate (SO_4^{2-})	$2.45 \cdot 10^2$	$2.512 - 2.718 \cdot 10^3$
Dissolved organic carbon (DOC)	$0.01 - 0.045 \cdot 10^2$	$0.0068 - 0.0071 \cdot 10^2$
pH	$6.6 - 8.8$	$7.8 - 8.3$

*   mean concentration in mg/l
**   mean value in mg/kg

Role of humic substances, oxygen, OH radicals and hydrogen
peroxide.

The O_2-concentration depends on the exchange with the atmosphere
and on oxygen producing and oxygen consuming assimilation
processes in water. In the upper water layer relevant for
photochemistry, however, oxygen concentrations near the saturation
limit are to be reckoned with even in eutrophic waters. With
regard to organic carbon, a distinction is made between the total
organic carbon content, which is mostly abbreviated as TOC, and
the dissolved organic carbon (DOC, particle size usually less than

the dissolved organic carbon (DOC, particle size usually less than o.4 μm). The TOC also comprises suspended matter and the biomass of plankton. However, this does not amount to more than roughly 5 %. The origin of the organic compounds varies. They are essentially metabolic products of organisms and degradation products of dead biomass.

Of the organic substances, the humins in particular are worth to be mentioned: they have a long-wave absorption extending into the visible spectrum. They form a major part of the dissolved or colloidally dissolved organic carbon, but are also present in particulate form.

Humic substances, including humic acids and fulvic acids, occur in almost every aquatic environment as well as in soil. They are not uniform compounds as defined in classical chemical terms. Humic acid, the main fraction of humic substances, is the proportion of post mortem, darkley stained organic substances in the soil that are soluble in diluted bases. They arise from rotting organic material in the course of humification. The definition is thus purely formal and is restricted to specification of the origin and the dissolution properties.

The chemistry of the humin is somewhat clarified by the determination of functional groups. The elements C, H and O are obligatory for the classification of substances like humins. The extent to which nitrogen (0.08 % to 5 %) is a necessary constituent of humins, remains an open question. The molecular weights in the humin fraction range between 1000 and 10000. Humins are thus of moderately high molecular weight (Stevenson, 1982; Choudhry, 1984), but are not high molecular weight substances with a partial particle mass far over 10000.

Aquatic humins have been found to sensitize phototransformation of pollutants in water exposed to sunlight (Zepp et al., 1985).

Our studies have shown that photolysis rates of environmental chemicals can be altered by the suspended and dissolved matter

present in various bodies of water. Zepp et al. (1985) have
examined the photochemical fate of humic acid solutions in order
to develop generalizations about the importance of indirect
photolysis. They demonstrated that humic substances are
responsible for the disappearance and the enhanced photooxidation
rate of organic compounds and possibly other compounds in natural
waters (Adrian et al., 1986). Draper and Crosby (1983) observed
that in sunlight waters aquatic humic acids act as sensitizers or
precursors for the production of photoreactants such as singlet
oxygen, humic derived peroxy radicals, hydrogen peroxide and OH
radicals. Depending on the water type, the season and the
pollution of the water, the photochemical generation of hydrogen
peroxide (H_2O_2) in natural water has been ascribed to the
macromolecules. In seawater, the concentration of hydrogen
peroxide is roughyl 14 to 290 nM, in surface freshwater 100 to
7000 nM and in sewage lagoons 100 to 33000 nM.

Superoxide anion ($O_2 \cdot ^-$) is known to be a precursor of hydrogen
peroxide according to the following reaction:

$$2 O_2 \cdot ^- + 2\ H^+ \longrightarrow H_2O_2 + O_2$$

According to the present state of knowledge as reported in the
literature (Mansour and Korte, 1986; Cooper and Zika, 1983), the
following reactions take place in aqueous solution under the
influence of UV light:

$$H_2O_2 \underline{\ \ \ \ \ hv\ \ } \longrightarrow\ \cdot OH + OH$$
$$OH + H_2O_2 \ \ ----\longrightarrow\ H_2O + HO_2 \cdot$$
$$HO_2 \cdot \ ---------\longrightarrow\ H^+ + O_2 \cdot ^-\ (pK\ \ 4.8)$$

Detection of the OH radicals formed in the aqueous photolysed
solution is carried out indirectly via product formation and the
kinetics of the reactive species (Mansour et al., 1986 b; Mansour,
1985). For this purpose, scavengers are used. Their kinetics in
reactions with OH radicals in aqueous solution are known. By
comparing the relative reaction rates of various scavengers,

inferences can be made about the nature of the reactive species. The kinetic data for the reaction of the scavengers with OH radicals are known from investigations in which these are produced radiolytically by electron-pulse irradiation of H_2O:

$$H_2O \;\text{------}hv\text{----}\text{>}\quad H_2,\; H_2O_2,\; H,\; \cdot OH$$

Properties of direct photolysis.

When a substance dissolved in a body of water absorbs light, it can return from the excited state to the ground state in various ways. However, it can also break down into reaction products with a certain quantum yield. A quantum yield is the ratio of the number of reacting molecules to the number of excited molecules. The photolysis is described by the simplified reaction equation:

$$A \;\text{-----}hv\text{----}\text{>}\; products$$

In monochromatic irradiation, the differential equation

$$\theta \;=\; \frac{R}{\displaystyle\int_{\lambda_1}^{\lambda_2} Ia(\lambda)\cdot d\lambda}$$

R   = rate of dissappearence (M · s^{-1})
Ia   = absorbed light intensity (einstein · l^{-1} · s^{-1})
λ_1, λ_2 = spectral range (overlap of the incident light and absorption spectrum of the compound)

applies to the reaction rate. However, this equation applies exactly only under quite specific conditions. In particular, the incident light of intensity I_o (reflection from the surface is neglected) must be not only monochromatic but also parellel. These conditions are not fulfilled by global radiation. However, since the global radiation is also the portion of sunlight which strikes a horizontal plane, it appears to be admissible to replace I_o by the corresponding fraction of global radiation. In agreement with this, the calculation is carried out as if the total global radiation struck the water surface perpendicularly. In addition it is assumed in the derivation of the equation that the incident light is only absorbed, not dispersed. This condition can also not be fulfilled exactly in natural waters. Since, on the other hand, measurements are available that document that the fraction of light scattered back in natural waters is only about 2 % of the incident light, the calculation is carried out as if the light was attenuated only by absorption. The pathlength 1 of the light will be chosen in accordance with the absorption of the water in such a way that 99 % of the light is absorbed within that distance. Equation 1 provides a mean value for the reaction rate, i. e. in accordance with equation 1 the value for the rate of photolysis is obtained. This is achieved if substance A is evenly distributed in the reaction volume at any time. The location of A is neglected in calculating the rate of photolysis.

If, as in natural water, irradiation is carried out with polychromatic light, the radiation intensity must be subdivided into suitable wavelength intervals. The resulting reaction rate is then a sum of the rates for all wavelength intervals. The quantum yield is treated as wavelength-independent. This procedure is in agreement with experimental experience.

The lifetime of those substance classes that undergo direct photolysis in water was expressed as: lifetime multiplied by 0.693. In the specification of the half-lives it is logical to take into account the fluctuation of meteorological conditions by specification of minimum, mean and maximum values per month. In the following tables (Table 5 and 6), the experimentally

Table 5   Photolysis of p-chlorophenol, chlorobenzene and pentachlorophenol

Compound	Water	Season (Year 1986)	Half-life (hr)	Quantum Yield
p-Chlorophenol	distilled	May – June	53	$4.2 \cdot 10^{-3}$
	Isar water	May – June	17	$3.5 \cdot 10^{-2}$
Chlorobenzene	distilled	May – June	19	$9.0 \cdot 10^{-2}$
	Isar water	May – June	11	0.1
Pentachloro-phenol	distilled	May – June	2.5	0.014
	Isar water	May – June	1.5	0.073

Table 6   Photolysis of chlorobenzene in water

Substance	Concentration (M)	Half-life Sunlight (Days)	Quantum Yield	Products (%)
Chlorobenzene	$2.10 \cdot 10^{-5}$	21	$2.90 \cdot 10^{-1}$	phenol/ benzene (trace)
DDE	$3.40 \cdot 10^{-5}$	1.5	$7.50 \cdot 10^{-1}$	photoisomer (Miller and Zepp, 1983)
1.4-Dichloro-benzene	$7.80 \cdot 10^{-4}$	25	$1.70 \cdot 10^{-2}$	chlorobenzene (Neely, 1978)
2-Chloro-biphenyl	$0.50 \cdot 10^{-4}$	31	$2.10 \cdot 10^{-2}$	2-hydroxy-biphenyl
4-Chloro-biphenyl	$0.35 \cdot 10^{-5}$	45	$0.25 \cdot 10^{3}$	4-hydroxybi-phenyl (trace)

determined and the calculated half-lives are listed for comparison. In all compounds investigated, there was good agreement of the values, thus verifying that the approach is correct.

In the investigations p-nitroanisol (Draper, 1985) or the system p-nitroanisol/pyridine, a well investigated sunlight actinometer, was included because it appeared to be an appropriate reference. In addition, the reaction quantum yield could be controlled by means of the pyridine concentration, so that it was possible to adapt the rate of reaction to the respective light conditions (Mansour et al., 1986 b).

As can be seen, the light attenuation is a very sensitive parameter. In the dye Direct Blue 15, e. g. the total half-life increases by a factor of 5 at a water depth of 1 m and by a factor of 20 at a water depth of 5 m. At a water depth of 10 m, there are indeed half-lives of 3 to 30 years. Haag and Mill (1987) studied the photostability of azo dyestuffs in water and they observed that these substances are photostable in pure and distilled water. They concluded that their half-lives in sunlight could generally be greater than two thousand hours. Direct Blue 15 is a diazodye with several sulfonic groups and can be transformed in aqueous solution in the presence of TiO_2 with simulated sunlight irradiation. Half-lives are 6 to 10 minutes (Fig. 5).

In order to examine the ecological significance of the occurrence, influence and fate of these substances in waters, it was first necessary to ascertain the rates at which these organic compounds are destroyed or transformed by light under natural conditions of irradiation. For these experiments we selected atrazine, parathion (Mansour et al., 1981), tetrachlorvinfos, diazinon, isoxaben and other pesticides (Fig. 6-10).

The photodegradation rate of the substances in the various water samples by natural sunlight decreased in the following order: aqueous acetone (20 ml/l), sea-water, river water and distilled water. The half-lives ranged from 8 hours to 1.5 weeks.

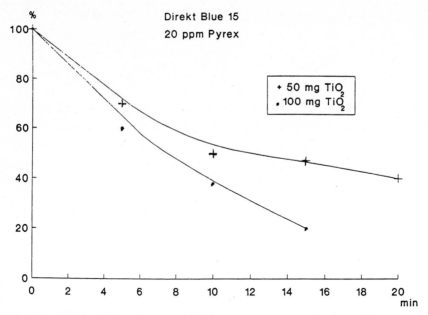

Fig 5. Photocatalytic degradation of Direct Blue 15 in water in the presence of TiO₂

Substance	Solvent	Irradiation λ [nm]	Quantum yield Φ	Products
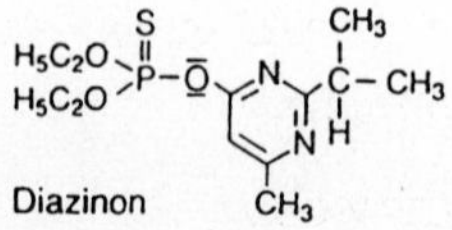 Diazinon	water pH 7.6	290	$1.8 \cdot 10^{-3}$	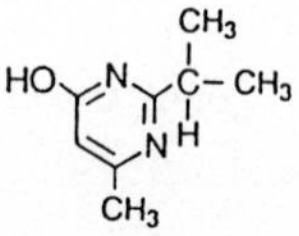
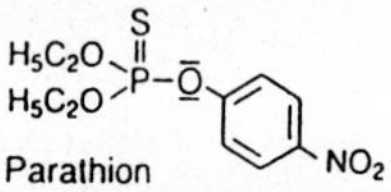 Parathion	water pH 7.3	290	$1.5 \cdot 10^{-3}$	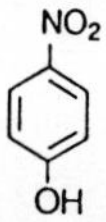 + Paraoxon
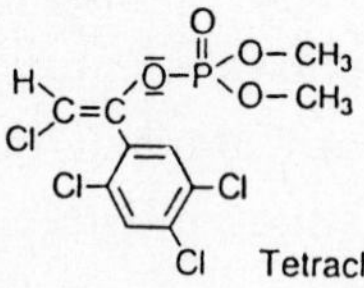 Tetrachlorvinfos	water pH 7.8	290	$1.7 \cdot 10^{-2}$	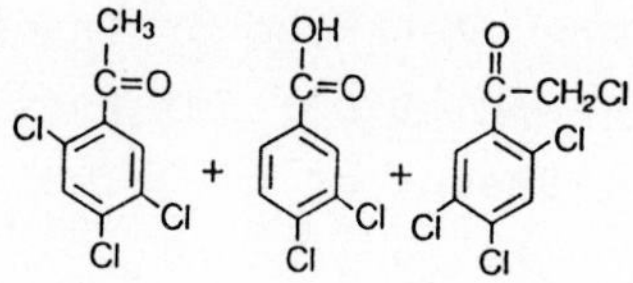
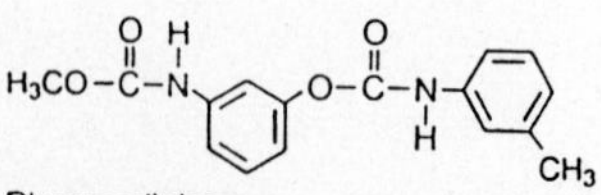 Phenmedipham	water pH 7.9	270	$1.3 \cdot 10^{-2}$	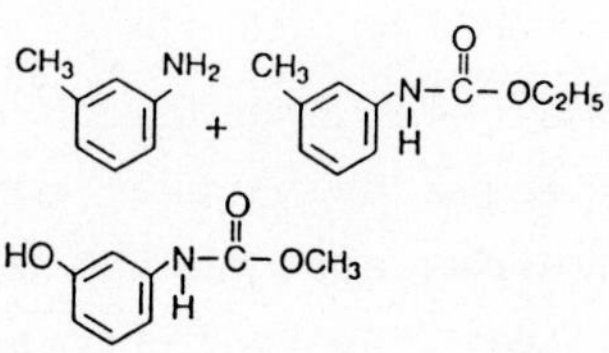

Fig. 6 Phototransformation of selected compounds in laboratory experiments (Mansour et al., 1981)

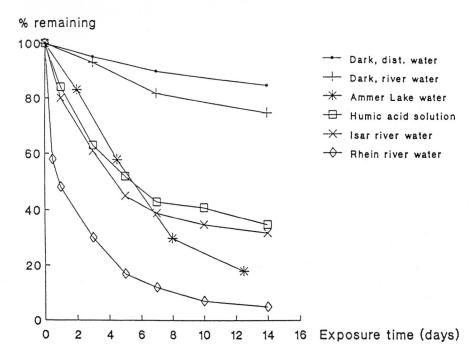

Fig. 7 Degradation rate of diazinon under sunlight

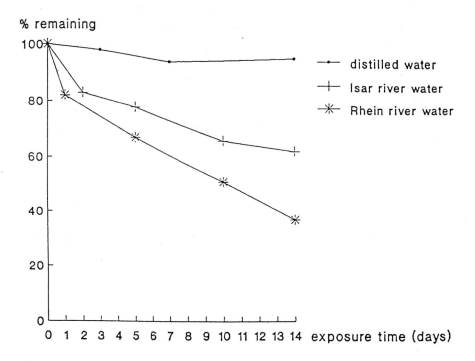

Fig. 8 Degradation rate of atrazine under sunlight

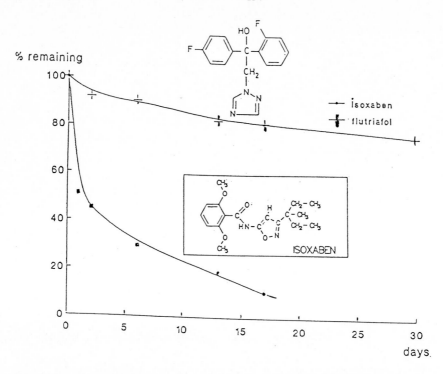

Fig. 9   Degradation of isoxaben and flutriafol under sunlight
          conditionin distilled water (summer 1991)

CONCLUSION

For the evaluation of ecotoxicological consequences of foreign
chemicals in the environment, it is not sufficient to investigate
the residues of the unchanged parent compounds alone. Every
chemical undergoes various alterations of its molecule, effected
by abiotic as well as by biotic processes. Although the type and
extent of alterations depend on the chemical structure of the
chemical and the reactivity of its functional groups, the absence
of any alteration cannot be postulated for any chemical. The same
applies to soil-bound residues. For a critical assessment of the
consequences of the presence of xenobiotics in soil and water,
above all mineralization must be considered. Abiotic or biotic
mineralization occurs for all compounds and its extent depends on
chemical stability.

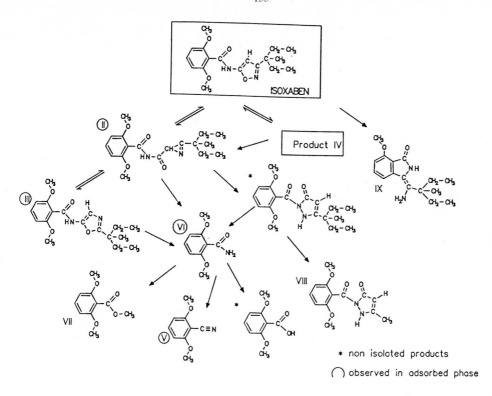

* non isolated products

◯ observed in adsorbed phase

Fig. 10 Proposed photodegradation pathways of isoxaben under
natural sunlight in aqueous and absorbed phase

REFERENCES

Adrian P, Andreux F, Metche M, Mansour M, Korte F (1986)
Autoxydation des ortho-diphénols catalysée par les ions Fe^{2+} et
Mn^{2+}: un modèle de formation des acides humiques. In: Comptes
rendues de l'Académie des Sciences, 303 II. Paris, p 1615

Adrian P, Lahaniatis ES, Andreux F, Mansour M, Scheunert I, Korte
F (1989) Reaction of the soil pollutant 4-chloroaniline with the
humic acid monomer catechol. Chemosphere 18: 1599-1609

Brooks GT (1972) Pathways of enzymatic degradation of pesticides. In: Coulston F, Korte F (eds) Environmental quality and safety 1. Thieme, Academic Press, Stuttgart New York London, p 106

Brunner W, Focht DD (1984) Three-half-order kinetic model for microbial degradation of added carbon substrates in soil. Appl Environ Microbiol 47: 167-172

Choudhry GG (1984) Humic substances: Structural aspects and photophysical, photochemical and free radical characteristics. In: Hutzinger O (ed) The handbook of environmental chemistry, Vol. 1, Part C. Springer, Berlin, p 1

Cooper WJ, Zika RG (1983) Photochemical formation of hydrogen peroxide in surface and ground waters exposed to sunlight. Science 220: 711-712

Draper WM (1985) Determination of wavelength-averaged, near UV quantum yields for environmental chemicals. Chemosphere 14: 1195-1290

Draper WM, Crosby DG (1983) The photochemical generation of hydrogen peroxide in natural water. Arch Environ Contam Toxicol 12: 121-126

Haag WR, Mill T (1987) Direct and indirect photolysis of water soluble azodyes: kinetic measurements and structure-activity relationships. Environ Toxicol Chem 6: 359-369

Mansour M (1985) Photochemical degradation testing of some aromatic compounds in dilute aqueous solutions containing hydrogen peroxide. In: ACS Abstract Book Vol. 25, 1 (Division of Environmental Chemistry). American Chemical Society, p 305

Mansour M, Korte F (1986) Abiotic degradation pathways of selected xenobiotic compounds in the environment. In: Pawlowski L, Alaerts G (eds) The studies on environmental science 29 - Chemistry for protection of the environment. Elsevier Science Publ Co Inc, Amsterdam, p 257

Mansour M, Korte F, Méallier P (1986 b) Evaluating the fate of some pesticides in aquatic media exposed to artificial light. Presented at the 6th International Congress of Pesticide Chemistry (IUPAC), Ottawa, 10.-15.8.1986

Mansour M, Scheunert I, Korte F (1985) Behaviour of some organochlorine compounds in terrestrial ecosystems and abiotic degradability in the environment. In: Compte rendu de la quinzième réunion du Groupe Français des Pesticides. Grignon, p 1

Mansour M, Scheunert I, Viswanathan R, Korte F (1986 a) Assessment of the persistence of hexachlorobenzene in the ecosphere. In: Morris CR, Cabral JRP (eds) Hexachlorobenzene: Proceedings of an International Symposium. IARC/WHO Scientific Publications, Lyon, p 53

Mansour M, Thaller S, Korte F (1981) Action of sunlight on parathion. Bull Environ Contam Toxicol 30: 358-364

Miller GC, Zepp RG (1983) Extrapolating photolysis rates from the laboratory to the environment. Residue Reviews 85: 88-110

Neely WB (1978) An integrated approach to assessing the potential of organic chemicals in the environment. Presented at the workshop on philosophy and implementation of hazard assessment procedures for chemical substances in the aquatic environment, Waterville Valley, New Hampshire, 14.-18.8.1978

Scheunert I (1992) Transformation and degradation of pesticides in soil. In: Ebing W (ed) Chemistry of plant protection, Vol. 8. Springer, Berlin Heidelberg New York, p 23

Scheunert I, Dörfler U, Adrian P (1990) Biomineralisierung von Umweltchemikalien in verschiedenen Böden in Abhängigkeit von jahreszeitlichen Bedingungen. In: Verband Deutscher Landwirtschaftlicher Untersuchungs- und Forschungsanstalten (ed) Landwirtschaft im Spannungsfeld von Belastungsfaktoren und gesellschaftlichen Ansprüchen, Kongreßband 1990 Berlin. VDLUFA Verlag, Darmstadt, p 633

Scheunert I, Vockel D, Schmitzer J, Korte F (1987) Biomineralization rates of ^{14}C-labelled organic chemicals in aerobic and anaerobic suspended soil. Chemosphere 16: 1031-1041

Scow KM, Simkins S, Alexander M (1986) Kinetics of mineralization of organic compounds at low concentrations in soil. Appl Environ Microbiol 51: 1028-1035

Simkins S, Alexander M (1984) Models for mineralization kinetics with the variables of substrate concentration and population density. Appl Environ Microbiol 47: 1299-1306

Stevenson FJ (1982) Humus chemistry. John Wiley Interscience Publication, New York

Sundstrom G, Ruzo LO (1978) Photochemical transformation of pollutants in water. In: Hutzinger O, van Lelyveld LH, Zoeteman BCJ (eds) Aquatic transformation and biological effects. Pergamon Press, New York, p 205

Zafiriou OC (1974) Sources and reactions of OH and daughter radicals in seawater. J Geophys Res 79: 4491-4497

Zepp RG, Schlotzhauer PF, Sink RM (1985) Photosensitized transformations involving electronic energy transfer in natural waters: role of humic substances. Environ Sci Technol 19: 74-81

Zepp RG, Wolfe NL, Gordon JA, Baughman GL (1975) Dynamics of 2,4-D esters in surface waters. Hydrolysis, photolysis, and vaporization. Environ Sci Technol 9: 1144-1150

Kinetics and Mechanisms of Environmentally Important Reactions on Soil Colloidal Surfaces

D.L. Sparks[1], S.E. Fendorf[1], P.C. Zhang[2], and L. Tang[1]
[1]Department of Plant and Soil Sciences
University of Delaware
Newark, Delaware 19717-1303

Without question, one of the pressing issues of our time is the preservation of the environment. Throughout the world, concerns have been expressed about an array of soil and water contaminants. These include: inorganic pollutants such as nitrates and phosphates, heavy metals such as arsenic, cadmium, chromium, copper, lead, mercury, and nickel, and radionuclides; and, organic contaminants such as pesticides, industrial chemicals, municipal sludges, and animal wastes. Intensive research efforts are being conducted by soil scientists, engineers, geochemists, and chemists to find ways to decontaminate polluted soils and ground and surface waters and to better predict the mobility and fate of contaminants in soils.

Much of the research on migration and retention of inorganic and organic contaminants in soils has been macroscopic in nature. The focus of most of these studies has been on the determination of partition coefficients and adsorption isotherm parameters and with development and utilization of purely equilibrium-based models. Sorption of inorganics and organics has been described using an array of empirical, semi-empirical, and surface complexation models including Freundlich, Langmuir, Temkin, constant capacitance, and triple layer. In some cases, these models equally well describe sorption data, and are often useful for describing sorbate reactions on an array of colloidal surfaces under various experimental conditions. However, many of these models have multiple adjustable parameters, and thus, it is not surprising that sorption data can be well described using them. In effect, use of sorption models to describe macroscopic data is often a curve-fitting exercise. Despite this, many

----

[2]Research Center, SUNY at Oswego, Oswego, NY 13126

NATO ASI Series, Vol. G 32
Migration and Fate of Pollutants in Soils and Subsoils
Edited by D. Petruzzelli and F. G. Helfferich
© Springer-Verlag Berlin Heidelberg 1993

investigators have employed them to make mechanistic interpretations about ion and molecular sorption on surfaces. However, some of these models can describe several different sorption mechanisms. Thus, conformity of material balance data to a certain model <u>does not</u> prove that a particular mechanism is operational.

The discussion above is in no way meant to imply that equilibrium-based approaches for modeling sorption reactions on colloidal surfaces are not meaningful and worthwhile. On the contrary, useful information can be gleaned from such studies. Our major point is that such studies provide no definitive information on reaction mechanisms or on the kinetics of the reactions. Moreover, such equilibrium studies are usually not particularly relevant to field conditions since soils are seldom, if ever, at equilibrium. To obtain definitive information on the mechanisms of inorganic and organic reactions on soils and soil components, one must conduct kinetic studies and employ surface spectroscopic and microscopic techniques.

In the past 10-15 years a number of important and enlightening studies have appeared in the literature on various aspects of kinetics of soil chemical processes. Some of these studies have dealt with mobility and retention of pollutants in soils. In this paper we should like to discuss a number of topics related to the kinetics of inorganic and organic contaminant reactions on soils and soil components such as clay minerals and metal oxides. Unfortunately, space does not allow for a comprehensive review of the topic, but the reader can consult a number of books and review chapters for more extensive details (Sparks, 1989; Sparks and Suarez, 1991).

To begin with, we shall discuss the application of chemical kinetics to heterogeneous systems such as soils and soil components. The pitfalls in applying empirical, various order chemical kinetics equations and other models to heterogeneous systems where chemical kinetics and transport processes are occurring simultaneously will be discussed. It will be argued that the kinetics of many soil chemical processes are transport-controlled and that a preferable way to describe these reactions is to employ diffusion models rather than chemical kinetics-based models. Other topics that will be discussed in this chapter include methods that can be used to measure rapid soil chemical reactions. The remainder of the chapter will discuss specific studies on the rates of inorganic and organic contaminant reactions on soil and soil colloids. These will include: adsorption/desorption of metals on oxide surfaces; ion exchange kinetics on clays; mechanisms of chromium(III) oxidation on manganese oxides; and kinetics of organic contaminant reactions on organo-clays and unmodified clay surfaces. In an effort to augment elucidation of reaction mechanisms, the application of in-situ surface spectroscopic and microscopic techniques for studying environmentally important reactions on soil colloidal surfaces will also be succinctly discussed.

Time Scales of Soil Chemical Processes

A variety of chemical reactions occur in soils and often in combination with one another. Reaction time scales can vary from microseconds for ion association, ion exchange, and sorption reactions to years for mineral-solution and mineral

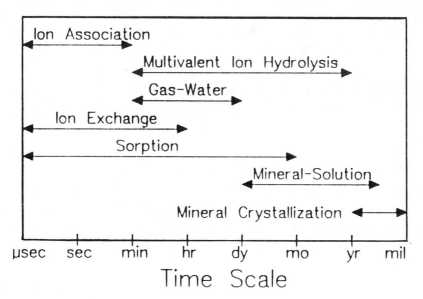

Figure 1. Time scales for reactions in soil environments (Amacher, 1991).

crystallization phenomena (Fig.1). Ion association reactions include ion pairing, inner- and outer-sphere complexation (ligand exchange and electrostatic attraction, respectively), and chelation in solution. Gas-water reactions involve gaseous exchange across the air/liquid interface. Ion exchange reactions occur when cations and anions are adsorbed and desorbed from soil surfaces via electrostatic attractive forces. Ion exchange reactions are reversible and stoichiometric. Sorption reactions can involve physical sorption, outer-sphere complexation, inner-sphere complexation, and surface precipitation. Mineral-solution reactions include precipitation/dissolution of minerals, and coprecipitation reactions whereby small constituents become a part of mineral structures (Amacher, 1991; Sparks, 1992).

The type of soil component can drastically affect the reaction rate. For example, metal sorption reactions are often rapid on clay minerals such as kaolinite and smectites and much slower on vermiculitic and micaceous minerals. This is in large part due to the availability of sites for sorption. For example, kaolinite has readily available planar external sites and smectites have primarily internal sites that are also quite available for retention of sorptives. Thus, sorption reactions on these soil constituents are often quite rapid, even occurring on time scales of seconds and milliseconds (Sparks, 1989). Metal sorption reactions on oxides, hydroxides, and humic substances depend on the type of surface and metal being studied, but are also generally rapid. For example, reaction rates of molybdate, sulfate, selenate, and selenite on goethite occurred on millisecond time scales (Zhang and Sparks, 1989, 1990a,b). Half-times for bivalent Pb, Cu, Zn sorption on peat ranged from 5 to 15 seconds (Bunzl et al., 1976).

On the other hand, vermiculite and micas have multiple sites for retention of metals and organics, including planar, edge, and interlayer sites, with some of the latter sites being partially to totally collapsed. Consequently, sorption and desorption reactions on these sites can be slow, tortuous mass transfer- controlled. Often, an apparent equilibrium may not be reached even after several days or weeks. Thus, with vermiculite and mica, sorption of inorganics and organics involves two to three different reaction rates- high rates on external sites, intermediate rates on edge sites, and low rates on interlayer sites (Jardine and Sparks, 1984; Sparks, 1991).

Application of Chemical Kinetics to Soil and Soil Colloidal Systems

The study of chemical kinetics, even in homogeneous systems, is complex and often arduous. When one attempts to study the kinetics of reactions in heterogeneous systems such as soils and even of soil components such as clay minerals, hydrous oxides, and humic substances, the difficulties are greatly magnified. This is largely due to the complexity of soils that are made up of a mixture of inorganic and organic components. These components often interact with each other and display different types of sites with various reactivities for inorganic and organic adsorptives. Moreover, the variety of particle sizes and porosities in soils further adds to their heterogeneity. In most cases, both chemical kinetics and multiple transport processes are occurring simultaneously. Thus, the determination of chemical kinetics, which can be defined as "the investigation of rates of chemical reactions and of the molecular processes by which reactions occur where transport is not limiting" (Gardiner, 1969) is extremely difficult, if not impossible, in soils and soil components. In these systems, one is studying kinetics, which is a generic term referring to time-dependent processes. Thus, apparent and not mechanistic rate laws and rate parameters are determined (Skopp, 1986; Sparks, 1989).

A number of transport processes can occur in soils and colloidal systems depending on the type of colloid and sites, as discussed previously. These include: transport in the soil solution, transport across a liquid film at the solid/liquid interface, transport in liquid-filled macropores, all of which are nonactivated diffusion processes; and diffusion of sorbate occluded in micropores, and diffusion processes in the bulk of the solid, both of which are activated diffusion processes (Fig. 2). Equations for and discussions of these phenomena can be found in Aharoni and Sparks (1991). Any one or more of these transport phenomena or chemical reaction could be rate-limiting.

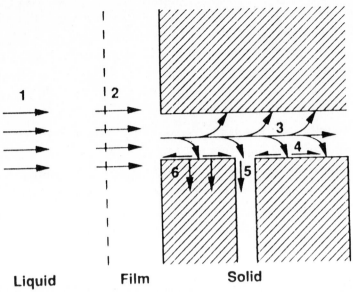

Liquid   Film   Solid

Figure 2.   Transport processes in solid-liquid soil reactions.
Nonactivated processes: (1) transport in soil solution; (2)
transport across liquid film at solid-liquid interface; and (3)
transport in liquid-filled micropore. Activated processes: (4)
diffusion of sorbate at surface of the solid; (5) diffusion of
sorbate occluded in micropore; and (6) diffusion in bulk of
solid (Aharoni and Sparks, 1991).

Analyses of Slow Chemical Reactions on Soil Colloids

An array of kinetic equations including zero-, first-, and
second-order, Elovich, fractional-power, and parabolic-diffusion
equations have been employed to analyze and describe soil
chemical rate data (Sparks, 1989). In some cases, a number of
these equations equally well describe time-dependent data if
simple correlation coefficients and standard errors of the
estimate are the indices that are employed to evaluate the data
(Chien and Clayton, 1980; Onken and Matheson, 1982; Sparks and
Jardine, 1984; Havlin et.al., 1985; Aharoni et al., 1991).

Unfortunately, there are many instances in the soil and
environmental sciences literature where mechanistic meanings
have been given to rate data solely on the fit of the data to
one of the equations mentioned previously. However, deduction of
reaction mechanisms based solely on the fit of the data to a
particular equation is incorrect (Sparks,1989). Despite the wide

applicability of one or more of these equations to various experimental results, there is often no consistent relation between the equation that gives the best fit and the physicochemical and mineralogical properties of the soil-sorbate system to which it is applied. Moreover, the kinetic equations that give the best fit are often empirical and the significance of the obtained rate parameters is not clear.

Aharoni and Sparks (1991) critically examined the inconsistencies and problems associated with the application of empirical equations to experimental rate data. These equations are approximations to which more general expressions reduce in certain limited time ranges. Aharoni and Sparks (1991) and Aharoni et al. (1991) showed that a generalized empirical equation can be derived by examining the applicability of power function, Elovich, and apparent first-order equations to experimental data.

Differentiating the power function, Elovich, and apparent first-order equations and writing them as explicit functions of the reciprocal of the rate $(Z) = (dq/dt)^{-1}$, one can show that the plot of Z vs. t for an experimental isotherm should be convex if the power-function equation is operational, linear if the Elovich equation is applicable, and concave if the apparent first-order equation is appropriate. However, Aharoni and Sparks (1991) have shown that Z vs. t plots for soil chemical reactions are usually S-shaped: convex at small t, concave at large t, and linear at some intermediate t range. These findings suggest that the kinetics follow some abstruse function that can be approximated by the power function equation at small t, by the Elovich equation at an intermediate t, and by an apparent first-order equation at large t. Thus the sigmoidal shape of the Z(t) plot means that the above equations are applicable, each at a limited time range of sorption.

One often assumes S-shaped behavior even when one of these equations seems to apply over the entire reaction period. The rate of sorption often decreases with the amount sorbed by many orders of magnitude before equilibrium is approached. However, experimental methods are usually designed for the measurements of rates varying within a shorter range, e.g., two to three orders of magnitude. There are cases, therefore, in which the measured data points represent only a limited part of the complete process. In these cases, the rate may seem to become zero before true equilibrium is approached, or the measured process may appear to be preceded by an instantaneous one. One of the simple equations mentioned earlier would be applicable during the entire experimental time if all the measured points are at times during which the assumptions underlying the equation are valid.

Appropriate kinetic models should lead to equations consistent with the above rule, i.e., a plot of Z vs. t that is S-shaped with a minimum slope at the inflection point. Models based on diffusion satisfy these conditions (Aharoni and Suzin, 1982a,b; Aharoni, 1984; Aharoni and Sparks, 1991; Aharoni et al., 1991), specifically models based on diffusion in either a homogeneous or heterogeneous medium. In homogeneous diffusion, equations yield S-shaped Z(t) plots in which the final and initial curved parts predominate whereas equations for heterogeneous diffusion give S-shaped Z(t) plots in which the intermediate linear part is dominant. These equations are derived, discussed, and applied to data for sorption of ions on soils in Aharoni and Sparks (1991) and Aharoni et al. (1991).

In general, one can conclude that for most slow soil chemical reactions it is likely that processes taking place at the solid phase are rate-determining and are transport-controlled. These transport-controlled reactions are most probably diffusion-controlled. The fact that diffusion-controlled reactions are dominant indicates that the kinetics of chemical processes cannot be considered separately from transport phenomena. Thus, such a combination of processes cannot be treated using first-order or other order chemical kinetics equations. When one states that a reaction between the species A and B is of first-order with respect to A, one assumes that the molecules of A have equal chances of participating in the reaction and therefore the rate is proportional to the concentration C_A. This reasoning can be extended to a reaction between an adsorbing surface and an adsorptive solute. The concentration C_A in this case, refers to the number of reactive sites per unit area, which corresponds to the number of unoccupied sites per unit area $(1-\theta_A)$. However, by using first-order kinetics (or other order kinetics), one assumes that all of the surface sites are potential reactants at any time, and that they have an opportunity of participating in the sorption processes. If one assumes that there are sites that cannot be reached directly from the fluid phase, but can be reached after the sorbate has undergone sorption and desorption at other sites, one cannot separate chemical kinetics from transport kinetics. This is certainly the case with most soils and soil colloids. Therefore, the overall kinetic process obeys a diffusion equation. However, the diffusion coefficient, which reflects the rate at which the sorbate jumps from one site to another, is determined by the rate of the chemical reactions by which the sorbent-sorbate bonds are created and destroyed. Also, the activation energy for diffusion is equal to the activation energy of the chemical reaction (Aharoni and Sparks, 1991).

Kinetics of Rapid Soil Chemical Reactions

Many reactions involving soils and soil components are rapid and occur on time scales of milliseconds and microseconds. These include certain sorption/desorption reactions, ion exchange processes, ion association, and other reactions involving inorganic and organic species (Fig. 1). For such reactions, batch and most flow techniques, with the exception of some stopped-flow methods, are not appropriate. For example, the fastest reactions one can measure with batch and stirred-flow techniques are about 15 seconds (Sparks, 1989; Amacher, 1991; Sparks and Zhang, 1991).

For rapid soil chemical reactions, chemical relaxation methods such as pressure-jump (p-jump), electric field pulse, temperature-jump, and concentration-jump can be employed. In this paper, only a brief discussion on the theory of chemical relaxation, and aspects of p-jump and stopped-flow techniques that have been developed and utilized in our laboratory will be discussed. For more information on rapid kinetic techniques, one can consult a number of sources (Eigen, 1954; Eigen and DeMaeyer, 1963; Takahashi and Alberty, 1969; Bernasconi, 1976; Sparks, 1989; Sparks and Zhang, 1991).

All of the chemical relaxation techniques are based on the principle that the equilibrium of a reaction mixture is rapidly perturbed by some external factor such as pressure, temperature, concentration, or electric field strength. Rate information can then be obtained by following the approach to the original equilibrium by measuring the relaxation time (the time it takes for the system to relax from the nonequilibrium state to the original equilibrium, after the perturbation pulse) via a particular detection system such as conductivity. The perturbation is small and because of this, all rate expressions are reduced to first-order equations regardless of reaction order or molecularity (Bernasconi, 1976). Therefore, the rate equations are linearized, simplifying determination of complex reaction mechanisms. A general linearized rate equation can be derived as:

$$\tau^{-1} = k_1(c_A + c_B) + k_{-1} \qquad [1]$$

where k_1 and k_{-1} are the forward and backward rate constants, respectively, and c_A and c_B are the concentrations of reactants A and B at equilibrium.

Pressure-jump relaxation is based on the principle that chemical equilibria depend on pressure as shown below (Bernasconi, 1976):

$$(\partial \ln K / \partial p)_T = \Delta V / RT \qquad [2]$$

where ΔV is the standard molar volume change of the reaction and p is pressure (MPa). For a small perturbation one can also write,

$$\Delta K / K = -(\Delta V \Delta p / RT) \qquad [3]$$

The p-jump occurs because of a quick pressure release or application. Then, the progression in a reaction is followed after the end of the pressure change. Other details on p-jump relaxation and experimental details can be found in a number of sources (Ljunggren and Lamm, 1958; Strehlow and Becker, 1959; Knoche, 1974; Knoche and Wiese, 1974; Bernasconi, 1976; Knoche and Strehlow, 1979; Sparks, 1989; Sparks and Zhang, 1991).

Recently, Fendorf et. al. (1992c) applied an electron

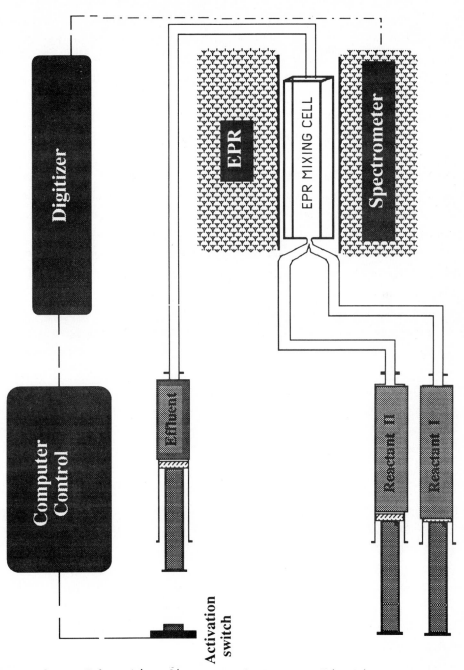

Figure 3. Schematic diagram of EPR-SF kinetic apparatus (Fendorf et al., 1992c).

Recently, Fendorf et. al. (1992c) applied an electron paramagnetic resonance stopped-flow (EPR-SF) technique to colloidal suspensions that is capable of measuring a reactant species _in-situ_ on a millisecond time scale. This technique, provided that one is measuring species that are EPR active in solution, has some advantages over most chemical relaxation approaches. With many of the latter, the reaction of interest must be fully reversible, reactant species are not directly measured, and rate constants are determined from linearized rate equations that are often dependent on equilibrium parameters. With this method, the stopped-flow mixing can be conducted in ≤10ms (and EPR signal digitized within a few μs).

A schematic diagram of the EPR-SF technique is shown in Fig.3. Dual 2mL in-port syringes feed an EPR-SF mixing cell. The mixing cell, located in the EPR spectrometer, allows for EPR detection of the cell contents. A single outflow port is fitted with a 2mL effluent collection syringe equipped with a triggering switch. The triggering switch activates the data acquisition system. Each run consists of filling the in-port syringes with the desired reactants, flushing the system with the reactants several times, and finally initiating and monitoring the reaction. Consecutive runs can be conducted with the same reactants simply by repeating the procedure.

Fendorf et.al.(1992c) used this method to study the kinetics of Mn(II) sorption on δ-MnO$_2$. The sorption reaction was essentially complete within 200ms (Fig.4). Data were taken every 50 μs and 100 points were averaged to give the time-dependent sorption of Mn(II) as shown. By maintaining a large excess of sorbent over sorbate, and monitoring the reaction far from equilibrium where the reverse desorption reaction rate was minimal, the overall reaction was pseudo first-order depending only on Mn(II) concentration- permitting only the forward rate constant, $k_{sorption}$, to be measured.

Kinetics of Environmentally Important Reactions on Soils and Soil Components

In this section, we should like to discuss the kinetics of some environmentally important reactions in soils. Since brevity

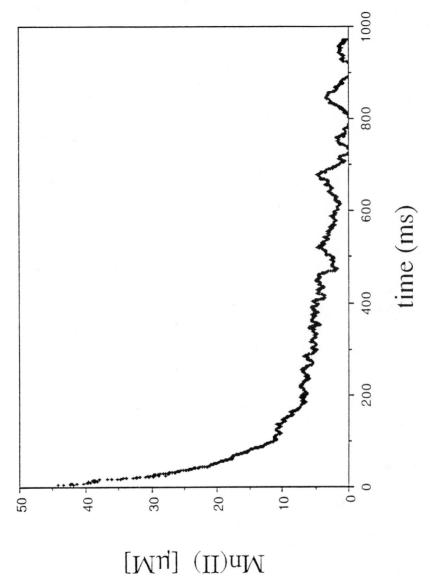

Figure 4. Typical curve of Mn(II) sorption on MnO_2 (Fendorf, 1992c).

is necessary we shall focus on some studies that have been investigated in our laboratories. We hope that the readers will excuse us for leaving out the research of others who have made significant contributions in the area of kinetics of soil chemical processes.

Anion Adsorption/Desorption Dynamics

Most anion adsorption/desorption reactions on soils and soil components are rapid. Recently, Zhang and Sparks (1989, 1990a,b) have successfully used pressure-jump relaxation to study the adsorption/desorption kinetics of metals including selenate, selenite, sulfate, and molybdate on goethite. Hayes and Leckie (1986) and Japanese workers (see e.g., Yasunaga and Ikeda, 1986 and references cited in Sparks, 1989) have also used p-jump relaxation to study cation and anion adsorption and desorption on metal oxides.

We shall summarize the results of the selenate and selenite studies of Zhang and Sparks (1990b). Selenate adsorption occurred mainly under acidic conditions. In the pH range studied, the dominant selenate species was SeO_4^{2-}, since the pK_2 for selenious acid is 2. As pH increased, SeO_4^{2-} adsorption rapidly decreased. At pH 2.98, the total percent of adsorption was 93. At pH>7.2, no adsorption occurred. Selenate adsorption was described very well with the triple layer model. With this model it was assumed that SeO_4^{2-} occurred via outer-sphere surface complexation (Zhang and Sparks, 1990b).

Zhang and Sparks (1990b) observed a single relaxation for selenate adsorption on goethite. Based on this and the equilibrium modeling, it was assumed that SeO_4^{2-} adsorption involved outer-sphere complexation and could be described as:

$$XOH + H^+ + SeO_4^{2-} \rightleftharpoons XOH_2^+ - SeO_4^{2-} \qquad [4]$$

where XOH represents 1 mol of reactive surface hydroxyl bound to a Fe ion in goethite.

A linearized relationship between reciprocal relaxation time and the concentration of species in suspension for the reaction in Eq.[4] was derived as:

$$\tau^{-1} = k_1 \, ([XOH][SeO_4^{2-}]+[XOH][H^+]+[SeO_4^{2-}][H^+])+k_{-1} \qquad [5]$$

where the terms in the brackets are the concentrations of species at equilibrium. Since the reaction is conducted at the solid/liquid interface, then the electrostatic effect has to be

considered in calculating the intrinsic rate constants (k^{int}).
Using the triple layer model to obtain electrostatic parameters,
a first-order equation is derived,

$$\tau^{-1} \exp\left(\frac{-F(\Psi\alpha-2\Psi\beta)}{2RT}\right) = k_1^{int}\left[\exp\left(\frac{-F(\Psi\alpha-2\Psi\beta)}{RT}\right)\right]\,([XOH][SeO_4^{2-}]$$

$$+ [XOH][H^+] + [SeO_4^{2-}][H^+]) + k_{-1}^{int} \quad [6]$$

A plot of this relationship was linear (Fig.5) indicating the
mechanism proposed in Eq.[4] may be appropriate. However, to
definitively confirm the mechanism, one would need to make
microscopic measurements (see later discussion).

Total selenite adsorption on goethite decreased with an
increase in pH. Using the modified triple layer model (TLM),
that assumes metals could be adsorbed in either the alpha
(inner-sphere complexes) or beta (outer-sphere complexes)
layers, it was found that in the pH range studied, selenite
adsorbed on goethite to form monovalent and bivalent selenite-Fe
complexes, since selenite existed as both SeO_3^{2-} and $HSeO_3^-$ in
suspension over the pH range studied. The amounts of both
complexes, $XHSeO_3^o$ and $XSeO_3^-$ were significant in suspension. The
amount of $XSeO_3^-$ dropped precipitously at about pH 8.3 ($pK_2=8.24$
for $HSeO_3^-$); the amount of $XSeO_3^-$ increased with pH until pH 8.3
and then dropped as the total adsorption decreased. The total
adsorption as predicted by the TLM matched the experimental data
very well.

Pressure-jump relaxation curves revealed a double
relaxation for selenite adsorption on goethite. Zhang and Sparks
(1990b) hypothesized a comprehensive two-step adsorption
mechanism:

Step 1   Step 2

Step 1 involves the formation of outer-sphere surface complexes
($XOH_2^+-HSeO_3^-$ and $XOH_2^+-SeO_3^{2-}$) in the β layer. Step 2 is a ligand-
exchange process, where the adsorbed selenite enters the α layer
and replaces a ligand from the goethite surface to form the
inner-sphere surface complexes ($XHSeO_3^o$ and $XSeO_3^-$). Two
protolytic equilibria (K_5 and K_6) are rapidly established

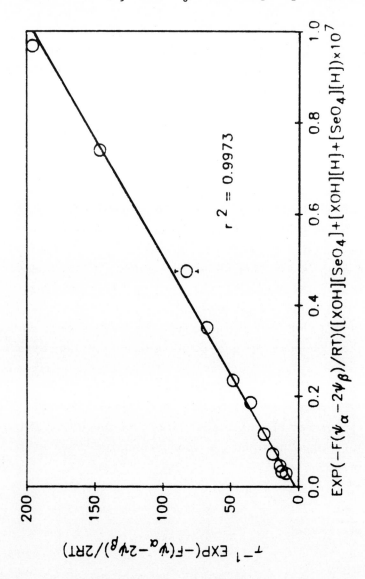

Figure 5. Plot of relationship between τ^{-1} with exponential and
concentration terms in Eq. [6] (Zhang and Sparks, 1990b).

compared to the complexation reactions.

To confirm the mechanisms in Eq.[7], Zhang and Sparks (1990b): (1) derived equations to show the relationships between the reciprocal relaxation times and the concentrations of species involved in the reaction; (2) solved the equations so that rate constants could be obtained for the expected mechanism; and (3) used the respective equation, observed values for $\tau^{-1}(\tau_{obsd}^{-1})$, and other concentration terms in the equation to calculate the four unknown intrinsic rate constants (k_{int}) at four of the pH levels studied. These rate constants were then inserted into the rate equation for other pH levels that were studied to calculate values of $\tau^{-1}(\tau_{calcd}^{-1})$. If the assumed mechanism is acceptable, the τ_{calcd}^{-1} values should match the τ_{obsd}^{-1} values. Also, the intrinsic equilibrium constants for formation of $XHSeO_3^{\circ}$ and $XSeO_3^{-}$ determined from the intrinsic rate constants $(K_{kinetic}^{int})$, which were calculated from the above derived equations, should be consistent with those obtained from equilibrium studies (K_{model}^{int}). As shown in Table 1 the $\tau_{1,obsd}^{-1}$ and $\tau_{2,obsd}^{-1}$ values agreed well with the $\tau_{1,calcd}^{-1}$ and $\tau_{2,calcd}^{-1}$ values, confirming the mechanism in Eq.[7]. The $K_{kinetics}^{int}$ and K_{model}^{int} values, although not shown, also compared well.

Table 1. Relaxation Data for Selenite Adsorption and Desorption on Goethite as a Function of pH at 298 K and an Ionic Strength of 0.02 M (Zhang and Sparks, 1990b).

pH	$\tau_{1,obsd,}^{-1}s^{-1}$	$\tau_{1,calcd,}^{-1}s^{-1}$	$\tau_{2,obsd,}^{-1}s^{-1}$	$\tau_{2,calcd,}^{-1}s^{-1}$
6.41	41.6	41.6	9.12	8.54
7.02	46.5	45.1	10.64	9.84
7.50	57.7	58.7	13.62	13.03
7.81	64.1	65.2	14.64	15.21
8.36	79.6	80.3	18.20	19.89
8.73	105.0	105.9	26.19	27.55
9.07	171.5	171.5	29.57	31.68
9.32	249.4	269.7	39.98	42.58
9.64	357.1	357.3	90.50	91.69

The conclusions of the study of Zhang and Sparks (1990b) that selenate and selenite are adsorbed on goethite as outer- and inner-sphere complexes, respectively, were previously directly confirmed with extended x-ray absorption fine structure spectroscopy (EXAFS) by Hayes et al. (1987).

Ion Exchange Kinetics on Clay Minerals

Since the pioneering research of J. Thomas Way in 1850 it has been known that metal ions can be exchanged on the surfaces of soil components. In Way's work (1850) it was shown that ion exchange on soil surfaces is rapid, in most cases instantaneous. Such reaction rates are too rapid to measure with batch and flow techniques. Recently, Tang and Sparks (1992) have used p-jump relaxation to study Ca-K and Ca-Na exchange on montmorillonite, an important clay mineral in soils.

A single relaxation (Tang and Sparks, 1992) was observed in both exchange systems. In these exchange systems, the following general reaction can be written:

$$2MX + M^{2+} \rightleftharpoons MX_2 + 2M^+ \qquad\qquad [8]$$

where M = Na or K and X denotes the montmorillonite surface. Based on this general reversible reaction, Tang and Sparks (1992) derived a linearized rate equation relating τ^{-1} to k_1 and k_{-1} values along with equilibrium concentration values for the ions. This equation is given below,

$$\tau^{-1} = k_1 \; ([\overline{MX}]^2 + 4[\overline{MX}][\overline{Ca}]) + k_{-1} \; ([\overline{M}]^2 + 4[\overline{CaX_2}][\overline{M}]) \qquad [9]$$

where [ ] with the overbar denotes equilibrium concentration.

Plots of Eq.[9] (rearranged) are shown in Fig. 6. The linearity of the plots confirms that Eq.[8] correctly describes the mechanism for Ca-K and Ca-Na exchange on montmorillonite. Thermodynamic equilibrium constants calculated kinetically (k_1/k_{-1}) agreed well with equilibrium constants calculated from static studies. This is an indication that with p-jump relaxation, chemical kinetics of ion exchange can be measured for Ca-K and Ca-Na exchange on montmorillonite since a kinetic approach cannot be employed to calculate equilibrium constants if mass transfer is occurring (Ogwada and Sparks, 1986).

Kinetics and Mechanisms of Cr(III) Oxidation on Mn-Oxide

Fendorf and Zasoski (1992) studied the rates of Cr(III) oxidation on Mn-oxide using a stirred-flow method. With such a

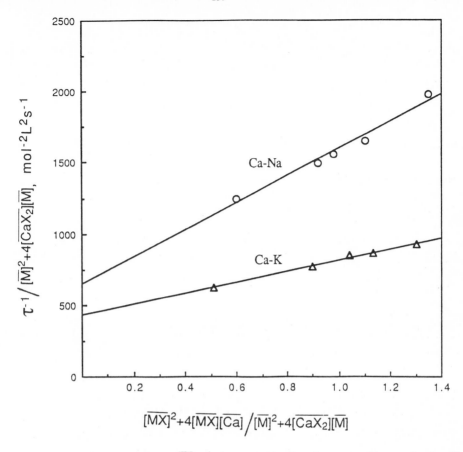

Figure 6. Plots of Eq. [9] (rearranged) for Ca-Na and Ca-K exchange on montmorillonite (Tang and Sparks, 1992).

method, relatively rapid reaction rates can be measured (>30 seconds), transport phenomena are minimized, and reactant concentrations are maintained and products are removed (Sparks, 1989; Amacher, 1991). The oxidation rate, followed by measuring the formation of Cr(VI), was too rapid at pH 5 to be measured with the stirred-flow method. Rates of reaction were lower at pH 3, and with a constant solution reactant influent, were dependent on the amount of surface consumed in the reaction. As the MnO_2 was consumed, the rate decreased, with the reaction continuing until the surface was decreased, viz., reaction rate was dependent on the amount of MnO_2. Rapid initial rates of

Cr(VI) formation at pH 5 were followed by a cessation in oxidation (Fendorf and Zasoski, 1992).

Based on these rate studies and other macroscopic investigations, the authors hypothesized that the inhibition of Cr(III) oxidation on Mn-oxide as Cr(III) concentration and pH increase was due to surface precipitation of chromium hydroxide. Of course, this conclusion was based strictly on macroscopic data. To be definitively confirmed, one needs to employ spectroscopic and microscopic techniques.

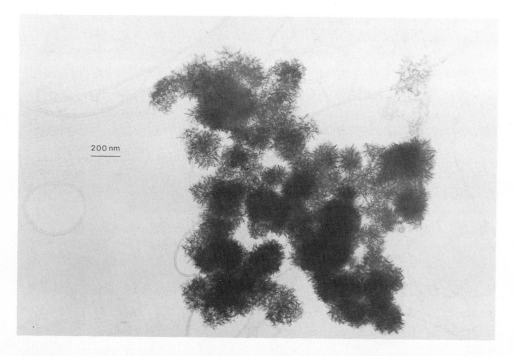

Figure 7.(a). Low magnification HRTEM image of unreacted δ-MnO$_2$ showing spherical clusters of needles characteristic of synthetic birnessite.

In a series of papers, Fendorf and coworkers (Fendorf et al., 1991, 1992a,b) employed high resolution transmission electron microscopy (HRTEM), energy dispersive x-ray spectroscopy (EDS), and electron energy loss spectroscopy (EELS) to ascertain the mechanism for inhibition of Cr(III) oxidation

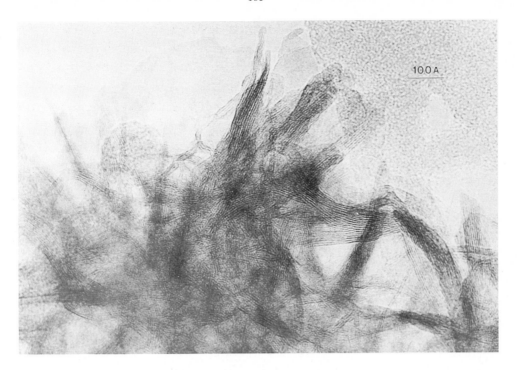

Figure 7.(b).   High   magnification   HRTEM   image   of δ-MnO$_2$
depicting   crystalline   needles   interspersed   with   amorphous
material (Fendorf et al., 1992a).

on MnO$_2$. High resolution TEM shows that the unreacted MnO$_2$ is
partially crystalline with spherical clusters of needles,
characteristic of synthetic birnessite (Fig. 7). After reaction
with Cr(III), a significant change in crystal structure results
(Fig.8). The birnessite needles are removed and the inner
material has fused, resulting in a convoluted mass (Fendorf et
al., 1992a). The oxide appears to have undergone dissolution,
with a new crystalline solid phase forming.

By also using EDS and EELS, Fendorf et al. (1992a)
confirmed that a surface precipitate, Cr(OH)$_3$, formed and was
responsible for the inhibition of Cr(III) oxidation on Mn-oxide
at higher Cr(III) concentrations and pH (greater than 4). This
was also independently confirmed through extended x-ray
absorption fine structure spectroscopy (EXAFS) studies (Fendorf
et al., unpublished data).

The above studies show the importance and desirability of employing microscopic and spectroscopic approaches for elucidating reaction mechanisms of ions and molecules on colloidal surfaces. The use of in-situ techniques that do not require high vacuums and that the sorbent be dried, such as EXAFS, are highly desirable for studying reactions on colloidal surfaces. We are presently investigating an array of metal interactions including chromium, copper, mercury, and lead on metal oxide surfaces using EXAFS.

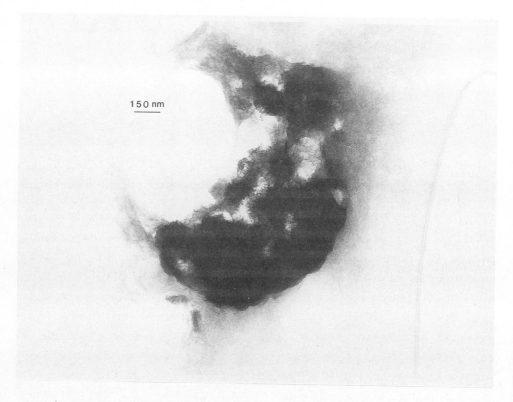

150 nm

Figure 8.(a). Low magnification HRTEM image of δ-MnO$_2$ after reaction with 1 m$\underline{M}$ Cr(III) pH 4.

Reactions of Organic Contaminants on Clay Surfaces

With the great concern about mobility of pesticides and industrial organics in soils, and their ultimate impact on water quality, it is imperative that rate studies be conducted that better simulate field reactions. Unfortunately, the literature in this area is sparse. In particular, desorption studies are lacking, especially ones that explore the effects of long-term reactions between soils and organics. Such data are needed if we are to minimize mobility of pollutants in soils and remediate contaminated soils. The need for long-term sorption and desorption studies with organics and indeed, inorganics, and soils and soil components was pointed out by Steinberg et al. (1987) who noted marked differences in the rates of 1,2-dibromoethane (EDB) reactions on freshly prepared and field "aged" samples.

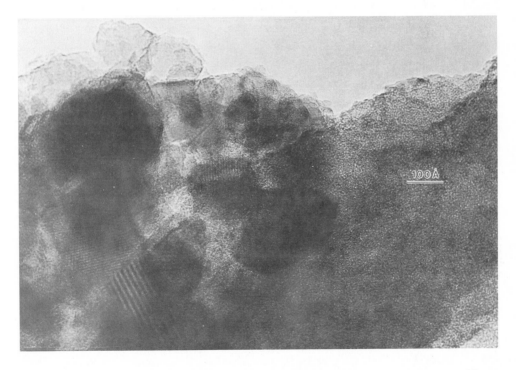

Figure 8.(b). High magnification HRTEM image of δ-MnO$_2$ after reaction with 1 m$\underline{M}$ Cr(III), pH 4 (Fendorf et al., 1992a).

Recently, Tang and Sparks (1992, unpublished data) have investigated the desorption of atrazine, a heavily used herbicide in crop production, from montmorillonite that was "aged" with atrazine. Desorption increased with "aging" time, indicating that slow diffusion processes were occurring. These results seem to corroborate the hypotheses of Steinberg et al. (1987) that organics slowly diffuse into and out of the micropores of soils and soil components.

Recently, studies have explored the use of organo-clays (clays that are amended with surfactants) for enhancing the sorption of organic pollutants (Mortland et al., 1986; Boyd et al., 1988). Such interactions increase the hydrophobicity of the clay surface and thus, increase organic pollutant retention. Organo-clays appear to have usefulness in settings where clay liners are employed to minimize the mobility of organic pollutants into ground water.

However, despite the potential use of organo-clays in minimizing soil and water pollution, many questions remain. For example, the kinetics of organo-clay and pollutant interactions have not been studied to any extent. In an effort to understand the dynamics of these reactions, Zhang and Sparks (1992) investigated the rates of phenol and aniline sorption/desorption from aqueous solutions on a hexadecyltrimethylammonium (HDTMA) clay. These kinetic studies were carried out using the stirred-flow method of Carski and Sparks (1985). Adsorption of the two organic pollutants was complete in about 40 minutes, while an apparent equilibrium in desorption was reached in about 30 minutes. There was marked hysteresis with phenol (about 42% of the sorbed phenol was subsequently desorbed) while about 70% of the adsorbed aniline was desorbed. Competitive sorption appeared to occur when the two organics were mixed and reacted with the organo-clay.

CONCLUSIONS

In this paper, we have attempted to provide an overview on various aspects of rates of environmentally important reactions on natural systems. This important area of research is only in its infancy and there are many unanswered questions. There is a veritable need to: combine in-situ spectroscopic techniques with rate studies to elucidate mechanisms of inorganic and organic pollutant reactions on soils; develop models that include kinetic parameters to better predict the mobility of pollutants in soils; conduct short- and long-term studies on pollutant desorption from soils in order to develop sound strategies for decontaminating soils and minimizing soil pollution; and, more fully understand the rates and mechanisms of humic substance interactions with pollutants.

REFERENCES

Aharoni C (1984) Kinetics of adsorption:The S-shaped z(t). Adsopt Sci Technol 1:1-29

Aharoni C, Sparks DL (1991) Kinetics of soil chemical reactions - A theoretical treatment. In: Sparks DL, Suarez DL (eds) Rates of soil chemical processes. Soil Sci Soc Am, Madison, WI

Aharoni C, Sparks DL, Levinson S, Ravina I (1991) Kinetics of soil chemical reactions: Relationships between empirical equations and diffusion models. Soil Sci Soc Am J 55:1307-1312

Aharoni C, Suzin Y (1982a) Application of the Elovich equation to the kinetics of occlusion: Part 1. Homogeneous microporosity. J Chem Soc Faraday Trans 1 78:2313-2320

Aharoni C, Suzin Y (1982b) Application of the Elovich equation to the kinetics of occlusion: Part 3. Heterogeneous microporosity. J Chem Soc Faraday Trans 1 78:2329-2336

Amacher MC (1991) Methods of obtaining and analyzing kinetic data. In: Sparks DL, Suarez DL (eds) Rates of soil chemicalprocesses. Soil Sci Soc Am, Madison WI, p 19-60

Bernasconi CF (1976) Relaxation kinetics, Academic Press, New York

Boyd SA, Mortland MM, Choiu CT (1988) Sorption characteristics of organic compounds on hexadecyltrimethylammonium-smectite. Soil Sci Soc Am J 52:652-657

Bunzl K, Schmidt W, Sansoni B (1976) Kinetics of ion exchange in soil organic matter. J Soil Sci 27:32-41

Carski TH, Sparks DL (1985) A modified miscible displacement technique for investigating adsorption-desorption phenomena in soils. Soil Sci Soc Am J 49:1114-1116

Chien SH, Clayton WR (1980) Application of Elovich equation to the kinetics of phosphate release and sorption in soils. Soil Sci Soc Am J 44:265-268

Eigen M (1954) Ionic reactions in aqueous solutions with half-times as short as 10^{-9} second. Applications to neutralization and hydrolysis reactions. Discuss Faraday Soc 17:194-205

Eigen M, DeMaeyer L (1963) Relaxation methods. Tech Org Chem 8(2):895-1054

Fendorf SE, Fendorf M, Sparks DL, Gronsky R (1991) Use of TEM for characterization of reactions of MnO_2 Cr(III) and Al(III). Proc 49th Annual Meeting of Electron Microscopy Society of America p 634-635

Fendorf SE, Zasoski RJ (1992) Chromium(III) oxidation by δ-MnO_2. 1. Characterization. Environ Sci Technol 26:79-85

Fendorf SE, Fendorf M, Sparks DL, Gronsky R (1992a) Inhibition mechanisms of Cr(III) oxidation of manganese oxides. J Colloid Interface Sci, In press

Fendorf SE, Sparks DL, Fendorf M, Gronsky R (1992b) Surface precipitation reactions on oxide surfaces. J Colloid Interface Sci 148:295-298

Fendorf SE, Sparks DL, Franz JA, Camaioni DM (1992c) An EPR stopped-flow kinetic study of rapid metal sorption reactions in colloidal suspensions. Soil Sci Soc Am J, In review

Gardiner WC Jr. (1969) Rates and mechanisms of chemical reactions. Wiley, New York

Havlin JL, Westfall DG, Olsen SR (1985) Mathematical models for potassium release kinetics in calcareous soils. Soil Sci Soc Am J 49:371-376

Hayes KF, Leckie JO (1986) Mechanism of lead ion adsorption at the goethite-water interace. ACS Symp Ser 323:114-141

Hayes KF, Roe HL, Brown GE, Hodgson KO, Leckie JO, Parks GA (1987) In situ X-ray adsorption study of surface complexes: Selenium oxyanions on α-FeOOH. Science 238:783-786

Jardine PM, Sparks DL (1984) Potassium-calcium exchange in a multireactive soil system: II. Thermodynamics. Soil Sci Soc 48:45-50

Knoche W (1974) Pressure-jump methods. In: Hammes GG (ed) Investigations of rates and mechanisms of reactions. Wiley, New York, NY

Knoche W, von Strehlow H (1979) Data capture and processing in chemical relaxation measurements. In: Gettins WJ, Wyn-Jones E (eds) Techniques and applications of fast reactions in solution. Reidel Publ., Dordrecht, The Netherlands

Knoche W, Wiese G (1974) An improved apparatus for pressure-jump relaxation measurements. Chem Instrum 5:91-98

Ljunggren S, Lamm (1958) A relaxation method for the determination of moderately rapid reaction rates near chemical equilibrium. Acta Chem Scand 12:1834-1850

Mortland MM, Shaobai S, Boyd SA (1986) Clay-organic complexes as adsorbents for phenol and chlorophenols. Clay Clay Miner 34:581-585

Ogwada RA, Sparks DL (1986) A critical evaluation on the use of kinetics for determining thermodynamics of ion exchange in soils. Soil Sci Soc Am J 50:300-305

Onken AB, Matheson RL (1982) Dissolution rate of EDTA extractable phosphate from soils. Soil Soc Sci Am J 46:276-279

Skopp J (1986) Analysis of time dependent chemical processes in soils. J Environ Qual 15:205-213

Sparks DL (1989) Kinetics of soil chemical processes. Academic Press, New York

Sparks DL (1991) Chemical kinetics and mass transfer processes in soil constituents. In: Bear J, Corapcioglu MY (eds) Transport processes in porous media, Kluwer Academic Publishers B V, Dordrecht, p 585-637

Sparks DL (1992) Soil kinetics. In: Nierenberg WA (ed) Encyclopedia of earth system science. Vol 4, Academic Press, New York, p 219-229 In press

Sparks DL, Jardine PM (1984) Comparison of kinetic equations to describe K-Ca exchange in pure and in mixed systems. Soil Sci Soc Am J 138:115-122

Sparks DL, Suarez DL (eds) (1991) Dynamics of soil chemical phenomena, Soil Sci Soc Am Spec Publ 27, Soil Sci Soc Am, Madison, WI

Sparks DL, Zhang P (1991) Relaxation methods for studying soil chemical processes. In: Sparks DL, Suarez DL (eds) Rates of soil chemical processes, Soil Sci Soc Am Spec Publ 27, Soil Sci Soc Am, Madison, WI

Steinberg SM, Pignatello JJ, Sawhney BL (1987) Persistence of 1,2- Dibromoethane in soils: Entrapment in intraparticle micropores. Environ Sci Technol 21:1201-1208

Strehlow von H, Becker M (1959) Ein Drucksprung-verfahren zur messung der Geschwindigkeit. Z Elektrochem 63:457-561

Takahashi MT, Alberty RA (1969) The pressure-jump methods. In: Kustin K (ed) Methods in enzymology. Academic Press, New York, NY

Tang L, Sparks DL (1992) Kinetics of calcium-potassium exchange on montmorillonite using pressure-jump relaxation. Soil Sci Soc Am J, In press

Yasunaga T, Ikeda L (1986) Adsorption-desorption kinetics at the metal-oxide-solution interface studied by relaxation methods. ACS Symp Ser 323:230-253

Way JT (1850) On the power of soils to absorb manure. J R Agric Soc Engl 11:313-379

Zhang PC, Sparks DL (1989) Kinetics and mechanisms of molybdate adsorption at the goethite/water interface using pressure-jump relaxation. Soil Sci Soc Am J 53:1028-1034

Zhang PC, Sparks DL (1990a) Kinetics and mechanisms of sulfate adsorption/desorption on goethite using pressure-jump relaxation. Soil Sci Soc Am J 54:1266-1273

Zhang PC, Sparks DL (1990b) Kinetics of selenate and selenite adsorption/desorption at the goethite/water interface. Environ Sci Technol 24:1848-1856

Zhang PC, Sparks DL (1992) Kinetics of phenol and aniline adsorption and desorption on an organo-clay (HDTMA-Montmorillonite) Soil Sci Soc Am J, In press

Adsorption-Desorption Methodologies and Selected Estimation Techniques for Transport-Modeling Parameters

W. R. Roy
Illinois State Geological Survey
615 East Peabody Drive
Champaign, Illinois 61820 USA

The physicochemical processes by which contaminants move through soil and groundwater are complex. The task of measuring or estimating numerical values that describe contaminant movement is essential in understanding the fate of potential groundwater pollutants. When chemicals are released from hazardous wastes sites, landfills, or chemical spills and enter the ground, the mass of the chemical may partition among the solid, liquid, and vapor/gas phases present in the subsurface. This chapter focuses on three topics; (1) the experimental measurement of adsorption and desorption of chemicals by soil, (2) techniques for estimating coefficients that are needed for computer-assisted modeling of volatilization and movement of organic-chemical vapors and (3), the biodegradation of trichloroethylene in groundwater.

Adsorption-Desorption at Solid-Liquid Interfaces

The migration rate and concentration of chemicals in groundwater determine how far the chemicals have moved, the extent of environmental degradation, and what remedial action can be implemented. Two of the most important geochemical processes affecting the rate of migration of chemicals is the adsorption to and desorption from

NATO ASI Series, Vol. G 32
Migration and Fate of Pollutants in Soils and Subsoils
Edited by D. Petruzzelli and F. G. Helfferich
© Springer-Verlag Berlin Heidelberg 1993

the solid phases (Olsen and Davis, 1990). The adsorption from solution at the solid-liquid interface is a complex and imperfectly understood phenomenon. The physicochemical forces responsible for adsorption can be broken down into eight categories, and have been discussed by Reinbold et al., 1979; Voice and Weber, 1983; Koskinen and Harper, 1990; and Roy et al., 1992a. These forces include London-van der Waals and Coulombic-electrostatic forces, hydrogen bonding, ligand exchange-anion coordination, chemisorption, dipole-dipole or orientation energy, induction or dipole-induced dipole, and the hydrophobic effect.

Estimating Adsorption Constants

The adsorption of chemicals from solution is often expressed as an adsorption isotherm. An adsorption isotherm is a graphical representation of the amount of solute adsorbed by an adsorbent as a function of the equilibrium concentration of the solute. The relationship is quantitatively defined by some type of partition function or adsorption equation that is statistically applied to the data to generalize the results. In studies concerned with the adsorption of gases by solids, more that 40 equations have been used to describe the data. Historically, only a few of the equations have been applicable to solid-liquid systems. Only one of the most commonly used of these equations will be discussed here--the Freundlich isotherm. This equation may not be appropriate for a given system. Kinniburgh (1986) discussed the applicability of other adsorption equations.

Probably the oldest, most widely used adsorption equation for solid-liquid systems, the Freundlich adsorption equation may be expressed as

$$x/m = K_f C^{1/n} \tag{1}$$

where x is the amount or concentration of the solute adsorbed, m is the mass of adsorbent, C is the equilibrium concentration of the solute, and K_f and n are system-specific constants.

When the adsorption isotherm is linear, n becomes unity, K_f reduces to a simple partition coefficient (sometimes written as Kd). This linear relationship is a desirable simplification for complex environmental models, as it allows simpler mathematical solutions of transport equations (Green and Karickhoff, 1990a). A linear adsorption constant (Kd) for hydrophobic solutes may be estimated. It has been known for some time that the amount of solute that disappears from solution when in contact with soil often correlates with the amount of organic matter in the soil material (Dzombak and Luthy, 1984). This observation has led to the advancement of the organic carbon-water partition coefficient (Koc), which is a soil-independent, compound-specific value. Koc values are experimentally derived from adsorption data using the relationship

$$Koc = Kd \ (100)/\%organic \ carbon \qquad (2)$$

The Koc values of many hydrophobic and organic solutes have been compiled and are given elsewhere (Kenaga, 1980; Kenaga and Goring, 1980, Banerjee et al., 1980; Hassett et al., 1983, Roy and Griffin, 1985, and Green and Karickhoff, 1990b). In order to estimate or characterize the extent of contaminant movement in site investigations, the organic carbon content of site samples is often measured, and eq. (2) is then used to estimate an adsorption constant. The constant is then incorporated into transport codes. It is, however, important to recognize three conditions or limitations that are inherent when applying this approach (Roy and Griffin, 1985):

1. K_f (or Kd) is an equilibrium constant, a value that is meant to describe solute partitioning between the adsorbent and solvent in a system that has attained chemical equilibrium or a steady state.

2. Koc values apply to relatively dilute systems or systems in which the 1/n term of the Freundlich adsorption equation is near 1.

3. Koc values only estimate adsorption as a function of the organic carbon content of the soil or soil materials; they do not take into account the contribution of inorganic hydrophobic sites of mineral matter.

The third limitation becomes especially significant in material with low organic-carbon contents, such as frequently the case in aquifers. It has been recognized that, below a certain level of organic carbon, mineral surfaces contribute to or dominate the extent of adsorption. It has been proposed that, when the organic carbon content of an adsorbent is below about 0.1%, the organic carbon fraction is not a valid predictor of the partitioning of nonpolar organic compounds (Stauffer and MacIntyre, 1986). Murphy et al. (1990) found, for example, that when adsorbents had an organic carbon content of less than about 0.5%, Koc values for the adsorption of carbazole, dibenzothiophene, and anthracene by hematite and kaolinite were sometimes higher, and sometimes lower than those predicted by the organic-carbon content. Schwarzenbach and Westall (1981) found that the adsorption of chlorobenzene was greater than predicted by the organic carbon content of the adsorbents below an organic carbon level of about 0.08%. The Koc of chlorobenzene is about 319 mL/g (Roy and Griffin, 1990), and the apparent Koc for a kaolin clay (0.06% OC) was measured as 1000. The Koc of trichloroethylene is about 106 mL/g (Garbarini and Lion, 1985). Pavlostathis and Jaglal (1991) reported apparent Koc values for trichloroethylene ranging from 344 to 9000 mL/g for clays and aquifer materials with an organic carbon content of less than or equal to 0.13%. Similarly, Hicken et al. (1991) indicated that the adsorption of atrazine by synthetic low organic-carbon clay mixtures did not correlate to organic carbon content.

Roy et al. (1992b) found that the adsorption of atrazine by low organic-carbon outwash sands did not correlate with the organic carbon content of the samples; the adsorption constants tended to be higher that those predicted by the application of eq. (2). They found that the extent of atrazine adsorption correlated more strongly with the pH of the sand-water suspensions, and that the use of only organic carbon would over-predict the mobility of atrazine. Clearly, the application of Koc values to predict the migration of organic solutes is a powerful tool, but the direct experimental measurement of adsorption may be required.

There is no *a priori* technique for estimating adsorption constants for ionic solutes such as polar organics or heavy metals. Hence, acquisition of adsorption data requires direct measurement.

Adsorption Measurements

The quantitative measurement of the extent of adsorption may be conducted by using three techniques; empirical field data, column studies, and batch or static reactor studies. The scope of this chapter will be confined to the application of batch-reactor procedures for measuring adsorption and desorption. Batch techniques have been used to measure adsorption for many years, but the experimental approach used warrants careful consideration in order to derive reproducible and unambiguous results. The batch-static technique consists of mixing a series of samples of the adsorbent with solutions in closed vessels containing the solute(s) under study for a period of time long enough to allow the attainment of adsorbate-solute equilibrium, and then the phases are separated. The solutions are then analyzed for the solute, and the difference between the initial amount in solution and the final concentration is taken as an indirect measurement of the amount that has partitioned to the solid phases. The apparent simplicity of this approach accounts for its wide use, but there are a number of important experimental factors that should be considered in order to generate meaningful data.

Effects of Adsorbent Preparation. The procedures used to prepare samples for laboratory investigations can directly influence adsorption results. Adsorbent samples (e.g., soils and clays) are usually dried so that they can homogenized, split and stored until needed. Studies (Ashton and Sheets, 1959; Van Lierop and MacKenzie, 1977; Dao and Lavy, 1978; Raveh and Avnimelech, 1978; Bartlett and James, 1980) have shown that the methods used to dry the sample may alter its chemical properties, which in turn can influence the results of batch-adsorption procedures. Oven-drying may increase the hydrophobicity of a soil, which in turn would increase the adsorbents's affinity for hydrophobic solutes. Research has established that, for example, forest fires can increase the hydrophobicity of soil materials near the ground surface. Bartlett and James (1980) found that a soil sample that had been oven-dried at 40° C adsorbed more phosphate than samples that were kept moist. Samples should be air-dried even though air-drying may take several days with large bulk samples. Air-drying anaerobic soils and sediments may require drying in an oxygen-free glove box or glove bag.

Effects of Temperature. Batch-adsorption measurements should be conducted under constant-temperature conditions. Kinniburgh and Jackson (1981) concluded that the effects of temperature on cation adsorption were small, but in some cases, temperature significantly influenced the adsorption data. For example, Kuo and Mikkelsen (1979) studied the adsorption of zinc by soils at a temperature range of 10° to 35° C and found that zinc adsorbed endothermically; increased adsorption was associated with higher temperature. The adsorption of phosphate by soils is often endothermic (Griffin and Jurinak, 1973; Singh and Jones, 1977; Taylor and Ellis, 1978; Roy et al., 1989).

Other studies (Hassett et al., 1983) found that, in contrast to the adsorption of ionic species, the adsorption of the nonpolar solutes phenanthrene and alpha-naphthol was largely unaffected by temperature variations from 15° to 35° C. The adsorption of 1,2-dichlorobenzene by a soil sample studied by Chiou et al. (1979) was not affected greatly by temperature differences between 3.5° and 20° C, but the adsorption of 1,1,1-trichloromethane was reduced at the lower temperature. Weber et al. (1983) found that the adsorption of Aroclor 1254 by river sediments was temperature-dependent; adsorption was reduced over a 10-degree range.

Because temperature variations may affect adsorption results, the temperature of the solutions should either be controlled by the use of a constant-temperature water bath or room. At the very least, the temperature of the solutions should be measured and reported with the adsorption data.

Stability of Solutes in Solution. In conducting batch adsorption measurements, investigators must consider the physicochemical stability of the solute in solution. Processes such as precipitation from a chemically saturated solution, photodegradation, volatilization, hydrolysis, and microbial degradation can potentially contribute to a decrease in solute concentration concomitantly with adsorption. For example, the photolytic half-life of hexachlorocyclopentadiene was found to be less than 5 minutes when exposed to sunlight (Chou and Griffin, 1983). Therefore, precautions should be taken to ensure that substances such as these are shielded from light, not only sunlight, but laboratory lights as well. The processes that can affect solute stability are solute-specific. Screening procedures for identifying stability problems are given in Roy et al. (1992a).

Effects of Solution pH. The adsorption behavior of soils in contact with ionic and ionizable inorganic and organic solutes is often influenced by the pH of the suspension. In general, the adsorption of inorganic cations increases with increasing pH (Kinniburgh and Jackson, 1981). The adsorption of inorganic anions is generally greater in acidic solutions. The adsorption of triazine compounds, for example, is greater at lower pH because triazine may be increasingly protonated which increases the magnitude of coulombic interaction with negatively charged sites on clay surfaces. The adsorption of neutral, nonpolar hydrophobic organic solutes appears to be largely unaffected by the pH of the soil-water system. Roy et al. (1992a) reported that the adsorption of the PCB Aroclor 1242 was not significantly influenced by pH in a pH range of 3 to 10.

Depending on the needs of the investigator, the pH of the solutions used to measure the extent of adsorption may need to be controlled by the addition of dilute acids or bases. At the very least, the pH of each solution should be measured, and reported with the adsorption data. The failure to measure and report pH data may render the adsorption data impossible to interpret.

Effects of Phase Separation and Method of Mixing. In batch adsorption studies in the literature, filtration was seldom used to separate the liquid from solid phases before analyzing the liquid phase. This is probably because the filter membrane can retain significant quantities of the solute, particularly organic solutes. Yaron and Saltzman (1972), and Griffin and Chou (1980), for example, initially used filter membranes in studies with organic solutes, but abandoned this approach in favor of centrifugation because of solute retention by the membranes. The investigator should always use centrifugation unless it can clearly be demonstrated that filtration does not significantly affect the adsorption results.

In batch studies, there is a need to agitate or mix the adsorbent-water suspensions. Mixing methods have included the use of hand shaking, rollers, mixers, reciprocating shakers, and rotating tumblers. Studies by Barrow and Shaw (1979) and Roy et al (1992a) have indicated that the method of mixing can influence adsorption data. Where a vigorous agitation was used, there may be the breakdown of soil particles, and the generation of "new" exposed sites for adsorption. Studies by Roy et at. (1992a) demonstrated that the end-over-end movement of rotating tumblers reduced the coefficient of variation in adsorption measurements in inter-laboratory comparisons.

Selection of a Solid:Liquid Ratio for Ionic Solutes. In batch studies, there is the need to select a ratio of adsorbent to liquid. If the solid:liquid ratio is too low, i.e, there is too much adsorbent or too little solution present, most of the solute may be adsorbed, calling for accurate measurements of small differences in low concentrations. If the ratio is too high, i.e., not enough adsorbent, the changes in the initial concentrations may be very small, which in turn may yield considerable scatter in the isotherms. A suitable ratio for ionic solutes cannot be determined *a priori*. A value of 10% to 30% adsorption for the highest solute concentration used is a useful criterion for selecting a suitable solid:liquid ratio. Experimentally, one may mix five or six different solid:liquid ratios (from 1:100 to 1:4) with the highest solute concentration for 24 hours. The solid:liquid ratio that results in about 10 to 30% adsorption should give satisfactory results for constructing an adsorption isotherm.

Selection of a Solid:Liquid Ratio for Nonionic Solutes. A simple calculation can be used to estimate a suitable solid:liquid ratio for nonionic solutes, particularly hydrophobic organic compounds. The derivation of this technique is given in Roy et al. (1992a). A suitable ratio (R) may be estimated from

$$R = 20/[(80)K_f] = 1/4K_f \qquad (3)$$

where $K_f = K_{oc}(\%OC)/100$.

For example, to obtain 20% adsorption of benzene by a soil containing 2.3% organic carbon, the estimated K_f would be

$$K_f \ (mL/g) = 83 \ mL/g(2.3\%)/100 = 1.91$$

The K_{oc} of benzene is 83 mL/g (Kenaga and Goring, 1980). Hence, R = 1/[4(1.91)] = 1:7.64, or simply, 1:8. Hence, a 1:8 solid:liquid ratio should yield a 20% reduction in the amount of benzene initially present. Twenty percent was chosen for illustration because it falls in the middle of the 10 to 30% window discussed above.

Constant and Variable Solid:Liquid Ratios. There are two experimental techniques used to generate batch-adsorption data (Roy et al., 1992a):

1. Constant solid:liquid ratio method: mixing a batch of aqueous solutions--each solution containing progressively decreasing solute concentrations--with adsorbent, keeping the amount of adsorbent (by weight) constant in all solutions.

2. Variable solid:liquid ratio method: mixing a batch of solutions (such as portions of the same leachate or extract of a solid waste) with progressively increasing amounts of adsorbent.

In the first technique, the initial (stock) solution--a solution prepared in the laboratory or a leachate collected in the field--is progressively diluted, forming a batch of dilutions that are added to the containers holding the same amount of adsorbent. The use of the second technique avoids diluting the initial solution. Such dilutions may result in unwanted side reactions (changes in pH or the reduction-oxidation potential) that will alter the chemical composition of the solution.

Determination of Equilibration Time. Adsorption at the solid-liquid interface is a thermodynamic phenomenon, and adsorption measurements are made when the system has equilibrated. Chemical equilibrium has been attained when the concentrations of the solutes cease to change with respect to time. Past studies have used equilibration times from 30 minutes to 6 days and longer. Adsorption is generally regarded as a fast reaction, and subsequent removal of solute from solution may be attributed to other processes. Karickhoff (1980) suggested that the adsorption process has a rapid component and a slower component, the rate of the latter depending largely on the movement of the solute to less-accessible adsorption sites. In some cases, equilibrium is never clearly attained. The ambiguity in the definition and measurement of equilibration intervals has been acknowledged as a major problem in adsorption studies (Anderson et al., 1981). For many systems involving complicated, heterogenous materials such as soils, it is very difficult to determine when adsorption processes dominate and when they become less important as other processes such as ion penetration, precipitation, or degradation, become significant.

It has been suggested that the equilibration time should be the minimum amount of time needed to establish a rate of change of solute concentration equal to or less than 5% in a 24-hour interval. This operational definition of adsorption equilibrium was evaluated by Roy et al. (1992a). When a suitable solid:liquid ratio was used (i.e., one that resulted in 10% to 30% adsorption), it was found that the rate of change in the concentration of both organic and inorganic solutes tended to be less than 5% after 24 hours of mixing. In some cases, the equilibration of inorganic anions required a 48-hour interval. Green and Karickhoff (1990a) summarized that, for most practical applications, equilibration intervals of 24 hours probably produce adsorption coefficients that represent macroscale equilibrium quite well, even though equilibrium may not be attained with internal adsorbent sites. It should be noted that, if the system appears to equilibrate within 24 hours, then the procedure for determining a solid:liquid ratio may yield data for constructing an adsorption isotherm, given that enough solutions, at least five, yield greater than 10% adsorption.

Desorption Measurements

Modelers often assume that adsorption is reversible. Once adsorbed, however, some chemicals may react further to become covalently and irreversibly bound, while others may become physically trapped in the soil matrix (Koskinen and Harper, 1990). The solutions generated by batch procedures can be used to measure desorption. The desorption behavior of adsorbates has not been measured as frequently as adsorption, and the methodologies for measuring desorption have not evolved as far as those for adsorption.

Desorption data have often been obtained by centrifuging the equilibrated solution (after the adsorption measurement), then removing a portion of the liquid, and replacing it with an equal volume of solution without the solute, equilibrating the system, then determining the amount of solute left in solution. Several such dilution steps would generate a desorption isotherm (Bowman, 1979). The dilution of the solution creates a concentration gradient away from the adsorbed phase, inducing the desorption of the adsorbate back into solution. This approach yields data points that may or may not conform to the adsorption isotherm. Hysteresis may, in some cases, be the result of experimental artifacts, lack of adsorption equilibrium, slow desorption kinetics, and solute/adsorbate instability. The causes of adsorption hysteresis are not well understood, and are beyond the scope of this lecture.

Volatilization and Solid-Vapor Partitioning of Organic Chemicals

Although adsorption may be a significant process in the environmental fate of some chemicals, many organic chemicals are highly volatile and may escape into the atmosphere from waste sites. Volatilization from the unsaturated zone and groundwater may be the most important fate of some organic chemicals. Henry's Law estimates the tendency of a chemical to partition between water and its gas or vapor phase. In general, when the Henry's Law constant (H) of a chemical is greater than about 10^{-4} atm-m^3/mol, then volatilization can be an important mechanism whereby a large fraction of the chemical is slowly removed from solution, depending on the mole fraction of the components in the aqueous phase, temperature gradients, and the extent of mixing. The Henry's Law constant of carbon tetrachloride, for example, is 3.0×10^{-2} atm-m^3/mol (Gossett, 1987). The magnitude of this value indicates that this compound would readily volatilize from water in the unsaturated zone. Bonazountas and Wagner (1984) modeled the fate of carbon tetrachloride in the unsaturated zone for a hypothetical scenario, and found that 99.3% of the CCl_4 present would volatilize and that 0.1% would remain as an adsorbed phase. About 0.6% was predicted to migrate down to the water table.

Henry's Law constants have been measured for some volatile organic chemicals, but data are lacking for most. An H value may be estimated as

$$Vp_{(t)}/S_{(t)} = H \text{ (atm-m}^3/\text{mol)} \tag{4}$$

where Vp is the vapor pressure of the chemical (atm) at temperature t, and S is the solubility of the chemical in water (mol/m^3) at the same temperature.

Organic vapors move through the unsaturated zone in response to either convection or diffusion. Several functional relationships have been developed between soil or porous media-diffusion coefficients and Fick's Law diffusion coefficients in terms of various soil properties. The expression of Millington and Quirk (1961), for example (eq. 5), has been used frequently to estimate a soil diffusion coefficient, and was the preferred method of Springer et al. (1984), but only after an empirical factor was added.

$$D_s = D_F n_a^{3.3}/n_t^2 \tag{5}$$

where D_s = soil diffusion coefficient

D_F = Fick's Law diffusion coefficient

n_a = air-filled porosity, and

n_t = the total porosity

There is no universally accepted method for estimating diffusion coefficients for soils (Roy and Griffin, 1990). The equation of Weeks et al. (1982) is another method to estimate a soil diffusion coefficient, and may be written as

$$D_s = Tn_a D_F/n_t + (n_t - n_a)p_w K_w + (1 - n_t)p_s K_s \tag{6}$$

where T = soil tortuosity factor (dimensionless term)

p_w = density of water (g/cm^3),

K_w = liquid-gas partition coefficient (volume [gas]/weight [solid]),

p_s = particle density of the porous media (g/cm^3), and

$K_s = K_w K_d$ where K_d is a liquid-solid partition

coefficient (volume[gas]/weight [solid]).

The application of eq. 6 requires several input parameters. There are very few data for the adsorption of vapor-phase organic chemicals by soil materials for the purposes of modeling vapor-phase movement. Moreover, there is no *a priori* method to predict adsorption constants for the partitioning of organic vapors onto porous media. Such constants for nonpolar chemicals may be estimated by using adsorption constants derived from saturated soil-water systems, if the relative humidity of the soil atmosphere is high. Chiou and Shoup (1985), Smith et al. (1990) and Pennell et al. (1992) have found that adsorption isotherms at high relative humidities were similar to those of liquid-solid systems. It is generally agreed that water molecules, because of their polar nature, effectively compete with nonpolar organic vapors for adsorption sites (Pennell et al., 1992). Because soil air usually has a relative humidity of 98 to 100%, first-order adsorption coefficients for the transport modeling of nonpolar vapors can be estimated from the liquid-solid literature (Roy and Griffin, 1990). Hence, the interaction of organic chemicals as vapors with soil materials may be separated into two processes; (1) gas-solid partitioning described by Henry's Law or an analogous relationship, and (2) solid-water partitioning described by a conventional adsorption constant.

Biodegradation of Organic Solutes: An Example with Trichloroethylene

Biodegradation may be another important fate of some organic compounds in groundwater. The selection of input parameters with which to model biodegradation without site-specific data may be difficult. There are no estimation techniques that can be used to predict biodegradation half-lives. Such values are often selected from the scientific literature. Trichloroethylene (TCE) is commonly found in groundwater at hazardous waste sites, and literature on TCE biodegradation will be summarized to illustrate the difficulty in selecting a biodegradation half-life for the purposes of transport modeling.

TCE does not biodegrade rapidly, but may do so slowly under anaerobic conditions. Dilling et al. (1975) reported that the half-life of TCE was about 10.7 months. This conclusion was based on a laboratory study in which water was purged with air just prior to the addition of TCE to the water. They concluded that most of the disappearance of TCE was probably due to oxidation. Given that groundwater is usually reduced (i.e., low in dissolved oxygen), selection of this half-life would probably underestimate TCE persistence.

Wilson et al. (1981) found that TCE was not significantly biodegraded under aerobic conditions in 45 days when a TCE solution was percolated through a low organic-carbon sandy soil. About 58 to 88% of the TCE volatilized from solution. Bouwer et al. (1981) used batch studies in which anaerobic conditions were created by seeding a methanogenic mixed culture in a laboratory-scale digester fed by activated sludge. The authors concluded that, after 16 weeks, TCE did not biodegrade significantly, and no estimate of a half-life was given.

In a comprehensive study on field data collected in the Netherlands, Zoetmann et al. (1981) estimated that the half-life of TCE in groundwater was about two years. It was not possible, however, to determine the removal mechanism (i.e., adsorption, volatilization, biodegradation or hydrolysis). Wilson et al. (1986), in batch studies using aquifer materials, measured the disappearance of TCE under methanogenic conditions. Their data suggest that half the initial TCE was gone after 193 days. Barrio-Lage et al. (1987) published data suggesting that the half-life of TCE was about 85 days. TCE depletion was studied in microcosms containing water and subsurface materials.

Howard et al. (1991) estimated that the half-life of TCE in groundwater was between 10.7 months and 4.5 years, based on the hydrolysis half-life (i.e., Dilling et al. (1975)), and anaerobic sediment sample data of Barrio-Lage et al. (1986).

Modelers are faced with considerable uncertainty in selecting biodegradation half-lives. It is likely that the great variation in half-lives is the result of the variation in microorganisms and the lack of standardized conditions under which the measurements were made. Most biodegradation studies document the possible extent of degradation, mechanisms, and degradation products. At present, the persistence of TCE is site-specific, depending on environmental conditions such as the reduction-oxidation potential, the presence of other organic compounds (anthropogenic and natural), and the biological activity and species of the microorganisms present. The modeler's only recourse is to use environmentally conservative estimates with caution. In the case of TCE, a half-life of 4.5 years may be the most conservative estimate available in the literature.

References

Anderson MA, Bauer C, Hansmann D, Loux N, Stanforth R (1981) Expectations and limitations for aqueous adsorption chemistry. In: Anderson MA, Rubin AJ (eds) Adsorption of Inorganics at Solid-Liquid Interfaces, Ann Arbor Publishers, Michigan, 327-347

Ashton FM, Sheets TJ (1959) The relationship of soil adsorption of EPTC to oats injury in various soil types. Weeds 7: 88-90

Banerjee S, Yalkowsky SH, Valvani SC (1980) Water solubility and octanol/water partition coefficients of organics: Limitations of the solubility-partition coefficient correlation. Environmental Science and Technology 14: 1227-1229

Barrio-Lage G, Parsons FZ, Nassar RS (1987) Kinetics of the depletion of trichloroethene. Environmental Science and Technology 21: 366-370

Barrio-Lage G, Parsons FZ, Nassar RS, Lorenzo PA (1986) Sequential dehalogenation of chlorinated ethenes. Environmental Science and Technology 20: 96-99

Bartlett R, James B (1980) Studying dried, stored soil samples--some pitfalls. Soil Science Society of America Journal 44: 721-724

Barrow NJ, Shaw TC (1979) Effects of soil:solution ratio and vigour of shaking on the rate of phosphate adsorption by soil. Journal of Soil Science 30: 67-76

Bonazountas M, Wagner J (1984) Modeling mobilization and fate of leachates below uncontrolled hazardous wastes sites. In: Fifth National Conference on Management of Uncontrolled Hazardous Wastes Sites, Washington, DC, Nov. 7-9, 1984, 97-102

Bouwer EJ, Rittman BE, McCarty PL (1981) Anaerobic degradation of halogenated 1- and 2-carbon organic compounds. Environmental Science and Technology 15: 596-599

Bowman BT (1979) Method of repeated additions for generating pesticide adsorption-desorption isotherm data. Canadian Journal of Soil Science 59: 435-437

Chiou CT, Peters LJ, Freed VH (1979) A physical concept of soil-water equilibria for nonionic organic compounds. Science 206: 831-832

Chiou CT, Shoup TD (1985) Soil sorption of organic vapors and effects of humidity on sorptive mechanism and capacity. Environmental Science and Technology 19: 1196-1200

Chou SFJ, Griffin RA (1983) Soil, clay, and caustic soda effects on solubility, sorption, and mobility of hexachlorocyclopentadiene. Environmental Geology Notes 104, Illinois State Geological Survey

Dao TH, Lavy TL (1978) Atrazine adsorption on soil as influenced by temperature, moisture content, and electrolyte concentration. Weed Science 26: 303-308

Dilling WL, Tefertiller NB, Kallos GJ (1975) Evaporation rates and reactivities of methylene chloride, chloroform, 1,1,1-trichloroethane, trichloroethylene, tetrachloroethylene, and other chlorinated compounds in dilute aqueous solutions. Environmental Science and Technology 9: 833-837

Dzombak DA, Luthy G (1984) Estimating adsorption of polycyclic aromatic hydrocarbons on soils. Soil Science 137: 292-308

Garbarini DR, Lion LW (1985) Evaluation of sorptive partitioning of nonionic pollutants in closed systems by headspace analysis. Environmental Science and Technology 19: 112-128

Gossett JM (1987) Measurement of Henry's Law constants for C1 and C2 chlorinated hydrocarbons. Environmental Science and Technology 21: 202-208

Green RE and Karickhoff SW (1990a) Sorption estimates for modeling. In: Pesticides in the Soil Environment. Soil Science Society of America Books Series, no. 2, Chapt. 4, 79-101

Green RE and Karickhoff SW (1990b) Estimating sorption coefficients for soils and sediments. In: DeCoursey DG (ed) Small watershed model (SWAM) for water, sediment, and chemical movement: Supporting documentation. U.S. Department of Agriculture, ARS-80

Griffin RA, Chou SFJ (1980) Attenuation of polybrominated biphenyls and hexachlorobenzene by earth materials. Environmental Geology Notes 87, Illinois State Geological Survey

Griffin RA, Jurinak JJ (1973) The interaction of phosphate with calcite. Soil Science Society of America Proceedings 37: 847-850

Hassett JJ, Banwart WL, Griffin RA (1983) Correlation of compound properties with sorption characteristics of nonpolar compounds by soils and sediments: Concepts and limitations. In: Francis CW, Auerback SI (eds) Characterization, Treatment, and Disposal, Environment and Solid Wastes. Butterworth Publishers, Mass., Chapter 15, 161-178

Hicken ST, Barr JS, McLean JE, Doucette WJ (1991) Sorption behavior of pesticides in soils containing low organic carbon. Agronomy Abstracts, 1991 Annual Meeting of the Soil Science Society of America, Denver, Colorado, October 27-November 1, 1991, 245

Howard PH, Boethling RS, Jarvis WF, Meylan WM, Michalenko EM (1991) Handbook of Environmental Degradation Rates. Lewis Publishers, Chelsea, Michigan

Karickhoff SW (1980) Sorption kinetics of hydrophobic pollutants in natural sediments. In: Baker RA (ed) Contaminants and sediments, v. 2, Ann Arbor Science Publishers, Ann Arbor MI, 193-205

Kenaga EE (1980) Predicted bioconcentration factors and soil sorption coefficients of pesticides and other chemicals. Ecotoxicology and Environmental Safety 4: 26-38

Kenaga EE, Goring AI (1980) Relationship between water solubility, soil sorption, octanol-water partitioning, and concentration of chemicals in biota. In: Eaton JG, Parrish PR, Hendricks AC (eds) Aquatic Toxicology. American Society for Testing and Materials, ASTM STP 707, 78-115

Kinniburgh DG (1986) General purpose adsorption isotherms. Environmental Science and Technology 20: 895-904

Kinniburgh DG, Jackson ML (1981) Cation adsorption by hydrous oxides and clays. In: Anderson MA, Rubin AJ (eds) Adsorption of Inorganics at Solid-Liquid Interfaces. Ann Arbor Publishers, Michigan, 91-160

Koskinen WC, Harper SS (1990) The Retention Process: Mechanisms. In: Pesticides in the Soil Environment. Soil Science Society of America Book Series, no. 2, Chapt. 3, 51-77

Kuo S, Mikkelsen DS (1979) Zinc adsorption by two alkaline soils. Soil Science 128: 274-279

Murphy EM, Zachara JM, Smith SC (1990) Influence of mineral-bound substances on the sorption of hydrophobic organic compounds. Environmental Science and Technology 24: 1507-1516

Millington RJ, Quirk JP (1961) Permeability of porous solids. Transactions of the Faraday Society 57: 120-127

Olsen RL, Davis A (1990) Predicting the fate and transport of organic compounds in groundwater. Hazardous Materials Control, v. 3, 38+ (various pages)

Pavlostathis SG, Jaglal K (1991) Desorptive behavior of trichloroethylene in contaminated soil. Environmental Science and Technology 25: 274-279

Pennell KD, Rhue RD, Rao PS, Johnson CT (1992) Vapor-phase sorption of p-xylene and water on soils and clay minerals. Environmental Science and Technology 26: 756-763

Raveh A, Avnimelech Y (1978) The effect of drying on the colloidal properties and stability of humic compounds. Plant and Soil 50: 545-552

Reinbold KA, Hassett JJ, Means JC, Banwart WL (1979) Adsorption of energy-related organic pollutants: a literature review. U.S. Environmental Protection Agency, Athens, GA, EPA-600/3-79-086

Roy WR, Griffin RA (1985) Mobility of organic solvents in water-saturated soil materials. Environmental Geology and Water Science 7: 241-247

Roy WR, Griffin RA (1990) Vapor-phase interactions and diffusion of organic solvents in the unsaturated zone. Environmental Geology and Water Sciences 15: 101-110

Roy WR, Hassett JJ, Griffin RA (1989) Quasi-thermodynamic basis of competitive-adsorption coefficients for anionic mixtures in soils. Journal of Soil Science 40: 9-15

Roy WR, Krapac IG, Chou SFJ, and Griffin RA (1992a). Batch-type procedures for estimating soil adsorption of chemicals. U.S. Environmental Protection Agency, Technical Resource Document, U.S. EPA/530/-SW-87-006-F

Roy WR, McKenna DP, Krapac IG, Chou SFJ, Mehnert E (1992b) Groundwater contributions to atrazine loadings in streams: First-year summary. In: Proceedings of the Second Annual Conference of the Illinois Groundwater Consortium, Springfield, IL, April 27-28, 1992

Schwarzenbach R P, Westall J (1981) Transport of non-polar organic compounds from surface water to groundwater: Laboratory sorption studies. Environmental Science and Technology 15: 1360-1367

Singh BB, Jones JP (1977) Phosphorus sorption isotherm for evaluating phosphorus requirements of lettuce at five temperature regimes. Plant and Soil 46: 31-44

Smith JA, Chiou CT, Kammer JA, Kile DE (1990) Effect of moisture on the sorption of trichloroethene vapor to vadose-zone at Picatinny Arsenal, New Jersey. Environmental Science and Technology 24: 676-683

Springer C, Lunnery PD, Valsaraj KT, Thibodeaux LJ (1984) Emissions of hazardous chemicals from surface and near-surface impoundments to air. U.S. Environmental Protection Agency Project Report 808161-02, Cincinnati, Ohio.

Stauffer T, MacIntyre WG (1986) Sorption of low-polarity organic compounds on oxide minerals and aquifer material. Environmental Toxicology and Chemistry 5: 949-955

Taylor RW, Ellis BG (1978) A mechanism of phosphate adsorption on soil and anion exchange resin surfaces. Soil Science Society of America Journal 42: 432-436

Van Lierop W, MacKenzie AF (1977) Soil pH measurement and its applications to organic soils. Canadian Journal of Soil Science 57: 55-64

Voice TC, Weber WJ (1983) Sorption of hydrophobic compounds by sediments, soils, and suspended solids. I. Theory and background. Water Research 17: 1433-1441

Weber WJ, Voice TC, Pirbazari M, Hunt GE, Ulanoff DM (1983) Sorption of hydrophobic compounds by sediments, soils, and suspended solids. II. Sorbent evaluation studies. Water Research 17: 1443-1452

Weeks EP, Earp DE, Thompson GM (1982) Use of atmospheric fluorocarbons F-11 and F-12 to determine the diffusion parameters of the unsaturated zone in the southern high plains of Texas. Water Resources Research 18: 1365-1378

Wilson JT, Enfield CG, Dunlap WJ, Cosby RL, Foster DA, Baskin LB (1981) Transport and fate of selected organic pollutants in a sandy soil. Journal of Environmental Quality 10: 501-506

Wilson BH, Smith GB, Rees JF (1986) Biotransformation of selected alkybenzenes and halogenated aliphatic hydrocarbons in methanogenic aquifer material: A microcosm study. Environmental Science and Technology 20: 997-1002

Yaron B, Saltzman S (1972) Influence of water and temperature on adsorption of parathion by soils. Soil Science Society of America Proceedings 36: 583-586

Zoetmann BCJ, De Greef E, Brinkmann FJJ (1981) Persistency of organic contaminants in groundwater, lessons from soil pollution incidents in the Netherlands. The Science of the Total Environment 21: 187-202

PART II. GLOBAL PROPAGATION PHENOMENA AND MODELING.

MULTIPHASE CONTAMINANTS IN NATURAL PERMEABLE MEDIA: VARIOUS MODELING APPROACHES

M. Yavuz Corapcioglu,
R. Lingam, and K.K.R. Kambham
Department of Civil Engineering
Texas A&M University
College Station, TX 77843-3136, USA

Sorab Panday
HydroGeoLogic, Inc.
1165 Herndon Parkway
Herndon, VA 22070, USA

Over the last decade, subsurface contamination by non-aqueous phase liquids (NAPL) has attracted considerable attention. With the increasing awareness of the potential of the mathematical modeling that offers understanding of the problems of subsurface flow and contaminant transport, a large number of papers have been published in the literature. This chapter summarizes the various approaches which have been employed to model the flow of groundwater/soil water, non-aqueous phase liquids, and soil air in a subsurface environment. These modeling studies can be classified in five categories. The first category, which is the most comprehensive, is the compositional approach. It incorporates the components comprising a NAPL. It has been demonstrated that various well-known equations of porous media flow such as Buckley-Leverett, Richards, black-oil, and solute transport are simplifications of the compositional model. The second category includes multiphase models with interphase mass transfer. The third category of models is those which incorporate capillary effects. The fourth group, sharp-interface models, do not consider any mass transfer or capillarity by assuming a piston-like idealization of the multiphase flow. The final category, which is the simplest of all, models the subsurface as a single cell. General modeling approaches in each category are discussed and solutions to practical problems are presented. The choice of modeling approach depends on the level of accuracy required and the availability of the information describing the complexities of the subsurface and fluids.

The research presented in this chapter has been partially supported by grants from National Science Foundation CEE-8401438, American Chemical Society/The Petroleum Research Fund PRF #15890-AC5, U.S. Department of Interior G-897/02, Gulf Coast Hazardous Substance Research Center, which is supported under cooperative agreement R 815197 with the U.S. EPA and the State of Texas as part of the program of the Texas Hazardous Waste Research Center. The contents do not necessarily reflect the views and policies of the U.S. EPA or the state of Texas, nor does the mention of trade names or commercial products constitute endorsement or recommendation for use.

Multiphase Multicomponent Models

Mass balance equations for transport processes in reservoirs and in porous media have been developed to various degrees of complexity by a number of researchers. Corapcioglu and Baehr (1987), Abriola and Pinder (1985), Kazemi *et al.* (1978), Ngheim *et al.* (1973), Crookston *et al.* (1979), Coats (1980), Young and Stephenson (1983), Youngren (1980), to name a few, have developed compositional simulators of a high degree of sophistication, as is necessary to model the multiphase, multicomponent flow. For a review of the subject, the reader is referred to Corapcioglu and Panday (1991a). A compositional simulator describes mass balances for components present in the reservoir fluids rather than the phases and, therefore, the complexity of the system is higher. In the definition of governing equations, our starting point is the three-dimensional macroscopic conservation of mass equations for each component in each phase, i.e., water, gas oil (NAPL) and soil solids.

The conservation of mass equations for any component (i) present in the water phase can be expressed as

$$
\frac{\partial}{\partial t} (C_w^i \rho_w n S_w) + \nabla \cdot (C_w^i \rho_w J_w - D_w^i \nabla \rho_w n S_w C_w^i)
$$
$$
- (R_{ow}^i - R_{wo}^i) + (R_{gw}^i - R_{wg}^i) + (R_{sw}^i - R_{ws}^i) + R_{source_w}^i - R_{sink_w}^i
$$

$$(1.1)$$

where C_w^i is the mass fraction of component i (i = 1,...,NC, where NC is the total number of components present in the system) in the water phase (w). A superscript denotes a component, while the subscript refers to a phase. ρ_w is the density of the water phase, S_w the degree of water saturation, n the porosity of the medium, J_w the volumetric flux of the water phase, and D_w^i is the hydrodynamic dispersion coefficient of component (i) in the water phase. Each component (i) in the water phase is transported by the convective flux of the water phase and by the diffusive and dispersive flux of the component (i) within the water phase. We assume that the solid grains (s) of the porous medium are non-reacting to other phases (i.e., $C_w^s = 0$) and, hence, in Eq. (1.1) a component term for solid grains is neglected. This implies that grains do not dissolve in either water or oil, or vaporize in the gas phase. However, in the presence of substances like hydrofluoric acids, sandstone particles can react (Hekim *et al.*, 1982). We must also note that in dealing with reservoirs like salt caverns, the reservoir solids might dissolve in water. All the terms on the right-hand side of Eq. (1.1) are either sources (the positive terms) or sinks (the negative terms) of mass in the water phase. R_{ow}^i is the rate of transfer of mass of (i) from the oil phase to the water phase and R_{wo}^i is the mass transfer rate of component component (i) from the water to the oil phase. Similarly, R_{gw}^i and R_{wg}^i denote mass transfer rates from gas to water, and water to gas phases, respectively. R_{sw}^i and R_{ws}^i are the desorption and adsorption terms which denote mass transfer rates of component (i) between the water phase and the soil solids. The last two terms are source and sink terms due to fluid injection or withdrawal or mass loss due to a degradation process (e.g., biodegradation). We must note that the summation of the equations for each component in the water phase would give the conservation of mass equation for the water phase.

The conservation of mass equation for component (i) (i = 1,...,NC) in the gas phase is

$$\frac{\partial}{\partial t}(C_g^i \rho_g n S_g) + \nabla \cdot (C_g^i \rho_g J_g - D_g^i \nabla \rho_g n S_g C_g^i)$$

$$= (R_{og}^i - R_{go}^i) + (R_{wg}^i - R_{gw}^i) + (R_{sg}^i - R_{gs}^i) + R_{source_g}^i - R_{sink_g}^i \tag{1.2}$$

where C_g^i is the mass fraction of component i in the gas phase, ρ_g is the density of the gas phase, S_g the gas saturation, J_g the volumetric flux of the gas phase, and D_g^i is the hydrodynamic dispersion coefficient of component (i) in the gas phase (g). R_{og}^i, R_{go}^i, R_{sg}^i, and R_{gs}^i are the mass transfer rates of (i) from oil to gas, from gas to oil, from soil solids to gas, and from gas onto soil solids, respectively. $R_{source_g}^i$ and $R_{sink_g}^i$ are source-sink terms for the gas phase, due to injection or withdrawal. Note that $C_g^s = 0$.

The conservation of mass equation for component (i) existing in oil phase is written as

$$\frac{\partial}{\partial t}(C_o^i \rho_o n S_o) + \nabla \cdot (C_o^i \rho_o J_o - D_o^i \nabla \rho_o n S_o C_o^i)$$

$$= (R_{wo}^i - R_{ow}^i) + (R_{go}^i - R_{og}^i) + (R_{so}^i - R_{os}^i) + R_{source_o}^i - R_{sink_o}^i \tag{1.3}$$

where C_o^i is the mass fraction of mass component (i) in the oil phase, ρ_o is the oil phase density, S_o the oil saturation, J_o the volumetric flux of the oil phase, and R_{os}^i and R_{so}^i the adsorption and desorption rates from the oil phase onto soil solids and vice-versa, respectively. D_o^i is the hydrodynamic dispersion coefficient of component (i) in the oil phase (o). $R_{source_o}^i$ and $R_{sink_o}^i$ are the source and sink terms for the oil phase. Note that $C_o^s = 0$. Finally, for a rigid porous medium, the mass balance equation of a component (i) in the solid phase (soil solids) is expressed as

$$\frac{\partial}{\partial t}(C_s^i \rho_s \{1 - n\}) = (R_{ws}^i - R_{sw}^i) + (R_{gs}^i - R_{sg}^i) + (R_{so}^i - R_{os}^i) \tag{1.4}$$

where C_s^i is the mass of component (i) per unit mass of soil solid particles (i.e., mass fraction), n is the porosity, and ρ_s is the density of the soil solids. In the absence of any adsorption and reaction $C_s^i = 1$. In writing Eq. (1.4), we neglected the diffusion and dispersion of components within the solid phase. This may not be true, in cases like activated carbon adsorption, and adsorbed components might diffuse within the secondary pores of carbon particles. Eqs. (1.1) to (1.4) have been discussed in greater detail by Corapcioglu and Baehr (1987) in relation to gasoline components in soil water and air.

The mass balance Eqs. (1.1) to (1.3) for the fluid phases are constrained by the total volume of void spaces existing within the matrix. Hence

$$S_w + S_g + S_o - 1 \tag{1.5}$$

Further, fully compositional models have mass fraction constraints which state that the total mass per unit volume of a phase is equal to the sum of the masses of all components comprising that phase per unit volume of the phase.

The volumetric flux of a fluid phase (f) in a rigid porous medium is obtained by Darcy's law as

$$J_f - - \frac{\rho_f g}{\mu_f} k_{r_f} k_o \nabla \left(\frac{p_f}{\rho_f g} - z \right) \qquad f - o, w, g \tag{1.6}$$

where g is the gravitational acceleration, μ_f the viscosity of the fluid, k_{r_f} the relative permeability to fluid f, k_o the absolute permeability, p_f the pressure in fluid (f), and z the vertical coordinate positive downwards (i.e., depth). In the literature, based on wettability concepts, it is generally assumed that the relative permeability to water is a function of water saturation only. Similarly, the gas relative permeability is a function of gas saturation only, and the oil relative permeability is a function of both gas and water saturations due to Eq. (1.5). Interactions that affect the relative permeability curves due to the other phases present are hence assumed to be small.

The pressures in each of the fluid phases present are related to each other by the capillary pressure which is the pressure difference across the interface of any two fluids and is taken to be positive when the pressure in a wetting phase is subtracted from the pressure in a nonwetting phase. Thus, we have the definition of capillary pressures as

$$p_{c_{aw}} - p_a - p_w \; ; \qquad p_{c_{ow}} - p_o - p_w \; ; \qquad p_{c_{ao}} - p_a - p_o - p_{c_{aw}} - p_{c_{ow}} \tag{1.7-9}$$

Note that Eq. (1.9) is not independent of Eqs. (1.7) and (1.8). $p_{c_{\alpha\beta}}$ is the capillary pressure across two phases, α and β, and can be represented as a function of any one of the phase saturations and the porosity. In many cases, it is assumed that p_c is fairly independent of porosity or the state of consolidation. This function, however, exhibits hysteresis during drainage and imbibition. The sum of Eqs. (1.1) to (1.4) for each component $i = 1, ..., NC$ over all the phases yields

$$\frac{\partial}{\partial t} (C_i^w \rho_w n S_w + C_g^i \rho_g n S_g + C_o^i \rho_o n S_o + C_s^i \rho_s \{1-n\}) + \nabla \cdot (C_w^i J_w \rho_w + C_g^i J_g \rho_g + C_o^i J_o \rho_o$$

$$\tag{1.10}$$

$$- D_g^i \nabla \rho_g n S_g C_g^i - D_o^i \nabla \rho_o n S_o C_o^i - D_w^i \nabla \rho_w n S_w C_w^i) - R_{source}^i - R_{sink}^i$$

where R_{source}^i and R_{sink}^i are the total source and sink terms for component (i) over all phases, and the relationships between the concentrations can be determined using equilibrium thermodynamics. Eq. (1.10) gives one equation for each component (i) distributed among all phases. The distribution between the phases is calculated efficiently, by assuming the ideal case laws to hold. For treatment of non-ideal cases occurring in compositional reservoirs, the reader is referred to Corapcioglu and Panday (1991a). Henry's law partitions masses between the liquid phases. Raoult's law depicts the equilibrium between the gas and aqueous phases, and an adsorption isotherm may be used to partition the mass onto the soil solids. The mass partition coefficients are functions of pressure, temperature and composition of each of the phases as discussed in the next section.

Application of the Multicomponent Approach to a Gasoline Spill Problem

Baehr and Corapcioglu (1987) have solved Eq. (1.10) to simulate a gasoline spill problem in an unsaturated soil. Gasoline, except for minute amounts of compounds containing sulfur, oxygen or nitrogen, is a mixture of hydrocarbons. There are several hundred different hydrocarbon components in various proportions in any single commercial gasoline. Certain components of gasoline have been shown to be hazardous. Benzene, for example, has been determined to be a human carcinogen. The partitioning of components among phases is achieved by assuming that the equilibrium approximation provides a meaningful working assumption. Jennings and Kirkner (1983) presented a criterion for evaluating the appropriateness of using the equilibrium assumption for solute transport, utilizing the Damköhler number which is the ratio of chemical reaction rates to the bulk flow rate. They defined a critical Damköhler number beyond which local equilibrium is valid.

Corapcioglu and Baehr (1987) employed the Raoult's law to quantify the ideal reference state for equilibrium between the oil and air phases. Raoult's law states that the ideal vapor pressure of a component solution is proportional to its volatility, as measured by the vapor pressure over the pure component, and its relative abundance in the oil phase as measured by the mole fraction of the component in the oil phase. By assuming the air phase behaves as an ideal gas, Corapcioglu and Baehr obtained a relationship between gas and oil phase concentrations

$$C_g^i \rho_g = H_{go}^i \gamma_i \left(\frac{1}{\omega_i \sum_{j=1}^{N} \frac{1}{\omega_j} C_o^i} \right) C_o^i \tag{1.11}$$

where ω_i is the molecular weight of the ith component, H_{go}^i is the equilibrium partition coefficient of the ith component between the gas and oil phases, and γ_i is the activity coefficient of the ith component adjusting for nonideality. Corapcioglu and Baehr (1987) applied Henry's law to express the equilibrium between the gas and water phases for the contaminant components. Henry's law is obeyed very well for sparingly soluble, nonelectrolyte constituents assuming an ideal gas phase. Henry's law states that the partial pressure of the ith component above the water phase is proportional to the concentration of the ith component in the water phase. This yields the relationship between the equilibrium concentrations of the oil and water phases,

$$C_w^i \rho_w = H_{wo}^i \gamma_i \left(\frac{1}{\omega_i \sum_{j=1}^{N} \frac{1}{\omega_j} C_o^i} \right) C_o^i \tag{1.12}$$

where H_{wo}^i is the partition coefficient of the ith component between the water and oil phases. Eqs. (1.11) and (1.12) suggest the following relationship between the equilibrium concentrations of the oil and adsorbed phases:

$$C_s^i = \left[H_{so}^i \gamma_i \left(\frac{1}{\omega_i \sum_{j=1}^{N} \frac{1}{\omega_j} C_o^i} \right) \right] C_o^i \tag{1.13}$$

where H_{si}^k is the partition coefficient of the ith component between the solid and the oil phases. Note that Eq. (1.13) holds if, and only if, $C_s^i = (H_{so}^i/H_{wo}^i)C_w^i \rho_w$. H_{so}^i depends on the nature of the solid surfaces and, unlike other partitioning coefficients, depends on the type of the porous media. At this point, it should be noted that fluid phase densities are functions of reservoir temperatures, fluid compositions as well as fluid pressures.

Baehr and Corapcioglu (1987) solved Eq. (1.10) for an immobilized gasoline column in unsaturated soil at residual saturations (i.e., $J_o = 0$; $S_o = S_{o_{res}}$). The gasoline was assumed to consist of eight hydrocarbon components, approximating the gasoline composition reported by Jamison et al. (1976). Benzene, toluene, 1-hexane, cyclohexane, and n-hexane were selected as individual components to analyze the fate of components. Aromatics such as benzene and toluene have relatively low air-water partition coefficients and thus will tend to favor the water phase, while alkanes such as n-hexane have higher air-water partition coefficients and favor the air phase. Cyclohexane (naphthene) and 1-hexene (an alkene) represent hydrocarbons of intermediate partition coefficient. Three composite component aromatics (besides benzene and toluene), alkanes (besides n-hexane), and "heavy ends" (any molecule with more than eight carbon atoms) completed the characterization of the gasoline. Values chosen for the thermodynamic properties of composite constituents were obtained by averaging the properties of individual hydrocarbons in each class. The transport of each component is coupled to the other components via the mole fraction only. Corapcioglu and Baehr (1987) also took into consideration the biodegradation of gasoline components in the water phase. This was achieved by relating the biodegradation rate (R^i_{sink}) to the oxygen concentration in the gas and water phases by a first-order kinetic equation. Baehr and Corapcioglu (1987) first obtained conservative predictions by neglecting the diffusion mechanism in the soil-gas phase.

Figure 1.1 illustrates the partitioning of the total hydrocarbon mass, obtained by summing over the eight hydrocarbons, into various fates (remaining in soil, leached down into the aquifer or biodegraded) as a function of time. With the high estimate of average recharge, $J_w = 105.6$ cm/yr, representative of precipitation rates of the Northeastern United States, one finds that at 100 years, 23% of the total initial hydrocarbon mass has entered the aquifer. Aerobic biodegradation at 100 years cannot account for more than 4% of the initial hydrocarbon mass trapped in the soil under these conditions. Figure 1.2 presents the rates at which the constituent mass enters the groundwater aquifer. Fluxes corresponding to cyclohexane, hexene, hexane, and heavy ends do not appear, since they remain less than 0.2 mg cm^{-2} yr^{-1}. Thus, nonaromatic components are essentially rendered immobile when movement is due to solute transport only. Figure 1.2 illustrates that approximately 82% of the total hydrocarbon flux into the aquifer is due to the three aromatic components, benzene, toluene, and other aromatics, which comprise 27% of the initial hydrocarbon mass trapped in the soil. The flux of total hydrocarbons experiences an undulating decline with time as more soluble components are leached out of the soil column. The points of inflection on the total curve correspond to the times when benzene and toluene are exhausted from the column.

Baehr and Corapcioglu (1987) also predicted the fate of gasoline components by allowing the diffusive transport in the air phase. Predictions obtained assuming nonzero values for soil diffusion coefficients allow for hydrocarbon mass to leave the soil column at the ground surface, entering the atmosphere. Also, oxygen recharge rates were increased as a diffusive component supplements the recharge due to dissolved oxygen in infiltrating water. This increase in available oxygen implies higher aerobic biodegradation rates. Thus, allowing for diffusive transport reduces the total mass available for leaching down to groundwater. Figure 1.3 illustrates the partitioning of hydrocarbon mass into various fates as a function of time. Referring to the total distribution of the initial hydrocarbon mass left the soil column after 70 years of which 17% entered the aquifer, 52% entered the atmosphere and 31% was biodegraded. Figure 1.4 presents the rates at which component mass enters the aquifer with $J_w = 105.6$ cm/yr. The three aromatic components comprise at least 99% of the total hydrocarbon mass entering

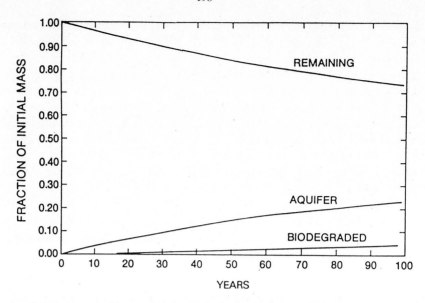

Figure 1.1   Total hydrocarbon distribution with no diffusive transport

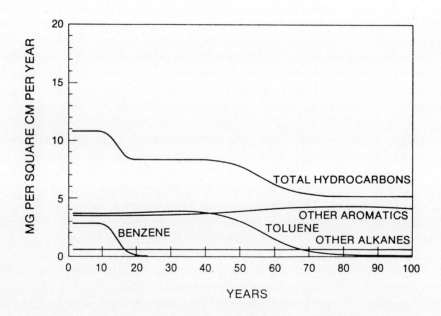

Figure 1.2   Contaminant flux into aquifer with no diffusive transport

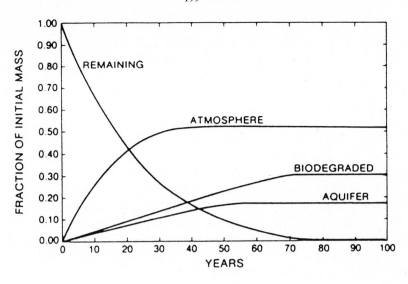

Figure 1.3   Total hydrocarbon distribution with a diffusive transport

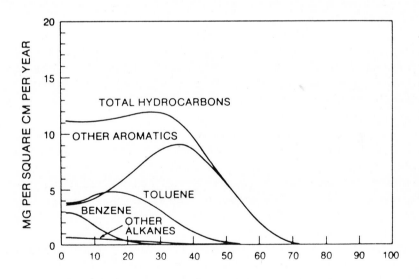

Figure 1.4   Contaminant fluxes into aquifer with a diffusive transport

the aquifer. Nonaromatic components (except for heavy ends), for the most part, ultimately enter the atmosphere.

Other Compositional Approaches

As noted in section 1.1, Corapcioglu and Baehr's (1987) model represented the transport of a NAPL phase with eight components. Abriola and Pinder (1985a,b) also had a model with an assumption of local equilibrium. The NAPL phase was composed of, at most, two distinct components, one of which may be volatile and slightly water soluble and the other is both nonvolatile and insoluble in water. Falta *et al.* (1992) presented a multicomponent model of steam injection for the removal of NAPL from the subsurface. Falta *et al.'s* model contained three mass components, three mass balance equations, and an energy balance equation. Corapcioglu and Panday (1991b) generalized Corapcioglu and Baehr's (1987) model for a non-isothermal system and applied it to predict the migration of petroleum components in frozen soils. In another paper, Panday and Corapcioglu (1991) applied the fundamentals of their formulation to study the solute rejection in freezing soils.

Multiphase Models with Mass Partitioning between Phases

Qualitative descriptions of the migration of immiscible plumes through the unsaturated zone, their spreading over the water table, and the entry of dissolved constituents into the saturated zone are provided by several researchers, including American Petroleum Institute (1972), Schwille (1981), Somers (1974) and Schiegg and Schwille (1991). Since petroleum products are less dense than water (e.g., specific density of gasoline, 0.7-0.8), the water table provides a barrier to the immiscible phase. van der Waarden *et al.* (1971, 1977) found evidence that water-extractable components of petroleum products are leached out of an oil zone by trickling water at a rate determined by partition coefficients of the components. Adsorption reduced concentrations and delayed transport. Hoffman (1970) has solved the classical one-dimensional dispersion equation for various point sources of oil contamination with an assumption of a constant exchange rate and contact area. Hoffman did not consider the immiscible phase in the transport equation. A similar approach has been provided by Fried *et al.* (1979). Pfannkuch (1984) described the theories developed to predict the mass transfer rates across interphase boundaries and has shown that these theories are limited to very simple systems. Despite the lack of published research in groundwater literature on hydrocarbon contamination until recent years, petroleum engineers have long recognized the importance of interphase mass transfer in oil reservoirs. Most oil reservoirs fall into the category of a black-oil reservoir. This is the one in which the water and gas compositions are assumed constant throughout the life of the reservoir, and the oil phase is composed of oil and dissolved gas, the latter being able to transfer mass between the oil and the gas phase depending on physical conditions. Primary oil production is modeled very efficiently using black-oil simulators. They have, in fact, become standard engineering tools in the petroleum industry, and several books are largely devoted to reservoir simulation [Peaceman (1977) and Aziz and Settari (1979)].

Black Oil Model as a Simplification of the Compositional Approach

The compositional model may be simplified by considering the gaseous phase to contain only gas, the water phase to contain only water, and the oil phase containing residual oil, and dissolved gas. Hence, there are a total of three components in this system of three fluid phases. This is the well-known **black-oil model** in reservoir engineering and, in many instances, this simplification may be applied to a reservoir simulation. Furthermore, the adsorption term is also neglected (i.e., $C_S^i = 0$, i = w, o, g).

The above assumptions lead to the constraints for the water component (i = w) as

$$C_w^w - 1 ; \qquad C_o^w - 0 ; \qquad C_g^w - 0 \qquad (2.1)$$

When incorporated into Eq. (1.10) with superscript i equal to w denoting the **component water** without any sink and source terms

$$\frac{\partial}{\partial t}(\rho_w n S_w) + \nabla \cdot (\rho_w J_w) - 0 \qquad (2.2)$$

In writing Eq. (2.2), we neglect both the dispersive flux and molecular diffusive flux due to spatial variations in the mass of water phase per unit water volume of the medium (bulk water phase density), (i.e., $\nabla \rho_w n S_w \sim 0$).

Eq. (2.2) can be obtained by using an alternative approach. If we sum the conservation of mass equation (1.1) for components in the water phase over all the components and note that $\sum_i C_w^i = 1$, we obtain

$$\frac{\partial}{\partial t}(\rho_w n S_w) + \nabla \cdot (\rho_w J_w) - \nabla \cdot \sum_i (D_w^i \nabla \rho_w n S_w C_w^i) - \sum_i \sum_j (R_{jw}^i - R_{wj}^i) + \sum_i (R_{source_w}^i - R_{sink_w}^i) \qquad (2.3)$$

For the case of a non-reacting water phase (i.e., $C_w^w = 1$), the terms on the right-hand side would vanish. The hydrodynamic dispersion coefficients of each compound in the water phase can be assumed as $D_w^w = D_w^o = D_w^g = D_w^i = D_w$. This assumption is justified by assuming that the velocity of each component in the water phase is the same as of the mass weighted velocity of the water phase. This way, the dispersion coefficient depends on the properties and the flow regime of the phase rather than on individual components. Hence the third term within the parentheses on the left hand side of Eq. (2.3) would reduce to $=(D_w \rho_w n \nabla S_w)$. As stated earlier, if we neglect the fluxes due to spatial variations in the mass of the water phase per unit volume of the porous medium, Eq. (2.3) would reduce to Eq. (2.2). Therefore, one does not have to assume that the nonadvective flux terms in the two species equation are equal and opposite in sign and thus cancel each others in the summation to obtain simple phase equations. Instead, gradients of the bulk water phase density ($\rho_w nS_w$) need to be neglected. Similar equations can be obtained for gas and oil phases by summing Eqs. (1.2) or (1.3), respectively, over all the components.

The density of the water in the reservoir is different from the density at surface conditions due to the pressure and temperature differences. To accommodate these changes a formation volume factor for water, B_w, is defined as

$$B_w = \frac{\rho_{ws}}{\rho_w} = \frac{V_w}{V_{ws}} \tag{2.4}$$

where ρ_{ws} is the density of water phase at surface conditions. V_{ws} and V_w are the volume of water at surface conditions and in the reservoir, respectively. Information for the volume factor for the water phase (e.g., Burcik, 1957) will help us to determine ρ_w in the reservoir. Hence, Eq. (2.2) may be written as

$$\frac{\partial}{\partial t}\left(nS_w \frac{\rho_{ws}}{B_w}\right) + \nabla \cdot \left(\frac{\rho_{ws}}{B_w} J_w\right) = 0 \tag{2.5}$$

The black oil assumptions when applied to the **oil component** in the reservoir give

$$C_o^o = \frac{M_o^o}{M_o^o + M_o^g} \; ; \qquad C_g^o = 0 \; ; \qquad C_w^o = 0 \tag{2.6}$$

where M_o^o is the mass of the oil phase, and M_o^g is the mass of gas dissolved in the oil phase. When Eqs. (2.6) are inserted into Eq. (1.10), we obtain

$$\frac{\partial}{\partial t}(C_o^o \rho_o nS_o) + \nabla \cdot (C_o^o \rho_o J_o) = R_{source}^o - R_{sink}^o \tag{2.7}$$

In writing Eq. (2.7) we neglect both the dispersive and diffusive fluxes due to spatial variations in bulk oil phase density. The formation volume factor for oil B_o is defined as

$$B_o = \frac{V_o}{V_{os}} \tag{2.8}$$

where V_o is the volume occupied by the oil phase in the reservoir and V_{os} is the volume of the oil under stock tank conditions (i.e., after the gas has been released due to pressure reduction). The density of oil phase ρ_o is defined as the mass of the oil phase divided by its volume under reservoir conditions. The density of the oil under stock tank conditions ρ_{os} is simply the mass of the oil stock (without the dissolved gas that escaped) divided by the volume of oil under stock tank conditions and, hence, we note that

$$\frac{\rho_{os}}{B_o} - \frac{\dfrac{M_o^o}{V_o^s}}{\dfrac{V_o}{V_o^s}} - \frac{M_o^o}{V_o} \qquad (2.9)$$

Note that also the term $C_o^o\rho_o$ (=mass of oil component per unit volume of oil phase, as used in groundwater hydrology to define concentration) can be expressed in terms of the basic variables of mass and volume as

$$C_o^o\rho_o - \frac{M_o^o}{M_o^o + M_o^g}\frac{M_o^o + M_o^g}{V_o} - \frac{M_o^o}{V_o} \qquad (2.10)$$

Equating Eqs. (2.9) and (2.10) and inserting into Eq. (2.7) gives

$$\frac{\partial}{\partial t}\left(\frac{\rho_{os}}{B_o}nS_o\right) + \nabla\cdot\left(\frac{\rho_{os}}{B_o}J_o\right) - R_{source}^o - R_{sink}^o \qquad (2.11)$$

Finally, the black-oil assumptions to simplify the equation for the **gas component** in the reservoir can be listed as

$$C_g^g - 1 ; \qquad C_o^g - \frac{M_o^g}{M_o^o + M_o^g} ; \qquad C_w^g - 0 \qquad (2.12)$$

which, when incorporated into Eq. (1.10), yield

$$\frac{\partial}{\partial t}(\rho_g nS_g + C_o^g nS_o) + \nabla\cdot(\rho_g J_g + C_o^g \rho_o J_o) - R_{source}^g - R_{sink}^g \qquad (2.13)$$

Also neglected are diffusive and dispersive fluxes due to spatial variations in bulk phase density. With the gas solubility 's' defined as volume of gas dissolved per volume of oil under stock tank conditions, i.e., $s = V_{gs}/V_{os}$ and the gas formation volume factor B_g given as the ratio of volumes occupied by a certain mass of gas under reservoir and at surface conditions, $B_g = \rho_{gs}/\rho_g = V_g/V_{gs}$, where again the differing values of V_g and V_{gs} are due to the temperature and pressure differences between the reservoir and the surface, and ρ_{gs} and ρ_g are the density of gas phase at surface conditions and in the reservoir, respectively. Similarly, V_{gs} and V_g are the volume of the gas phase at surface conditions and in the reservoir, respectively. Then we can rewrite Eq. (2.13) as

$$\frac{\partial}{\partial t}\left(\frac{\rho_{gs}}{B_g}nS_g + \frac{s\rho_{gs}}{B_o}nS_o\right) + \nabla\cdot\left(\frac{\rho_{gs}}{B_g}J_g + \frac{s\rho_{gs}}{B_o}J_o\right) = 0 \qquad (2.14)$$

Depending on the level of modeling desired, the mass balance Eqs. (1.1) to (1.4) may be simplified to Eq. (1.10) which may be further simplified to the set of **Eqs. (2.5), (2.11), and (2.14)** which describe a **black oil** simulation in reservoir engineering by neglecting adsorption.

Other Approaches

Pinder and Abriola (1986) applied the model of Abriola and Pinder (1985) to study the migration of a single component NAPL (trichloroethylene) in groundwater. Local equilibrium assumptions provided the experimental relationships of mass partition between the water phase, TCE and soil solids. A two-dimensional finite difference solution was given as an example problem. Sleep and Sykes (1989) developed a model for a similar problem for predicting the fate of immobilized volatile organic compounds in variably saturated media. Variably saturated water flow, density dependent gas flow, and water and gas phase transport are included. Water and gas phase flow are partially decoupled by assuming that capillary effects between the water and gas phases are negligible. The capability to account for nonequilibrium conditions with respect to interphase mass transfer is also implemented. The model is tested on a simple hypothetical field problem involving the transport of trichloroethylene (TCE).

Mendoza and Frind (1990a, b) focused on the mechanisms by which organic vapors can spread within the aerated pore space of the porous medium. As noted by Mendoza and Frind, because vapor diffusion rates tend to be high and substances such as pesticides are dilute, the emphasis in most previous research has been on diffusive transport (e.g., Corapcioglu and Baehr, 1987). However, in the case of dense vapors, e.g., those of chlorinated solvents, density-driven advection can also play an important role. Another possible transport mechanism is advection due to the vapor mass flux caused by vaporization at the interface between the residual NAPL phase and the gas phase. Mendoza and Frind's model includes and distinguishes between the processes of vaporization at the source. Phase partitioning to the soil moisture was represented as an equilibrium retardation process that both slows plume development and makes contaminants available for transport by infiltrating water. Their formulation is analogous to that for density-dependent transport in the saturated zone. Axisymmetric coordinates were used in order to represent localized residual-saturation solvent sources. Application to a laboratory experiment shows that inclusion of density-driven advection provides a better match with the observations. They have demonstrated the effectiveness of the advective transport mechanism by a simulation of the migration of 1,1,1-

trichloroethane vapors in a highly permeable coarse sand. Falta *et al.* (1989) also presented a density-driven gas flow model.

Another group of studies that incorporate the interphase mass transfer are transport models of a NAPL vapor. Studies that assumed equilibrium partitioning between phases include Wilson *et al.* (1987, 1988), Forsyth (1988) and Shoemaker *et al.* (1990), and non-equilibrium models include Gierke *et al.* (1990, 1992), Crittenden *et al.* (1986) and Roberts *et al.* (1987). Reviews of multiphase models are also given by Abriola and Reeves (1990) and Allen (1987).

Multiphase Models with Capillary Effects

A further level of simplifications achieved for a model incorporating three simple fluid phases, by neglecting any form of mass transfer between phases, i.e., oil, water and gas exist in their respective phases only. When applied to Eq. (1.10), this yields Eq. (2.2) or its equivalent Eq. (2.5) or the water phase, and Eq. (2.7) with $C^o_o = 1$. For the gas phase, we obtain from Eq. (2.14) by taking $C^g_o = 0$, and neglecting the source and sink terms

$$\frac{\partial}{\partial t} (\rho_g n S_g) + \nabla \cdot (\rho_g J_g) - 0 \qquad (3.1)$$

The well-known equation of water flow in unsaturated soils is a classic example of multiphase flow with capillary effects. The rigorous formulation of the simultaneous flow of gas and water phases would include Eq. (3.1), Eq. (2.2), and the saturation constraint, saturation-dependent specific retention curve and permeability expressions. However, as discussed in the following section, the system of equations are further simplified to a single equation.

Water Flow in Unsaturated Soils - Richards' Equation

If we neglect the effect of the gas phase on the nonreacting water flow (i.e., immobile gas under atmospheric conditions), and taking ρ_w = constant, with no sink or source terms, Eq. (2.2) or (2.3) reduces to the well-known **Richards'** (1931) equation for unsaturated soils occupied by two phases (water and air). This can be achieved by rewriting Darcy's law for a rigid medium and incompressible water. The functional relation between the suction pressure and the volumetric water content θ_w ($n = n S_w$) is expressed by the specific retention curve. When the air is at atmospheric conditions, i.e., $p_a = 0$, the Richards' equation is expressed in the vertical direction (z) as

$$\frac{\partial \phi}{\partial t} + \frac{\partial}{\partial z}\left[K_w \frac{\partial \phi}{\partial \theta_w} + K_w\right] - 0 \tag{3.2}$$

In Eq. (3.2), the suction pressure is expressed in terms of suction head $\phi(=p_{c_{aw}}/\rho_w g)$. The hydraulic conductivity K_w is defined as $K_w = k_{rw} k_o \rho_w g / \mu_w$. The term $(K_w d\phi/d\theta_w)$ is known as the soil water diffusivity $D(\theta_w)$ in the infiltration literature. $d\theta_w/d\phi$ is known as the specific water capacity. ϕ-based equation Eq. (3.2) can be stated as a θ_w-based equation

$$\frac{\partial \theta_w}{\partial t} - \frac{\partial}{\partial z}\left[D(\theta_w) \frac{\partial \theta_w}{\partial z}\right] - \frac{dK_w}{d\theta_w} \frac{\partial \theta}{\partial z} - 0 \tag{3.3}$$

Fractional Flow Equations of Buckley-Leverett (1942)

Another simplification to the phase equations yields the fractional flow equations used in petroleum reservoir engineering. Assuming the densities of the phases to be constant, the porous medium to be rigid and neglecting the source and sink terms, these equations can be written in general form as

$$n \frac{\partial S_f}{\partial t} + \nabla \cdot (Jf_f) - 0 \tag{3.4}$$

for any phase f = w, o, g, where fractional flow is expressed by $f_f = J_f/J$ and the total flux J is defined as $J = J_w + J_o + J_g$. Substituting Darcy's law and neglecting capillarity, gravity and thermal gradients, we obtain an expression for the fractional flow as

$$f_f - \frac{k_{r_f}/\mu_f}{\sum_f (k_{r_f}/\mu_f)} \tag{3.5}$$

when oil and water are the only two phases present. Eq. (3.4), represented in a one-dimensional (horizontal) form, reduces to the Buckley-Leverett (1942) equation used to simulate water flooding operations. Corapcioglu and Hossain (1990) extended the fractional flow approach to two dimensions to study the migration of a non-dissolving TCE plume in groundwater where $S_g = 0$. In Buckley-Leverett's approach, we neglect capillary pressure differences. In a two-dimensional space of x (horizontal) and z (vertical), fractional water flow expressions f_{wx} and f_{wz} are functions of degree of water saturation S_w only. The viscosities are assumed to be constant at isothermal soil conditions.

3.3   Three-phase Flow with Capillary Effects

In addition to two-phase flow of an immiscible and water phases, there are a number of studies in the literature solving the transport of three phases, i.e., oil, water and air. Hochmuth and Sunada (1985) developed a two-dimensional finite element model to simulate the movement of a hydrocarbon and water in groundwater. Faust (1985) solved the governing equations for the simultaneous flow of water and an immiscible fluid under saturated and unsaturated conditions. Osborne and Sykes (1986) presented a two-phase finite element model to study immiscible organic transport at a landfill. Osborne and Sykes studied the effects of heterogeneities and anisotropy of the medium and concluded that the extent of migration is very much dependent on these parameters. Faust included an immobile air phase to model transport in the unsaturated zone, while Osborne and Sykes' model is applicable only for saturated regions below the water table. Hochmuth and Sunada, Faust, and Osborne and Sykes assume no mass transfer between phases and thus do not include the transport of solubilized and vaporized constituents in the water and air phases, respectively. Kuppasamy et al. (1987) and Kaluarachchi and Parker (1989) presented finite element models analogous to Faust's (1985) model. Later, Faust et al. (1989) extended the solution of his model to three-dimensions. Three phase equations include 16 dependent variables, p_w, p_o, p_g, S_w, S_o, S_g, ρ_w, ρ_o, ρ_g, μ_w, μ_o, μ_g, n, k_{rw}, k_{ro}, k_{rg}. The saturation constraint, six density and viscosity expressions as functions of phase pressures, three relative permeability expressions as functions of saturations, a porosity expression as a function of pressure, and two capillary pressure expressions (p_g - p_o) and (p_o - p_w) as functions of saturations. Three-phase relative permeabilities are estimated based on data for two-phase permeabilities (Stone, 1970, 1973). For example, Lin et al. (1982) obtained relative permeability data for trichloroethylene (TCE) imbibition in a TCE-water system. Corapcioglu and Hossain (1990) fitted a third order expression for S_w to the Lin et al. (1982) data. Curves of similar forms were also employed by Faust (1985). As an example of a multiphase flow problem, we illustrate the solution of Faust (1985) in Figure 3.1. It compares the analytical solution of Buckley-Leverett (1942) for linear water flood of an oil reservoir with the finite element solution of Faust.

Solution of the Infiltration Equation

Soil water flow in the unsaturated zone is caused by a driving force resulting from the suction gradient. As noted in soil physics text books soil suction is due to the physical affinity of water to the soil particle surfaces and capillary voids. Under conditions prevailing in the field, the soil water flow is highly complex and variant due to soil properties, and boundary and initial conditions. Therefore a rigorous solution of the problem includes the coupled solution of two-phase (i.e., water and air) flow equations. Due to nonlinearity of the problem, the solution of unsaturated flow problems require the use of numerical techniques. However, the use of numerical techniques in the soil physics did not become wide spread till 1970s. Instead, analytical solutions of the Richards' (1931) equation were generally employed to analyze the problems based on various assumptions. The major assumption was the assumption of an air phase at atmospheric conditions, i.e., $p_g = 0$. A review of these techniques was provided by Philip (1969, 1970) and Parlange (1980).

Morel-Seytoux and Billica (1985a,b) have presented a two-phase approach and demonstrated the need for such an approach instead of one-phase formulation for a couple of illustrated situations. Morel-Seytoux and Billica (1985a) have shown that the two-phase approach is needed when the mobile air in the semi-infinite column of soil cannot be confined. They have shown that water content profiles are convex only at the toe of the wetting front of one-phase water flow. Behind the front, the water content is very close to saturation. However, two-phase flow approach shows a convex shape at high water contents. Morel-Seytoux and Billica (1985a) noted that a transition zone near the surface of a semi-infinite column as well as behind the wetting zone agree more closely with experimental data. Similar conclusions were reached for a soil column with an impervious bottom. Morel-Seytoux and Billica (1985b) noted that a comparison of results by the two approaches demonstrates a measure of error resulting from the assumption of $p_g = 0$. The use of two phase approach reduces the capacity of infiltration rate significantly after ponding of soil surface as well as the ponding times due to increased resistance resulting from air-phase flow. Ponding occurs while the soil is very dry in a two-phase approach. However, if one-phase approach is used, ponding takes place when the soil is almost fully saturated.

Sharp Interface Models

Abriola and Reeves (1990) noted that although sharp interface models have some limitations, such as negligence of capillary forces in comparison to pressure and gravity forces, they "may prove useful as screening or site assessment tools because of their relative simplicity." They may be employed for source identification in an oil contaminated groundwater. Especially, analytical expressions obtained by solving governing equations are quite attractive in routine exposure assessment studies. Therefore it is the objective of this section to present an analytical solution to calculate the variation of free product thickness and volume of free product in a two-pump recovery operation by using a sharp interface model.

Recovery of oil accumulated on water table can be accomplished by creating a cone of depression with a pumping recovery well. Free product is induced to flow into the recovery well by the influence of the water table gradient established by pumping. Once the well is in place and the cone of depression is formed, a scavenger oil recovery unit is placed in the well on top of the oil layer and the oil is pumped to a recovery tank.

Although sharp interface models have been investigated by a number of researchers, analytical solutions have been attempted in very few of them. Lateral spreading of free product on water table has been studied by van Dam (1967), Mull (1971, 1978), Dracos (1978), Greulich

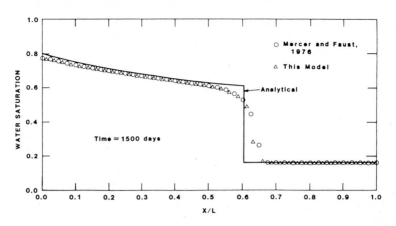

Figure 3.1   Saturation profile of linear waterflood: saturation versus distance from water injection showing analytical solution (Buckley and Leverett, 1942), results using finite-element of oil-water flow (Mercer and Faust, 1976), and results of Faust (1975) (After Faust, 1975)

Table 5-1. Calculation for the Single Cell Method

	Spill 1 Benzene	Spill 2 1,1,1 Trichloroethane	Spill 3 1,1,2 Trichlorethylene
$m_o = F[z - z_b]$ $= F(211)$	7.4 g/cm^2	11.4 g/cm^2	12.2 g/cm^2
$t_c = m_o/RS_o$	39.3 yrs	147.8 yrs	105.4 yrs
$J = RS_o$	1.88×10^{-1} g/cm^2 yr	7.71×10^{-1} g/cm^2 yr	1.16×10^{-1} g/cm^2 yr
$J_T = JA$ $= J(211^2)$	8.4×10^3 g/yr for 39.3 yrs	3.4×10^3 g/yr for 147.8 yrs	5.2×10^3 g/yr for 105.4 yrs

and Kaergaard (1984), Greulich (1985), Holzer(1976), Levy *et al.*, (1990), Schiegg (1977) and Hochmuth and Sunada (1985). van Dam (1967) obtained an expression to estimate the areal extent of an oil lens at equilibrium by assuming a constant oil saturation within the lens equal to half residual saturation and a stationary water table. Mull (1971, 1978) presented an analysis by assuming the oil lens to cease spreading when the lens thickness at the spill site matches the capillary rise. Greulich (1985) [also presented in Greulich and Kaergaard (1984)] represented the oil lens as a combination of a cylinder at the center and a thin circular disc outside. Holzer (1976) applied Hantush's (1968) theory by making an analogy between the decay of an oil lens on a horizontal water table and the movement of a freshwater lens in an unconfined saline aquifer. Dracos (1978) has proposed an approach to estimate the spread of an oil slick in soils and presented expressions to calculate the maximum spread of the oil-polluted area. Levy *et al.* (1990), too, used Hantush's methodology to estimate the rate and volume of oil leakage from a tank. Hochmuth and Sunada (1985) developed a two-dimensional numerical finite-element model to simulate the movement of a hydrocarbon and groundwater by solving the Boussinesq equation stated for each phase. Schiegg (1977) [also presented in Schiegg and Schwille (1990)] presents a numerical lateral spreading solution by using a semi-analytical vertical infiltration expression as an initial condition. Schiegg assumes that spreading starts after the infiltration of oil ceases.

Formulation of the Sharp Interface Approach

The first step in the mathematical formulation of any transport phenomena involves the balance equations. The lateral spreading of oil is simulated by considering the migration of oil and water phases simultaneously. Since both phases are mobile, the starting point is the mass balance equations of oil and water phases in a porous medium [Eqs. (2.2) and (2.7) with $C_o^\circ=1$ and without any sink and source terms].

Governing equations to estimate oil thickness L, and water elevation η, as a function of space and time, can be obtained by averaging the oil and water phase equations along the vertical in their respective regions, by applying appropriate boundary conditions on top and bottom surfaces. By averaging the oil phase equation along the vertical and substituting the appropriate boundary conditions, Corapcioglu *et al.* (1992) obtained a linearized equation for the oil thickness L and water table elevation η.

$$\nabla^2 L + \nabla^2 \eta = \pm \frac{R_o}{K_o L_o} \delta(x-\xi)\delta(y-\zeta) + \frac{n\,[S_{oo}-S_{oun}]}{K_o L_o}\frac{\partial L}{\partial t} - \frac{n\,[S_{oun}-S_{ow}]}{K_o L_o}\frac{\partial \eta}{\partial t} \quad (4.1)$$

Similarly, water phase equation is derived by using a similar procedure followed in the derivation of oil phase equation. The mass balance equation for water is integrated between impervious bed rock and water table and appropriate boundary conditions are applied.

Substitution of approximate boundary conditions and averaging rules, yields an equation in L and η as

$$\nabla^2 \eta + \nabla L^2 \left(\frac{\rho_o}{\rho_w}\right) = \pm \frac{Q_w}{K_w \eta_o} \delta(x-\xi') \delta(x-\zeta') + \frac{n[S_{ww}-S_{wo}]}{K_w \eta_o} \frac{\partial \eta}{\partial t} \tag{4.2}$$

In order to obtain an expression for free product thickness, the oil phase equation is solved uncoupled. Uncoupling of the oil phase equation is done by solving unconfined ground water flow equation independently for η and then substituting it into the oil phase equation. The unconfined ground water flow equation, known as Boussinesq's equation, can be expressed as

$$\frac{\partial^2 \eta}{\partial r^2} + \frac{1}{r} \frac{\partial \eta}{\partial r} - \frac{S_y}{K_w \eta_o} \frac{\partial \eta}{\partial t} \tag{4.3}$$

where S_y is the specific yield of the phreatic aquifer and r is the radial distance from the well. The solution of Eq. (4.3) for an aquifer of infinite extent is given by the Theis solution as

$$\eta_o - \eta(r,t) = \frac{Q_w}{4\pi K_w \eta_o} W(u) ; \quad where \quad u = \frac{r^2 S_y}{4 K_w \eta_o t} \tag{4.4}$$

W(u) is an exponential integral known as the Well function in groundwater hydrology. Substitution of Eqs. (4.4) in Eq. (4.1) written in axis-symmetric cylindrical coordinates yields

$$\frac{\partial^2 L(r,t)}{\partial r^2} + \frac{1}{r} \frac{\partial L(r,t)}{\partial r} - \frac{n(S_{oo}-S_{oun})}{K_o L_o} \frac{\partial L}{\partial t} - \frac{Q_w}{4\pi K_w \eta_o}\left[\frac{S_y}{K_w \eta_o} + \frac{n(S_{oun}-S_{ow})}{K_w L_o}\right]\frac{e^{-\frac{r^2 S_y}{4K_w \eta_o t}}}{t} \tag{4.5}$$

Recovery of An Established Oil Mound

In this section, we seek a solution to the recovery problem af an established oil mound on an initially horizontal water table. The initial condition can be expressed as

$$L(r,0) = \frac{L_o}{\ln\left(\frac{r_o}{r_w}\right)} \ln\left(\frac{r_o}{r}\right) \tag{4.6}$$

where r_w and r_o are the radii of the recovery well and the oil lens, respectively. In a two-pump recovery system, one pump creates a cone of depression by pumping groundwater and it draws groundwater, and the free product floating on it, into the recovery well. A second pump at the water table skims off the free product. The condition at the recovery well is given by

$$\lim_{r \to 0} r \frac{\partial L(r,t)}{\partial r} - \frac{1}{2\pi}\left(\frac{Q_o}{K_o L_o} - \frac{Q_w}{K_w \eta_o}\right) \tag{4.7}$$

where Q_o and Q_w are oil and water pumpage rates, respectively. At the outer edge of the oil lens a no-flux boundary condition is defined. Solution of Eq. (4.5) using the initial and boundary conditions is obtained by Laplace transformation technique. It can be expressed as

$$L(r,t) - C_1^* I_0\left(r\sqrt{\frac{A}{2t}}\right) + \frac{C_2^*}{2} W(u_1) + \frac{2B}{(a-A)}\frac{1}{2} W(u_2) + \frac{L_o}{\ln\left(\frac{r_o}{r_w}\right)} \ln\left(\frac{r_o}{r}\right) \tag{4.8}$$

where

$$C_1^* - \left(-\frac{Q_w}{2\pi K_w \eta_o} + \frac{2B}{(a-A)}\right)\sqrt{\frac{a}{A}}\frac{K_1\left(r_o\sqrt{\frac{a}{2t}}\right)}{I_1\left(r_o\sqrt{\frac{A}{2t}}\right)} + C_2\frac{K_1\left(r_o\sqrt{\frac{A}{2t}}\right)}{I_1\left(r_o\sqrt{\frac{A}{2t}}\right)} + \frac{L_o}{r_o\ln\left(\frac{r_o}{r_w}\right)\sqrt{\frac{A}{2t}}I_1\left(r_o\sqrt{\frac{A}{2t}}\right)} \tag{4.9}$$

$$C_2^* - -\left[\frac{1}{2\pi}\left(\frac{Q_o}{L_o K_o} - \frac{Q_w}{\eta_o K_w}\right) + \frac{L_o}{\ln\left(\frac{r_o}{r_w}\right)} + \frac{2B}{(a-A)}\right] \; ; \; where \; A - \frac{n(S_{oo} - S_{oun})}{K_o L_o} \tag{4.10}$$

$$B - \frac{Q_w}{4\pi K_w \eta_o}\left[\frac{S_y}{K_w \eta_o} + \frac{n(S_{oun} - S_{ow})}{K_o L_o}\right] \; ; \; a - \frac{S_y}{K_w \eta_o} \; ; \; u_1 - \frac{r^2 A}{4t} \; ; \; u_2 - \frac{r^2 a}{4t} \tag{4.11}$$

The variation of free product thickness with time is illustrated in Figure 4.1. As seen in this figure, an initially 1 m thick oil lens at the recovery well is reduced to 0.17 m after a month of oil pumping at a rate of 4.25 m³/day. Figure 4.2 shows that, in 20 days, 90 percent of the initial volume of oil is recovered. Figure 4.3 illustrates a graphical procedure to estimate the recovery efficiencies based on the readings at the monitoring wells. For example, a 20 cm free product thickness in an observation well 10 m from the recovery well indicates 40 percent removal efficiency.

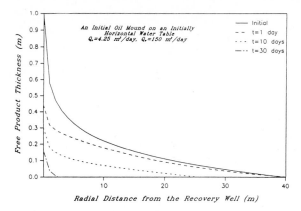

Figure 4.1   Variation of free product thickness

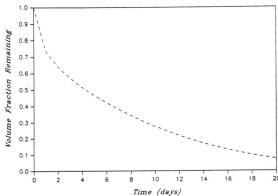

Figure 4.2   Efficiency of the free product recovery operation

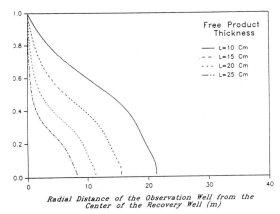

Figure 4.3   Estimation of recovery efficiency based on readings at the monitoring wells

The Single Cell Model

The multicomponent multiphase modeling approach leads to a system of nonlinear partial differential equations as given by Eq. (1.10). In a regional study of nitrate and chloride pollution, Mercado (1976) cited occasional lack of data required to justify the use of the standard diffusion-advection equation. Instead, Mercado proposed the use of a simplified single-cell model based on a macroscopic interpretation of the conservation of mass principle. A similar approach is adopted by Corapcioglu and Baehr (1985) to provide a conservative estimate of the rate at which the solubilized fraction of an immobilized, immiscible contaminant plume in the unsaturated zone is leached down entering an underlying aquifer. As a first approximation, this model makes use of available or readily accessible data and is conservative in the sense that assumptions made should tend to overestimate the flux of the solubilized fraction into the aquifer. Conceivably, this estimate could provide the top boundary condition in a solute transport model for the aquifer. The single-cell model of Corapcioglu and Baehr (1985) is a simplified version of the multicomponent model. The single cell represents a portion of the aquifer volume within the investigated area. The whole system is represented by a single element. The conservation of mass within one element is described by a simplified version of the system of governing partial differential equations.

The discussion given here closely follows Corapcioglu and Baehr (1985). First, let us consider a homogeneous (= one component, i=1) immiscible contaminant. Then, R is the average recharge rate for water, F is the field capacity of immiscible fluid in natural soil, $C_w(x,y,z,t)$ is the concentration of component dissolved in water, $J(x,y,t)$ is the flux of dissolved component into the water table, J_T is the areal flux of dissolved component into water table, $Z_T(x,y)$ and Z_B are the top and bottom of the immiscible plume, respectively, and $A_1(x)$ and $A_2(x)$ are the curves defining the cross-section defined by the bottom of the immiscible plume or the intersection of the immiscible plume and the water table at $z - z_B$ = constant.

In developing the model, Corapcioglu and Baehr (1985) made various assumptions to simplify the physics, chemistry and biology of the phenomenon. Incorporation of various assumptions in Eq. (1.10) leads to

$$\frac{\partial}{\partial t}\left\{ \int_{z_B}^{z_T(x,y)} [C\theta_{ww} + \rho_o\theta_o]\, dz \right\} - - \int_{z_B}^{z_T} \frac{\partial}{\partial z} [J_w C_w]\, dz - RC \mid_{z_T} - RC \mid_{z_B} \tag{5.1}$$

where $q_w = -R$ in Eq. (1.10). Then, we define

$$m = \int_{z_B}^{z_T(x,y)} \left[C_w \theta_w + \rho_o \theta_o \right] dz = \text{mass of contaminant in column of unit cross-sectional area}$$

The assumptions of equilibrium chemistry ($Cl_{z_B} = S_o^*$) and clean water injection at top yield

$$\frac{dm}{dt} = -RS_o^* \qquad t < t_c ; \qquad m = 0 \quad \text{otherwise} \tag{5.2}$$

The initial condition is $m_o = m(x,y,0) = F[z_T(x,y) - z_b]$. Then, the integration of Eq. (5.2) gives

$$m = m_o - RS_o^* t \qquad t < t_c ; \qquad m = 0 \quad \text{otherwise} \tag{5.3}$$

Then, the contaminant flux is obtained by

$$J = -\frac{dm}{dt} = RS_o^* \qquad t < t_c ; \qquad J = 0 \quad \text{otherwise} \tag{5.4}$$

The total areal flux J_T of solubilized contaminant entering the underlying aquifer is obtained by integrating J over the region A bounded by A_1 and A_2. Eqs. (5.3) and (5.4) constitute the proposed single-cell model. The applicability of this model is restricted to estimate the fate of one-component immiscible contaminants less dense than water (e.g., benzene), which do not penetrate the water table no matter how extensive the spill. It is only applicable to incidents involving contaminants denser than water when the immiscible fluid introduced to the soil environment is immobilized before reaching the water table.

As we noted earlier, the contaminant is represented by a single component. Hence, the application of this single-cell model would give a very conservative estimate for each component by ignoring the presence of the other components. Corapcioglu and Baehr (1985) applied the single-cell model individually to three different hypothetical spills of benzene, 1,1,1 trichloroethane, and 1,1,2 trichloroethylene (TCE) assuming a leak of 100 gallons of each component from a storage tank. We assume that, for simplicity, the volumetric field capacity for each substance is given as 40 liters of contaminant per cubic meter of soil or 4.0×10^{-2} cm^3 of contaminant/cm^3 of soil (sand at field capacity with water for a light oil). This field capacity implies approximately 9.5 m^3 ($= 100$ gallons/40 liters/m^3) of contaminated soil distributed in a cube. Corapcioglu and Baehr (1985) assumed as cube of soil (211x211x211 cm) contaminated right below the ground. Calculations are given in Table 1. These conservative estimates for the mass fluxes, based on realizable parameters, suggest that unless the saturated zone could provide a large diluting effect, a spill as small as 100 gallons could threaten a water supply.

Corapcioglu and Baehr (1985) have presented the comparisons of results given in Table 5.1 with the results of an application of the model of Eq. (1.10). They have shown that the single- cell model yields very conservative estimates as much as two to four times of the generalized model and sometimes higher. This is due to the fact that the single-cell model neglected the vaporization in the air phase and considered only the advective transport in the water. These components of the transport processes are realistically incorporated into the generalized model of Eq. (1.10).

Conclusions

We presented a summary of predictive models of multiphase fluid flow in subsurface environments. The mathematical models reviewed were categorized in five groups. Each group of models require a different set of input parameters describing the subsurface and fluid(s). The level of complexity of the information is directly related to the level of sophistication of the model. The choice of modeling approach depends on the level of accuracy required and the availability of data. Usually the availability of information limits the choice of the model. However, since models provide relatively inexpensive and useful tools to design and test effective remediation schemes, they are critical in our clean up efforts. Understanding and prediction of hydrocarbon migration in subsurface environments will advance effective recovery and clean up techniques of these contaminants, as well as help to protect existing water resources.

References

Abriola, L.M. and Reeves, H.W. (1990) Slightly miscible organic chemical migration in porous media: Present and future directions in modeling, *Proc. Env. Res. Conf. Groundwater Quality and Waste Disposal*, edited by I.P. Murarka and S.S. Cordle, EPRI, Palo Alto, CA, 15.1-15.24.

Abriola, L.M. and Pinder, G.F. (1985a) A multiphase approach to the modeling of porous media contamination by organic compounds, 1, Equation development, *Water Resour. Res.*, 21(1), 11-18.

Abriola, L.M. and Pinder, G.F. (1985b) A multiphase approach to the modeling of porous media contamination by organic compounds, 2, Numerical simulation, *Water Resour. Res.*, 21(1), 19-26.

Allen, M.B., III. (1987) Numerical modeling of multiphase flow in porous media, in *Advances in Transport Phenomena in Porous Media*, ed. by J. Bear and M.Y. Corapcioglu, Martinus Nijhoff Publications, Dordrecht, The Netherlands.

American Petroleum Institute (API). (1972) The migration of petroleum products in soils and groundwater, Principles and counter-measures, *Publ. 4149*, API, Washington, DC.

Aziz, K. and Settari, A. (1979) *Petroleum Reservoir Simulation*, Applied Science Publishers, London.

Baehr, A.L. and Corapcioglu, M.Y. (1987) A compositional multiphase model for groundwater contamination by petroleum products, 2, Numerical solution, *Water Resour. Res.*, (23)(1), 201-213.

Burcik, E.J. (1957) *Properties of Petroleum Reservoir Fluids*, Wiley, New York.

Buckley, S.E. and Leverett, M.C. (1942) Mechanism of fluid displacement in sands, *Pet. Trans. AIME*, 146, 107.

Coats, K.H. (1980) In situ combustion model, *Soc. Petrol. Eng. J.*, 533-554.

Corapcioglu, M.Y. and Panday, S. (1991a) Compositional Multiphase Flow Models, in M.Y. Corapcioglu (ed.), *Advances in Porous Media*, Vol. 1, Elsevier, Amsterdam, The Netherlands, 1-60.

Corapcioglu, M.Y. and Panday, S. (1991b) Soil and groundwater contamination by petroleum products in frozen soils, *Proc. Third International Symp. on Cold Regions Heat Transfer*, June 11-14, University of Alaska, 303-310.

Corapcioglu, M.Y. and Hossain, M.A. (1990) Ground water contamination by high-density immiscible hydrocarbon slugs in gravity-driven gravel aquifers, *Ground Water*, 28, 403-412.

Corapcioglu, M.Y. and Baehr, A.L. (1987) A compositional multiphase model for groundwater contamination by petroleum products, 1, Theoretical considerations, *Water Resour. Res.*, 23(1), 191-200.

Corapcioglu, M.Y. and Baehr, A.L. (1985) Immiscible contaminant transport in soils and groundwater with an emphasis on petroleum hydrocarbons: System of differential equations vs. single-cell model, *Wat. Sci. Tech.*, 17, 23-37.

Corapcioglu, M.Y., Lingam, R., and Heisler, V.K. (1992) Analytical prediction of gasoline thickness on the water table, Proceedings of the Water Forum, ASCE, Baltimore, Maryland, August 2-5.

Crittenden, J.C., Hutzler, N.J. Geyer, D.G., Oravitz, J.L. and Friedman, G. (1986) Transport of organic compounds with saturated groundwater flow: Model development and parameter sensitivity, *Water Resour. Res.*, 22(3), 271-284.

Crookston, R.B., Culham, W.E. and Chen, W.H. (1970) A numerical simulation model for thermal recovery processes, *Soc. Petrol. Eng. J.*, 37-58.

Dracos, T. (1978) Theoretical considerations and practical implications in the filtration of hydrocarbons in aquifers, in *Proc. International Symp. on Ground Water Pollution by Oil Hydrocarbons*, Stavebn'i Geologie Praha, Prague, Czechoslovakia, 127-137.

Falta, R.W., Pruess, K., Javandel, I. and Witherspoon, P.A. (1992) Numerical modeling of steam injection for the removal of nonaqueous phase liquids from the subsurface, Numerical formulation, *Water Resour. Res.*, 28, 433-449.

Falta, R.W., Javendel, I., Pruess, K. and Witherspoon, P.A. (1989) Density-driven flow of gas in the unsaturated zone due to the evaporation of volatile organic compounds, *Water Resour. Res.*, 25(10), 2159-2169.

Faust, C.R. (1985) Transport of immiscible fluids within and below the unsaturated zone: A numerical model, *Water Resour. Res.*, 21(4), 587-596.

Faust, C.R., Guswa, J.H. and Mercer, J.W. (1989) Simulation of three-dimensional flow of immiscible fluids within and below the unsaturated zone, *Water Resour. Res.*, 25(12), 2449-2464.

Forsyth, P.A. (1988) Simulation of nonaqueous phase groundwater contamination, *Adv. Water Resour.* 11(6), 74-83.

Fried, J.J., Muntzer, P. and Zilliox, L. (1979) Groundwater pollution by transfer of oil hydrocarbons, *Ground Water*, 17(6), 586-594.

Gierke, J.S., Hutzler, N.J. and McKenzie, D.B. (1992) Vapor transport unsaturated soil columns: Implications for vapor extraction, *Water Resour. Res.*, 323-335.

Gierke, J.S., Hutzler, N.J. and Crittenden, J.C. (1990) Modeling the movement of volatile organic chemicals in columns of unsaturated soil, *Water Resour. Res.*, 26(7), 1529-1547.

Greulich, R.H. (1985) Groundwater contamination from oil: Behavior, control and treatment, *Water Supply*, 3, 233-242.

Greulich, R. and Kaergaard, H.. (1984) The movement of a continuously growing body of oil on the groundwater table, *Nordic Hyd.*, 15, 265-272.

Hantush, M.S. (1968) Unsteady movement of fresh water in thick unconfined saline aquifers, *Bull. Internat. Assoc. Sci. Hydrology*, 13, 40-60.

Hekim, Y., Fogler, H.S. and McCune, C.C. (1982) The radial movement of permeability fronts and multiple reaction zones in porous media, *Soc. Petrol. Eng. J.*, 99-107.

Hochmuth, D.P. and Sunada, D.K. (1985) Ground water model of two-phase immiscible flow in coarse material, *Ground Water*, 23, 617-626.

Hoffman, B. (1970) Dispersion of soluble hydrocarbons in groundwater stream, in *Advances in Water Pollution Research*, ed. by S.H. Jenkins, Pergamon, HA/7(b)/1-8.

Holzer, T.L. (1976) Application of groundwater flow theory to a subsurface oil spill, *Ground Water*, 14, 138-145.

Jamison, V.W., Raymond, R.L. and Hudson, J.O. (1976) Biodegradation of high octane gasoline, in *Proceedings, Third International Biodegradation Symposium*, ed. by J.M. Sharpley and A.M. Kaplan, Applied Science Publishers, Englewood, NJ, 187-196.

Jennings, A.A. and Kirkner, D.J. (1983) Criteria for selecting equilibrium or kinetic sorption descriptions in groundwater quality models, in *Frontiers in Hydraulic Engineering*, ed. by H.T. Shen, American Society of Civil Engineers, 42-47.

Kaluarachchi, J.J. and Parker, J.C. (1989) An efficient finite element method for modeling multiphase flow, *Water Resour. Res.*, 25(1), 43-54.

Kazemi, H., Vestal, C.R. and Shank, G.D. (1978) An efficient multicomponent numerical simulator, *Soc. Petrol. Eng. J.*, 355-368.

Kuppusamy, T., Sheng, J.J., Parker, J.C. and Lenhard, R.J. (1987) Finite element analysis of multiphase immiscible flow through soils, *Water Resour. Res.*, 23, 625-631.

Levy, B.S., Riordan, P.J. and Schreiber, R.P. (1990) Estimation of leak rates from underground storage tanks, *Ground Water*, 28, 378-384.

Lin, C., Pinder, G.F. and Wood, E.F. (1982) Water and trichlorethylene on immiscible fluids in porous media, *Water Resources Progress Report, 83-WR-2*, Princeton University.

Mendoza, C.A. and Frind, E.O. (1990a) Advective-dispersive transport of dense organic vapors in the unsaturated zone, 1, Model development, *Water Resour. Res.*, 26(3), 379-387.

Mendoza, C.A. and Frind, E.O. (1990b) Advective-dispersive transport of dense organic vapors in the unsaturated zone, 2, Sensitivity analysis, *Water Resour. Res.*, 26(3), 388-398.

Mercado, A. (1976) Nitrate and chloride pollution in aquifers: A regional study with the aid of a single-cell model, *Water Resour. Res.*, 12(4), 731-747.

Mercer, J.W. and Faust, C.R. (1976) The application of finite-element techniques to immiscible flow in porous media, *Finite Elements in Water Resources*, ed. W.G.Fray, G.F.Pinder, and C.A.Brebbia, Pentech, London, 1.21-1.57.

Morel-Seytoux, H.J. and Billica, J.A. (1985a) A two-phase numerical model for prediction of infiltration: Applications to a semi-infinite soil column, *Water Resour. Res.*, 21(4), 607-615.

Morel-Seytoux, H.J. and Billica, J.A. (1985b) A two-phase numerical model for prediction of infiltration: Case of an impervious bottom, *Water Resour. Res.*, 21(9), 1389-1396.

Mull, R. (1971) Migration of oil products in the subsoil with regard to groundwater pollution by oil, *Advances in Water Pollution Research*, S.H. Jenkins (ed.), Pergamon Press, Oxford, 2Ha-7a, 1-8.

Mull, R. (1978) Calculations and experimental investigations of the migration of oil products on natural soils, *Proc. Int. Symp. on Groundwater Pollution by Oil Hydrocarbons*, Prague, June 5-9, Int. Assoc. Hydrogeol., 167-181.

Nghiem, L.X., Fong, D.K. and Azis, K. (1973) Compositional modeling with an equation of state, *Soc. Petrol. Eng. J.*, 687-698.

Osborne, M. and Sykes, J. (1986) Numerical modeling of immiscible organic transport at the Hyde Park landfill, *Water Resour. Res.*, 22(1), 25-33.

Panday, S. and Corapcioglu, M.Y. (1991) Solute rejection in freezing soils, *Water Resour. Res.*, 27, 99-108.

Parlange, J.-Y. (1980) Water Transport in Soils, *Ann. Rev. Fluid Mech.*, 12, 77-102.

Peaceman, D.W. (1977) *Fundamentals of Numerical Reservoir Simulation*, Elsevier, Amsterdam.

Pfannkuch, H.O. (1984) Determination of contaminant source strength from mass exchange processes at the petroleum/groundwater interface in shallow aquifer systems, in *Petroleum Hydrocarbons and Organic Chemicals in Groundwater*, National Water Well Association, 111-129.

Philip, J.R. (1969) Theory of infiltration, *Adv. in Hydrosciences*, ed. V.T.Chow, vol. 5, Academic Press, New York, 215-297.

Philip, J.R. (1970) Flow in porous media, *Ann. Rev. Fluid Mech.*, 2, 177-204.

Pinder, G.F. and Abriola, L.M. (1986) On the simulation of nonaqueous phase organic compounds in the subsurface, *Water Resour. Res.*, 22(9), 109S-119S.

Richards, L.A. (1931) Capillary conduction of liquids through porous mediums, *Physics*, 1, 318-333.

Roberts, P.V., Goltz, M.N., Summers, R.S., Crittenden, J.C. and Nkedi-Kizzi, P. (1987) The influence of mass transfer on solute transport in column experiments with an aggregated soil, *Contam. Hydrol.*, 1, 375-393.

Schiegg, H.O. and Schwille, F. (1991) Hydrocarbons in porous media, in *Transport Processes in Porous Media*, eds. J. Bear and M.Y. Corapcioglu, Kluwer, 69-202.

Schiegg, H.O. (1977) Methode zur abschatzung der ausbreitung von erdol derivaten in mit wasser and luft erfullten boden, *Mitteilung der Versuchanstalt fur Wasserbau, Hydrologie and Glaziologie an der ETH Zurich*, 22.

Schwille, F. (1981) Groundwater pollution in porous media by fluids immiscible with water, in *Quality of Groundwater*, ed. by W. Van Duijvenbooden, P. Glasbergen and H. van Lelyveld, Elsevier Science, New York, 451-463.

Shoemaker, C.A., Culver, T.B., Lion, L.W. and Peterson, M.G. (1990) Analytical models of the impact of two-phase sorption on subsurface transport of volatile chemicals, *Water Resour. Res.*, 26(4), 745-758.s

Sleep, B.E. and Sykes, J.F. (1989) Modeling the transport of volatile organics in variably saturated media, *Water Resour. Res.*, 25(1), 81-92.

Somers, J.A. (1974) The fate of spilled oil in the soil, *Hydrol. Sci. Bull.*, 19(4), 501-521.

Stone, H.L. (1970) Probability for estimating three-phase relative permeability, *Trans. AIME*, 229, 214.

Stone, H.L. (1973) Estimation of three phase relative permeability and residual oil data, *J. Can. Pet. Tech.*, (Oct.-Dec.), 53.

van Dam, J. (1967) The migration of hydrocarbons in a water-bearing stratum, *Joint Problems of the Oil and Water Industries*, edited by P. Hepple, the Inst. of Petroleum, London, 55-96.

Van der Waarden, M., Birdie, A.L.A. and Groenewood, W.M. (1971) Transport of mineral oil components to groundwater, *I, Water Res.*, 5, 213-226.

Van der Waarden, M., Birdie, A.L.A. and Groenewood, W.M. (1977) Transport of mineral oil components to groundwater, II, *Water Res.*, 11, 359-365.

Wilson, D.E., Montgomery, R.W. and Sheller, M.R. (1987) A mathematical model for removing volatile subsurface hydrocarbons by miscible displacement, *Water Air Soil Pollut*, 33, 231-255.

Wilson, D.J., Clarke, A.N. and Clarke, J.H. (1988) Soil clean up by in-situ aeration, I. Mathematical modeling, *Sep. Sci. Technol.*, 23, 991-1037.

Young, L.C. and Stephenson, R.E. (1983) A genealized compositional approach for reservoir simulation, *Soc. Petrol. Eng. J.*, 727-742.

Youngren, G.K. (1980) Development and application of an in situ combustion reservoir simulator, *Soc. Petrol. Eng. J.*, 39-51.

Transport of Linearly Reactive Solutes in Porous Media. Basic Models and Concepts.

D. Schweich

Laboratoire des Sciences du Génie Chimique, CNRS

ENSIC, INPL

1 rue Grandville, BP 451

F - 54001 Nancy cedex

France

Modeling the transport of reactive solutes in natural porous media requires a multidisciplinary approach which involves chemistry, biochemistry, fluid mechanics, system dynamics, thermodynamics, kinetics, etc. Some concepts originating from these sciences reveal themselves to be powerful tools for modeling purposes, and for understanding what is occuring in the porous medium. For the sake of simplicity, we will essentially focus on the main tools necessary to model the transient behavior of linearly interacting solutes, and will exclude biochemical phenomena.

We will focus on the transient behavior of a reactive or nonreactive solute at various length scales. The longest length scale is a field between an injection and a detection well (or piezometer), or a laboratory column (Figure 1 left). Intermediate scales are defined either by the flowing carrier fluid at the exclusion of the porous matrix, or by an "elementary aggregate" of the sorbing phase (Figure 1 middle). The elementary aggregate is defined as a continuous piece of stationary phase surrounded by mobile water. It can be composed of various solids and immobile water.

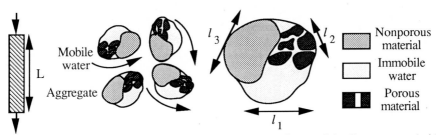

Figure 1: Schematic of flow system (left), of elementary volume of the flow system (middle), and of sorbing aggregate (right). l_i are chararcteristic lengths.

NATO ASI Series, Vol. G 32
Migration and Fate of Pollutants in Soils and Subsoils
Edited by D. Petruzzelli and F. G. Helfferich
© Springer-Verlag Berlin Heidelberg 1993

The smallest length scales define the internal structure of the sorbing aggregate (Figure 1 right). Predicting the behavior of the solute requires that each of the length scales be modeled as easily as possible. Linear system dynamics provide an efficient solution.

Let $x_i(t)$ be the concentration input to a flow system, and $y_i(t)$ the associated time response. The system is said to be linear when the response to a linear combination of inputs $x_i(t)$ is the same linear combination of the individual responses $y_i(t)$. Laplace transformation is especially appropriate for the study of transient linear systems. The Laplace transform of f(t) is defined by

$$\overline{f}(s) = \int_0^\infty f(t)\ e^{-st}\ dt \tag{1}$$

where s is the Laplace variable. The properties of a linear dynamic system are completely determined by the transfer function G(s) defined by

$$G(s) = \frac{\overline{y}(s)}{\overline{x}(s)} \tag{2}$$

where $\overline{y}$ and $\overline{x}$ are the Laplace transforms of the output and input variables respectively. The impulse response of the system is obtained when x(t) is a Dirac $\delta(t)$-function of Laplace transform $\overline{x}(s) = 1$. Consequently, G(s) is the Laplace transform of the impulse response. The goal of Section 1 is to provide a method for deriving models for the transfer functions involved at various scales. As soon as the transfer functions at small scales are available, new transfer functions at a larger scale can be established by use of simple theorems. Four of which will be useful.

1 - Subsystems are said to be in series when the output variable of one system is the input variable to the next. If $G_j(s)$ is the transfer function of subsystem j, then the transfer function of the whole system is given by

$$G(s) = \prod_j G_j(s) \tag{3}$$

2 - When a system reaches steady state, the following equality holds

$$\frac{y(\text{at steady state})}{x(\text{at steady state})} = G(0) \tag{4}$$

3 - The normalized kth-order moment of the impulse response h(t) is defined by

$$\mu_k = \frac{\int_0^\infty t^k\ h(t)\ dt}{\int_0^\infty h(t)\ dt} \tag{5}$$

If G(s) is the Laplace transform of h(t), then

$$\mu_k = \frac{(-1)^k}{G(0)} \left(\frac{\partial^k G}{\partial s^k} \right)_{s=0} \tag{6}$$

4 - The Laplace transform of the response to a Heaviside step function is

$$\overline{y}(s) = \frac{G(s)}{s} \tag{7}$$

Residence time distribution of mobile water

We will show in Section 3 that the behavior of a reactive solute is governed by the flow pattern in the voids between the aggregates and by the dynamics of the sorption process. Consequently, we will first need to describe the mobile water phase, assuming that the solute cannot penetrate the internal pores of the solid matrix and eventually any immobile water.

Consider a flow system, and assume that the volumetric flow rate, Q, is constant. Provided that there is no dispersion at the inlet and outlet cross-sections, the Residence Time Distribution (RTD), $E_m(t)$, can be defined as the normalized impulse response to an injection of a nonreactive solute (Sardin et al., 1991; Villermaux, 1982; Wen and Fan, 1975). Let $x(t)$ and $y(t)$ be the flux averaged inlet and outlet concentrations and n_I the amount of solute injected, then

$$x(t) = \frac{n_I}{Q}\delta(t) \tag{8-1}$$

$$E_m(t) = \frac{Qy(t)}{\displaystyle\int_0^\infty Qy(t)\,dt} = y(t)\frac{Q}{n_I} \tag{8-2}$$

The flux averaged concentrations $x(t)$ and $y(t)$ are defined with respect to a possible velocity profile at the inlet and outlet cross-sections respectively. When the velocity profile is flat, the flux-averaged and resident concentrations coincide. The expressions for the residence time distribution can be obtained by solving the mass balance equation for a nonreactive solute in the transient state and using equations (8). For example, if one-dimensional plug flow takes place

$$v\frac{\partial y}{\partial z} + \frac{\partial y}{\partial t} = 0, \quad y = x \text{ at } z=0 \qquad E_m(t) = \delta(t - t_m) \qquad t_m = L/v \tag{9}$$

where L is the length of the system, v the pore fluid velocity, and t_m the mean residence time of mobile water. If the system is a perfect mixing cell of uniform composition and of volume V, then

$$Qx = Qy + V\frac{dy}{dt} \qquad E_m(t) = \frac{1}{t_m}\exp(-t/t_m)H(t) \qquad t_m = V/Q \tag{10}$$

where H(t) is the Heaviside step function. For more complex flow patterns, the expression for $E_m(t)$ is either complex (involving series, integrals, special functions, etc.) or not available, as illustrated by the numerous publications devoted to the simple convective-dispersive flow. Fortunately, in these complex situations, it is generally simple to derive the Laplace transform, $G_m(s)$, of $E_m(t)$. For the sake of simplicity we will deal exclusively with a one-dimensional convective-dispersive flow. The mass balance equation and the appropriate boundary conditions which give the residence time distribution are (Sardin et al., 1991):

$$-D\frac{\partial^2 C_m}{\partial z^2} + v\frac{\partial C_m}{\partial z} + \frac{\partial C_m}{\partial t} = 0 \tag{11-1}$$

$$v\,x = v\,C_m(0^+,t) - D\frac{\partial C_m(0^+,t)}{\partial z}, \qquad \frac{\partial C_m(L,t)}{\partial z} = 0, \qquad x = \delta(t), \qquad y = C_m(L,t) \tag{11-2}$$

The Laplace transform of the residence time distribution is:

$$G_m(s) = \frac{4q\,\exp[P(1-q)/2]}{(1+q)^2 - (1-q)^2\exp(-Pq)}, \qquad q = \left\{1 + 4\,\frac{s\,t_m}{P}\right\}^{1/2} \tag{11-3}$$

where P is the Péclet number defined by

$$P = \frac{vL}{D} \tag{12}$$

Other expressions for the Laplace transform of the impulse response, which depend on the chosen boundary conditions for the mass balance equation, are available in the literature (van Genuchten and Parker, 1984; van Genuchten and Wierenga, 1986; van Genuchten, 1981). Although useful, these expressions do not define the residence time distribution which requires that convection prevails at inlet and outlet (Sardin et al., 1991; Villermaux, 1982). We will further show that the mathematically different expressions yield similar results in most situations.

Complex flow patterns can also be represented by discrete models. They are obtained by assuming that the flow system is composed of a suitable network of mixing cells. For example, the one-dimensional convection-dispersion equation and boundary conditions (11-1,2) can be replaced by

$$\left.\begin{aligned} QC_{m,k-1} &= QC_{m,k} + \frac{V_m}{J}\frac{dC_{m,k}}{dt} \qquad k = 1 \text{ to } J \\ x &= C_{m,0}, \quad y = C_{m,J} \end{aligned}\right\} \tag{13}$$

where k is the cell index, V_m the total volume of mobile water, and J the total number of mixing cells in series. From (13) and (3), the corresponding transfer function is

$$G_m(s) = \left[1 + \frac{s\,t_m}{J}\right]^{-J} \qquad t_m = V_m/Q \tag{14}$$

Remark that (14) allows one to use noninteger J. Using equation (6), the moments of the residence time distribution can be deduced from knowledge of $G_m(s)$. It can be shown that the first-order moment, or mean residence time, μ_{1m}, of the residence time distribution is the ratio of the volume accessible to the fluid to the volumetric flow rate whatever the flow pattern (Villermaux, 1982). We strongly emphasize that the first-order moment of the residence time distribution is independent of any dispersion and kinetic parameters. In equations (9), (10), (11), and (14), $\mu_{1m} = t_m$. For some boundary conditions to the convection-dispersion equation, a dispersion contribution to μ_{1m} is found. As already mentionned, these boundary conditions do not define the residence time distribution of the system.

The second-order centered moment, or variance, measures the width of the residence time distribution. It is defined by

$$\sigma_m^2 = \mu_{2m} - (\mu_{1m})^2 = \int_0^\infty (t - t_m)^2 \, E_m(t) \, dt \qquad (15)$$

The following expressions are obtained for the one-dimensional convection-dispersion problem

$$\frac{\sigma_m^2}{t_m^2} = \frac{2}{P} - \frac{2}{P^2}(1 - e^{-P}) \approx \frac{2}{P} = \frac{2D}{vL} \qquad \text{continuous model} \qquad (16\text{-}1)$$

$$\frac{\sigma_m^2}{t_m^2} = \frac{1}{J} \qquad\qquad \text{discrete model} \qquad (16\text{-}2)$$

In most situations, P is greater than 20. Thus, the second contribution to the middle expression of (16-1) becomes negligible, and the reduced variance is close to 2/P. Matching the variances of the continuous and discrete models yields

$$P \approx 2J \qquad (17)$$

Villermaux (1981, 1982, 1987) has shown that equation (17) makes the residence time distributions corresponding to the two models so close that the differences are generally less than experimental errors (except when P is smaller than 20). Moreover, these residence time distributions are very close to the impulse responses deduced from continuous models based on various boundary conditions. This means that, provided that P is greater than 20, the boundary condition problem is secondary, and is no more than an artificial mathematical problem. Similar results have been published by van Genuchten and Parker (1984).

From a numerical point of view, equation (13) is a first-order backwards (i.e., implicit) expansion of equation (9) with respect to z. It is known that this expansion involves "numerical dispersion" owing to the finite size of the elementary cell. Equation (17) renders this numerical dispersion as close as possible to the physical dispersion. In other words, the discrete model, or an appropriate expansion of the spatial derivatives of a model where dispersion is neglected, can be used to simulate the physical dispersion process.

The equivalence between the continuous and discrete models is further supported by the physical interpretations of the Péclet number

$$P = \frac{vL}{D} = \frac{L/v}{D/v^2} = \frac{t_m}{t_D} = \frac{L}{D/v} = \frac{L}{\xi_D} \qquad (18)$$

where t_D is a characteristic dispersion time, and ξ_D a characteristic dispersion length or dispersivity. For sufficiently high fluid velocity, molecular diffusion can be neglected, and the dispersion process is essentially due to the random structure of the porous matrix. This regime is called "statistical dispersion".

One must first wonder whether Fick's law is able to describe this regime. To answer this question, let us represent the porous matrix by a 3-dimensional lattice of channels. The dispersion process is due to the random walk of water globules in the lattice. Let us first consider the simplest case of a one-dimensional lattice. A uniform and symmetrical one-dimensional random walk is made of successive jumps over a length Δ chosen randomly, in either the positive or negative direction, with equal probabilities at each time step τ (Derrida et al, 1988; Montroll and Weiss, 1962; Hughes and Prager, 1980). The (discrete) probability density function (symmetrical binomial distribution) of the walker becomes nearly gaussian in space when the number of time steps is larger than about 20. In the latter case, Fick's law holds in the continuous limit when Δ and τ tend toward zero and $\Delta^2/\tau = 2D$ is constant. In the porous matrix, water globules are the walkers which jump from node to node every t_D, whereas the time required for a globule to move over a distance L is t_m. The corresponding number of elementary time steps is thus $t_m/t_D = P$. Consequently, Fick's law can be used to represent the statistical dispersion process provided that P is larger than about 20. As a result, the different impulse responses obtained according to the boundary conditions when P is lower than 20 are physically questionnable because Fick's law should no longer be used.

The random change of direction of water globules on the lattice is a key phenomenon which explains the dispersion process. To estimate t_D, we must first estimate the average length of a straight path, l_p, along a given direction on the lattice. Let p be the probability that the globule remains on the same straight path at a node. The average length l_p is $\Delta(1 + p + p^2 + p^3 + ...) = \Delta/(1 - p)$. Assuming that there are two identical downstream channels leaving a node, p is close to 0.5, and $l_p \approx 2\Delta$. The length of a channel is then of the order of the aggregate size d_p, and thus $l_p \approx 2d_p$. The time spent by the globule along the straight path is finally given by

$$t_D = t_{Ds} = \frac{l_p}{v} \approx 2\frac{d_p}{v} \qquad (19)$$

where t_{Ds} is the statistical dispersion time. Substituting (19) into (18) gives:

$$P \approx 0.5\frac{L}{d_p} \qquad P_g = \frac{v\,d_p}{D} = \frac{d_p}{\xi_{Ds}} \approx 0.5 \qquad (20)$$

where P_g is the aggregate Péclet number and ξ_{Ds} the statistical dispersion length.

For an extremely slow fluid velocity, molecular diffusion becomes responsible for dispersion. In this case the dispersion coefficient D is close to the molecular diffusivity $\mathcal{D}_m$ of water. The threshold between the molecular (low velocity) and the statistical (high velocity) regimes is defined by the molecular Péclet number at the aggregate scale:

$$P_{gm} = \tau\frac{v d_p}{\mathcal{D}_m} = \frac{d_p}{\xi_{Dm}}, \qquad \xi_{Dm} = \frac{\mathcal{D}_m}{\tau\,v} \approx \frac{\mathcal{D}_m}{2\,v} \qquad (21)$$

where τ is a tortuosity factor about 2 accounting for the interaggregate paths which are not parallel to the main flow direction. When P_{gm} is much less than unity, the water globule loses its integrity because molecular diffusion becomes significant compared with convection over a length smaller than d_p ($\xi_{Dm} \gg d_p$ or $t_{Dm} = \tau d_p^2/\mathcal{D}_m \ll t_{Ds}$). Consequently, the random change of direction at the nodes becomes secondary, and molecular diffusion prevails. Conversely, when P_{gm} is greater than unity, the globule keeps its integrity over the length d_p

and statistical dispersion prevails (i.e., (20) holds). More generally, one may model both regimes by adding the contributions of the statistical and molecular diffusion processes. Using equations (20) and (21), one obtains

$$\xi_D = \xi_{Ds} + \xi_{Dm} \qquad D \approx 2 v \, d_p + \mathcal{D}_m/\tau \qquad (22)$$

In terms of the mixing cell model one has:

$$\xi_D = \frac{L}{2 J} \qquad J \approx \frac{L}{4 \, d_p + \mathcal{D}_m/(\tau v)} \qquad (23)$$

The discussion above shows that dispersion models, either continuous or discrete, can be used provided that the characteristic dispersion length ξ_D exists and is much smaller than the largest scale, L, of the system. Equivalently, t_D must be smaller than t_m. ξ_D is smaller than L in most column experiments because the packing is made of a disturbed (i.e., repacked) soil. In field experiments, ξ_D is generally ill-defined and it can be of the order of L because of cracks and fissures at various scales. In the latter case, measuring the residence time distributions over increasing sizes L often yields an ever increasing dispersion coefficient.

There is no efficient and simple method to model flow systems which exhibit a scale-dependent dispersion length, except when the system is self-similar. We will briefly outline some properties of residence time distributions in self-similar structures, and we refer the reader to Villermaux and Schweich (1992) for technical details. A self-similar flow system is obtained by a recurrent procedure illustrated in Figure 2A. At each step, the network has the same volume V.

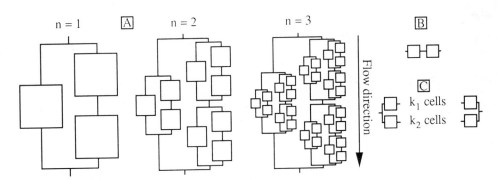

Figure 2: Schematic of a self-similar flow system. A: n = 1 defines generating pattern; n = 2 and 3 define two successive generations. B: generating pattern for mixing cells in series. C: generating patterns made of two parallel lines containing k_1 and k_2 mixing cells respectively. Generating patterns are made of identical elementary systems (i.e., mixing cell, plug flow).

A generating pattern made of identical elementary flow systems is first defined by n=1. In the special case of Figure 2A, it is made of three identical mixing cells. At step n=2, each mixing cells are replaced by a contracted generating pattern. The contraction factor (1/3 in Figure 2A) is chosen so as to keep the volume of the system constant. The procedure is then repeated as many times as desired. Only three steps are shown in Figure 2A. The same recurrent method can be applied whatever the generating pattern (see Figure 2B and 2C) and whatever the elementary flow system involved in the generating pattern.

The problem consists of finding the limiting residence time distribution at infinite generation. Let us first consider the generating pattern of Figure 2B. At step n, the system is composed of $J = 2n$ mixing cells in series, and it is equivalent to a one-dimensional convective-dispersive flow with $P = 4n$. When n goes to infinity the limiting system is equivalent to ideal plug flow. In this system, all water molecules moves along a single path and the residence time distribution is a delayed Dirac $\delta(t)$-function (See eqn. (9)). Conversely, in cases 2A and 2C, water molecules can move along different paths because of the presence of parallel branches. One easily guesses from Figure 2A that there is no characteristic length smaller than L at infinite generation because there are branches with any number of mixing cells in series of vanishingly small volume, starting from a single cell in the left branch and ending with an infinite number of cells in the right branch. Consequently, it would be surprising that a standard dispersion model could be applied to 2A and 2C.

Flow rate partioning between the branches of the generating pattern can be arbitrary. However, for the sake of simplicity, we will assume equal flow rates downstream a node. With this assumption, it can be shown (Villermaux and Schweich, 1992) that: (1) there is a unique limiting residence time distribution , (2) the moments of any order are bounded, (3) the limiting residence time distribution is independent of the elementary flow system involved in the generating pattern, and (4) the Laplace transform of the residence time distribution is the solution of a *nonlocal* algebraic fixed point equation. For the pattern of Figure 2A, the equation is (Villermaux and Schweich, 1992):

$$G_m(s) = \frac{1}{2}\left\{\left[G_m\left(\frac{2s}{3}\right)\right]^2 + G_m\left(\frac{2s}{3}\right)\right\} \qquad (24\text{-}1)$$

There is no analytical expression for $G_m(s)$ and only numerical procedures (based on the recurrent method) can be used to evaluate $G_m(s)$ and eventually $E_m(t)$ (by Fast Fourier Transform). Figure 3A illustrates the residence time distributions for various n, and Figure 3B compares the limiting residence time distribution ($n \geq 50$) with the best fit obtained with the mixing cell in series model. When n is greater than about 10 the network reaches the asymptotic behavior, and the limiting residence time distribution is somewhat asymetrical. The best fit with the mixing cell in series model is qualitatively satisfactory and yields $J = 2.5$ or $P \approx 5$. Although the curves of Figure 3B have similar shapes, the low equivalent Péclet number clearly indicates that the dispersion model is not appropriate from a physical point of view.

The inadequacy of the dispersion model is dramatically illustrated in Figure 4 for an elementary pattern given by Figure 2C with $k_1 = 2$ and $k_2 = 4$ which induce two successive maxima in the limiting residence time distribution . When there is a single path in the network (Figure 2B), the limiting residence time distribution is classical. When there are different paths in the generating pattern (Figures 1A and 1C), the self-similar networks exhibit an "anomalous dispersion" (nongaussian in space) behavior. According to the distribution of path lengths (i.e., the generating pattern), the residence time distribution can be "regular" (i.e., it qualitatively fits a standard dispersion model), or "irregular" (i.e., standard dispersion fails to model the residence time distribution). Let us finally remark that the fixed point equation which defines $G_m(s)$ is independent of t_m (See eqn. (24-1)). This means that the limiting G_m and E_m distributions are scale-invariant. This is illustrated by the reduced variance resulting from the pattern of Figure 2C

$$\frac{\sigma_m^2}{t_m^2} = \frac{(k_1 - k_2)^2}{(k_1 + k_2)(k_1 + k_2 - 2)} \tag{24-2}$$

which is independent of L contrary to a classical Péclet number (Compare (24-2) with (16-1)).

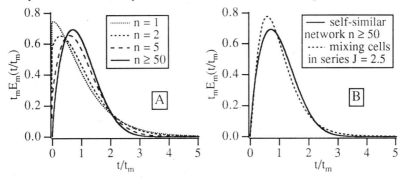

Figure 3: RTDs in self-similar structure. A: RTDs of patterns of Figure 2A at various n. B: best fit of RTD n ≥ 50 by mixing cell in series model.

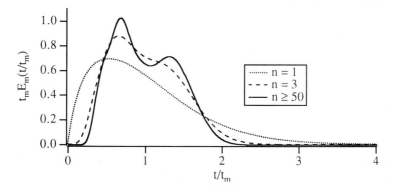

Figure 4: RTDs resulting from generating pattern of Figure 2C with $k_1 = 2$ and $k_2 = 4$.

Little is known concerning the properties of these residence time distributions for finite n. When flow rates are not equi-partitioned downstream a node, the properties of the limiting residence time distribution (if any) have not been investigated in detail. Moreover, in the latter case, the moments of the residence time distribution can diverge at some order depending on the structure of the generating pattern. There is little doubt that the properties of these self-similar networks could be useful in the near future to account for some unusual scaling behaviors of fitted dispersion coefficients reported in the literature.

Dynamics of a linearly sorbing aggregate

The aggregate is a dynamic system which exchanges solute with its environment. The problem is to derive the transfer function which relates the concentration of sorbed solute averaged over the aggregate volume to the concentration in the environment. We will assume that the immediate vicinity of an aggregate is of uniform composition. This means that the characteristic length scale of the concentration gradient in the environment of the aggregate is larger than the aggregate dimension. This assumption, which is implicit in most transport models of the literature, deserves a few comments.

Let us assume that the external fluid is the mobile water phase. Because of the dispersion process, C_m is nearly constant over the distance ξ_D. In the statistical regime $\xi_D = 2d_p$ and the environment of the aggregate can be assumed to be of uniform composition. However, close the injection point, the length, z, of the flow system can be smaller than $20\xi_D$, or $P(z) = vz/D$ can be smaller than 20. In this case, the concentration gradients are governed by the injection signal instead of the dispersion process, and the aggregate can be in contact with a solution of nonuniform composition. When the flow system has a size L much larger than ξ_D, the region where the uniformity assumption is violated is small and we *hope* that it can be neglected. Let us remark that we are facing again the boundary condition problem dealt with in Section 1.

Figure 5 schematically illustrates the four aggregate structures which will be considered. In Figure 5, the white area above the dashed line is the mobile fluid, the dashed line indicates the viscous boundary layer, and the shaded areas are the sorbing aggregate.

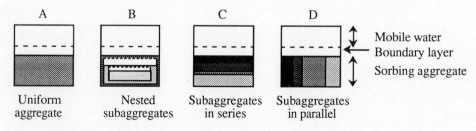

A	B	C	D	
Uniform aggregate	Nested subaggregates	Subaggregates in series	Subaggregates in parallel	Mobile water Boundary layer Sorbing aggregate

Figure 5: Schematic of aggregate structures.

The simplest situation (Figure 5A) assumes that the aggregate behaves as an effective medium with uniform properties. In 5B the aggregate is composed of nested subaggreggates, each of which can have different properties. In this latter case, we will speak of nested sorbing sites. This situation will account for multiple porosity levels as encountered in a macroporous particle made up of numerous microporous subparticles (a macroporous aggregate made of microporous clay crystals for example). In 5C the aggregate consists of concentric shells of different properties. In this case, we will speak of sorbing sites in series. This could be used to describe sorbing aggregates made of a weathered shell which surrounds an unaltered core. Finally, in 5D the aggregate is made of a population of sorbing subaggregates placed in parallel and locally in contact with the same solution. Here, we will speak of sorbing sites in parallel. This case could be used to account for a distribution of aggregate sizes.

Let $<C_p>$ be the concentration of the solute averaged over the aggregate volume. The problem consists of deriving the transfer function

$$L(s) = \frac{<\overline{C_p}>}{\overline{C_m}} \tag{25}$$

The uniform aggregate

We assume that the solute penetrates the aggregate by diffusion. The solute molecules have first to diffuse across the viscous bounadry layer. This mass transfer process is accounted for by the linear driving force approximation for diffusion. Second, the solute molecules penetrates the bulk of the aggregate by molecular diffusion along tortuous paths defined by the geometry of the pore network. This second mass transfer process is modeled by Fick's law using an effective diffusivity. Third, the solute molecules adsorb on the solid matrix. The adsorption rate is assumed to be first-order.

We will only give the result of the model, and we refer the reader to Sardin et al. (1991) and Villermaux (1981) for the derivations. Assuming that the adsorbed solute is motionless (i.e., no surface diffusion), one finds

$$L(s) = \frac{<\overline{C_p}>}{\overline{C_m}} = \frac{\alpha}{\dfrac{1}{B(s)\, H^*[sB(s)]} + st_e} \tag{26-1}$$

$$B(s) = \frac{\alpha - \beta + \beta st_{des}}{\alpha - \beta + \alpha st_{des}} \tag{26-2}$$

where H^* and B account for the internal diffusion process in the pore fluid and the adsorption process on the solid respectively. t_e and t_{des} are the characteristic external mass transfer time and desorption time respectively. They are given by

$$t_e = \frac{\alpha l}{k_e} \qquad (27)$$

$$t_{des} = \frac{\alpha - \beta}{\alpha k_{des}} \qquad (28)$$

l is the characteristic aggregate size defined as the ratio of the aggregate volume to its external surface area. β is the fraction of immobile water in the internal pores relative to the aggregate volume. k_e is the external mass transfer coefficient which can be estimated from empirical correlations (Crittenden et al., 1986; Nicoud and Schweich, 1989). Finally, k_{des} is a first-order desorption constant which accounts for the dynamics of the adsorption from the pore liquid to the solid matrix. Knowing that $L(0)$ is the ratio of $<C_p>$ to C_m at steady state (see (4)), and that the aggregate is in equilibrium with the mobile phase at steady state, $L(0) = \alpha$ is the equilibrium partition factor, where

$$\alpha = \beta + \rho_p K_A \qquad (29)$$

In (29), K_A is the adsorption constant (litre of water per kg of solid) and ρ_p the bulk density of the aggregate.

H*(s) describes the internal diffusion process, and its expression is given in Table 1 as a function of the aggregate shape. D'_e is the apparent effective diffusivity of the solute in the aggregate

$$D'_e = \frac{\beta \mathcal{D}_m}{\alpha \tau_p} = \frac{D_e}{\alpha}, \qquad D_e = \mathcal{D}_m \frac{\beta}{\tau_p} \qquad (30)$$

where τ_p is the tortuosity factor of the internal pores and D_e the standard effective diffusivity.

Most of the dynamic features are governed by characteristic times such as t_e and t_{des}. New characteristic times can be defined by the first-order moments of the inverse of a transfer functions. Using (6), the internal diffusion time is defined as

$$t_i = - \left(\frac{\partial H^*}{\partial s} \right)_{s=0} = \mu \frac{l^2}{D'_e} \qquad (31)$$

where μ is a shape factor given in Table 1. Then, the characteristic time of the sorption process at the aggregate level becomes

$$t_M = - \frac{1}{L(0)} \left(\frac{\partial L}{\partial s} \right)_{s=0} = t_e + t_i + t_{des} \qquad (32)$$

Equations (26) and (32) make it easy to obtain the description of degenerate cases. When one of the three contributions to t_M is negligible, it can be then set equal to zero and $L(s)$ simplifies. For example when t_{des} is negligible, then $B(s) = 1$ and

$$L(s) = \frac{\alpha}{\dfrac{1}{H^*(s)} + s t_e} \qquad (33)$$

Furthermore, if one wants to describe solute transfer to immobile water trapped in an aggregate where there is no adsorption on the solid matrix, then $\alpha = \beta$. These examples show that equations (26) cover a wide range of situations.

Table 1: Standard transfer functions $H^*(s)$ for various aggregate shapes. I_0 and I_1 are zero- and first-order modified Bessel functions of the first kind. For nested, parallel or series sites parameters l and D'_e depend on site index k.

$\lambda = l \left(\dfrac{s}{D'_e} \right)^{1/2}$			
Aggregate shape	Characteristic length l	$H^*(s)$	Shape factor f
Slab, thickness $2l$	l	$\dfrac{\tanh(\lambda)}{\lambda}$	1/3
Infinite cylinder, radius r	$r/2$	$\dfrac{I_1(2\lambda)}{\lambda I_0(2\lambda)}$	1/2
Sphere, radius r	$r/3$	$\dfrac{\cotanh(3\lambda)}{\lambda} - \dfrac{1}{3\lambda^2}$	3/5

Nested sorbing sites

The aggregate is now considered to be a nest of n shells. The aggregate is made of macrograins and macropores. The macrograins are made of micrograins and micropores, and so on. The innermost level is referred to by the subscript 1, and the outermost by n. Any internal level k is considered pointwise when seen from the outer level k+1. The innermost level is assumed to be a uniform medium with no substructure. Every inner level is in contact with a uniform environment, locally defined by the immediately outer level. Each level has its own associated properties (l_k, k_{ek}, D'_{ek}, τ_{pk}) and characteristic times t_{ek} and t_{ik}. β_k is defined as the fraction of pore water in pores at level k, the pores at inner levels being excluded, relative to the volume of level k. Villermaux et al. (1992) have shown that

$$L(s) = \frac{<\overline{C_p}>}{\overline{C_m}} = L_n(s) \qquad\qquad L_k(s) = \frac{\alpha_k}{\dfrac{1}{H_k(s)} + s t_{ek}} \qquad (34\text{-}1,2)$$

$$H_k(s) = B_k(s)\, H^*[s B_k(s)] \qquad\qquad B_k(s) = \frac{\beta_k + (1-\beta_k)L_{k-1}(s)}{\alpha_k} \qquad (34\text{-}3,4)$$

$$\alpha_k = \beta_k + (1-\beta_k)\alpha_{k-1} \qquad\qquad\qquad (34\text{-}5)$$

with the initial transfer function L_1 given by (26-1). In (34-3), H^* is calculated with the appropriate size l_k and diffusivity D_{ek}. The characteristic sorption time is given by

$$t_M = t_{Mn}, \qquad t_{Mk} = t_{ek} + \frac{\alpha_k - \beta_k}{\alpha_k} t_{Mk-1} + t_{ik} \qquad (35)$$

where t_{ik} and t_{ek} are given by (31) and (27) respectively with the appropriate subscripts k on μ, l, k_e, α and D'_e. Finally, t_{M1} is given by (32). One easily recognizes that (34-2) to (34-4) generalize (26-1) (See also Table 2). Furthermore, comparing (35) with (32) shows that a desorption time t_{des} at a macrolevel can be interpreted in terms of a more detailed sorption process at a deeper level through the characteristic time $(\alpha_k-\beta_k)t_{M,k-1}/\alpha_k$. The external mass transfer time, t_{en}, at the outermost level is the characteristic time of the mass transfer process through the boundary layer. At inner levels, t_{ek} is no longer given by the classical correlations for this mass transfer process. t_{ek} accounts for a possible external resistance, and can be estimated provided that physical explanations and models for this resistance are available. As in Section 2.1, degenerate transfer functions can be obtained when some of the characteristic times are set to zero:

when $t_{ik} = 0$ set $H^*[s\ B_{k-1,k}(s)] = 1$

when $t_{Mk} = 0$ set $L_k(s) = \alpha_k$

In most situations, physical and chemical considerations show that two or three levels are sufficient, and that at most two characteristic times are important.

Sorbing sites in series

This problem has been first presented by Villermaux (1981). A more complete description has been given by Schweich (1991). The aggregate is assumed to have a continuous distribution of properties along a space variable ξ ranging from 0 (center of the aggregate) to l (external surface of the aggregate). The distribution coefficient $\alpha(\xi)$, the fraction of pore water $\beta(\xi)$, the tortuosity factor $\tau_p(\xi)$, and the effective diffusivity $D_e(\xi)$ defined by

$$D_e(\xi) = \frac{\beta(\xi)}{\tau_p(\xi)}\ \mathcal{D}_m(\xi) = \alpha(\xi)\ D'_e(\xi) \tag{36}$$

depend on the location inside the aggregate. Let us define the local average of $f(\xi)$ by

$$<f>_\xi = \frac{p}{\xi^p} \int_0^\xi f(u)\ u^{p-1}du \tag{37}$$

where p is a geometric factor (p=1 for a slab, 2 for a cylinder, and 3 for a sphere). L(s) is now given by

$$L(s) = \frac{<\widetilde{C_p}>}{\widetilde{C_m}} = \frac{<\alpha>_l}{\dfrac{<\alpha>_l}{<\alpha(\xi)B(\xi,s)>_l\ H(l,s)} + s t_e} \tag{38-1}$$

where H(l,s) is the solution at $\xi= l$ of the differential equation

$$\frac{1}{p\xi^{p-1}}\frac{\partial \xi^p <\alpha B>_\xi H}{\partial \xi} + \frac{s\xi^2(<\alpha B>_\xi H)^2}{D_e(\xi)} = \alpha(\xi)B(\xi,s) , \qquad H(0,s) = 1 \qquad (38\text{-}2)$$

Note that $<\alpha>_l$ is the *average equilibrium partition factor* (The same as α in Sections 2.1 and 2.2). Depending on the structure assumed for each of possible sublevels, $B(\xi,s)$ is given by $B(\xi,s) = 1$ when there are no sublevels and adsorption is instantneous, (26-2) with $t_{des} = t_{des}(\xi)$ when there is a finite rate of adsorption, and by

$$B(\xi,s) = \frac{\beta(\xi) + [1-\beta(\xi)]L_{n-1}(s)}{\alpha(\xi)} \qquad (38\text{-}3)$$

when there are nested dynamic sublevels of transfer functions $L_{n-1}(s)$. Note that L_{n-1} does not depend on ξ since nested sublevels are considered to be pointwise at the scale of the aggregate. The characteristic sorption time is given by

$$t_M = t_e + t_i + t_B \qquad (39\text{-}1)$$

$$t_i = -\left(\frac{\partial H(l,s)}{\partial s}\right)_{s=0} = \frac{p}{l^p<\alpha>_l}\int_0^l \frac{\xi^{p+1}}{\alpha(\xi)D'_e(\xi)}(<\alpha>_\xi)^2\,d\xi = \frac{p}{l^p<\alpha>_l}\int_0^l \frac{\xi^{p+1}}{D_e(\xi)}(<\alpha>_\xi)^2\,d\xi \qquad (39\text{-}2)$$

$$t_B = \frac{-1}{<\alpha>_l}<\alpha\left(\frac{\partial B}{\partial s}\right)_{s=0}>_l \qquad (39\text{-}3)$$

(39-2) reduces to (31) when α and D_e are independent of ξ. When adsorption is instantaneous, $t_B = 0$, whereas

$$t_B = \frac{<\alpha t_{des}>_l}{<\alpha>_l} \qquad \text{for a first-order adsorption} \qquad (40\text{-}1)$$

$$t_B = t_{Mn-1}\frac{<\alpha - \beta>_l}{<\alpha>_l} \qquad \text{for nested sublevels} \qquad (40\text{-}2)$$

Table 2 shows that equations (39) and (40) are generalized forms of (32) and (35-1).

Sorbing sites in parallel

This problem has been investigated by Villermaux (1981) and Rasmuson (1985). Here, the aggegate is a population of subaggregates with various properties, referred to by subscript k. These subaggregates are assumed to be in contact with the same solution of concentration C_m. Let $<C_{pk}>$ be the average concentration of sorbed solute on site k, and

$$L_{pk}(s) = \frac{<\overline{C_{pk}}>}{C_m} \qquad (41)$$

the transfer function of the kth site. L_{pk} can be one of the transfer functions, $L(s)$, given by (26-1), (34-1), (38-1) and presented in Sections 2.1 to 2.3. If w_{pk} is the volume fraction of the kth site with respect to the total aggregate volume, then

$$L(s) = \frac{<\overline{C_p}>}{\overline{C_m}} = \sum_k w_{pk} \, L_{pk}(s) \tag{42}$$

where $<C_p>$ is the concentration of sorbed solute averaged over the site population. Let α_{pk} be the equilibrium partition factor on the kth site . We then have

$$L(s) = \alpha \sum_k \omega_{pk} \, L_{pk}(s)/\alpha_{pk} \tag{43-1}$$

$$\omega_{pk} = \alpha_{pk} \, w_{pk}/\alpha \tag{43-2}$$

$$\alpha = \sum_k w_{pk} \, \alpha_{pk} \tag{43-4}$$

where α is the equilibrium partition factor averaged over the population, ω_k the *fraction of solute sorbed at equilibrium* on the kth site, and $L_{pk}(s)/\alpha_{pk}$ the Laplace transform of the *normalized* impulse response of the kth site.

The characteristic sorption time is given by

$$t_M = \sum_k \omega_{pk} \, t_{Mpk} \tag{44}$$

where t_{Mpk} is the sorption time associated to $L_{pk}(s)$.

Note that L_{pk}/α_{pk} does not explicitly depend on α_{pk} since L_{pk} is proportional to α_{pk} (see (26-1), (34-2), (38-1)). This means that the unique equilibrium parameter which explicitely appears in $L(s)$ is the average partition factor α. Conversely, L_{pk} generally involves several characteristic times. There is no convenient continuous counterpart to (43-1) in the general case. However, when L_{pk} depends on a single characteristic time τ, the latter can be chosen as the parameter which defines the population. In this case

$$L(s) = \alpha \int_0^\infty \frac{L_p(\tau,s)}{\alpha(\tau)} \, f(\tau) \, d\tau \tag{45}$$

where $f(\tau)d\tau$ is the Transfer Time Distribution, or the *fraction of solute sorbed at equilibrium* on sites which have a characteristic time between τ and $\tau+d\tau$. From equation (6), the characteristic sorption time becomes

$$t_M = \int_0^\infty \tau \, f(\tau) \, d\tau \tag{46}$$

Degenerate cases are easily obtained from the general equations above. When internal diffusion at a single level is the dominating process in a slab, one then obtains from Table 1

$$L(s) = \alpha \int_0^\infty \frac{\tanh(\lambda)}{\lambda} \, f(\tau) \, d\tau \, , \quad \lambda^2 = 3\tau s \, , \quad \tau = t_i = \frac{l^2}{3 \, D'_e} \tag{47}$$

The last equality in (47) shows that $f(\tau)$ can represent a distribution of either l or D'_e.

Table 2: Expressions for characteristic sorption time according to aggregate structure and underlying mass transfer and sorption processes.

Homogeneous aggregate	$t_M = t_e + t_i + t_{des}$ $\quad$ $t_e = \dfrac{\alpha l}{k_e}$ $t_i = \mu \dfrac{l^2}{D'_e}$ $\quad$ $t_{des} = \dfrac{\alpha - \beta}{\alpha k_{des}}$
Nested sorbing sites	$t_M = t_{Mn}$ $\quad$ $t_{Mk} = t_{ek} + \dfrac{\alpha_k - \beta_k}{\alpha_k} t_{Mk-1} + t_{ik}$ $\quad$ $t_{ek} = \dfrac{\alpha_k l_k}{k_{ek}}$ $t_{ik} = \mu_k \dfrac{l_k^2}{D'_{ek}}$ $\quad$ $t_{M1} = t_{e1} + t_{i1} + t_{des1}$ $\quad$ $t_{des1} = \dfrac{\alpha_1 - \beta_1}{\alpha_1 k_{des}}$
Sorbing sites in series	$t_M = t_e + t_i + t_B$ $\quad$ $t_e = \dfrac{l <\alpha>_l}{k_e}$ $\quad$ $t_i = \dfrac{p}{l^p <\alpha>_l} \displaystyle\int_0^l \dfrac{\xi^{p+1}(<\alpha>_\xi)^2}{D_e(\xi)} d\xi$ $t_B = \dfrac{<\alpha t_{des}>_l}{<\alpha>_l} = \dfrac{p}{l^p <\alpha>_l} \displaystyle\int_0^l \dfrac{\alpha(\xi) - \beta(\xi)}{\alpha(\xi) k_{des}(\xi)} d\xi$ $\quad$ for finite adsorption rate and no sublevels. $t_B = t_{Mn-1} \dfrac{<\alpha - \beta>_l}{<\alpha>_l}$ $\quad$ for sublevels.
Sorbing sites in parallel	$t_M = \displaystyle\sum_k \omega_{pk} t_{Mpk}$ $\quad$ t_{Mpk} given by one of the t_M's above. $t_M = \displaystyle\int_0^\infty \tau f(\tau) d\tau$ $\quad$ when there is a single dominating kinetic process.

The first-order approximation

In spite of the different mathematical expressions for $L(s)$, most cases are closely approximated by

$$L_N(s) = \frac{\alpha}{1 + st_M} \qquad (48)$$

where α is the average partition factor and t_M the average sorption time defined in Table 2.

Applications

For the sake of simplicity we will assume throughout this section that internal diffusion with instantaneous adsorption on the solid surface is the dominating kinetic process. We will further assume that the aggregate is a slab in which α is uniform.

Let us consider a laboratory batch experiment where solute-free aggregates are in contact with a solution of constant concentration C_m. The average solute concentration $<C_p(t)>$ is the step response of the aggregates. According to (7), $<C_p(t)>$ is the inverse of $L(s)/s$. This type of experiment is often used to fit the equilibrium parameter α from the ratio $<C_p>/C_m$ at steady state, and the kinetic parameters from the transient part of the experimental curve. The sorption time t_M is uniquely defined by the experiment, and it can be determined from the experimental curve by

$$t_M = \int_0^\infty [1 - <C_p(t)>/<C_p(\infty)>] \, dt$$

The problem is now to interpret t_M in terms of D_e. If one interprets t_M with various physical models (uniform aggregate, nested or parallel or series sites), then one can find as many physical parameters (D_e, D_{ek}) as there are models. This means that the determination of a physical kinetic parameter from an experimental curve is an ill-defined problem, even when the governing process (i.e., internal diffusion in this example) is clearly identified. If one uses the uniform aggregate model, one then obtains an average effective diffusivity $<D_e>$ defined by

$$t_M = t_i = \frac{\alpha l^2}{3 <D_e>} \tag{49}$$

If one uses the model of sites in series or in parallel, the diffusivity obeys (see Table 2)

$$t_M = t_i = \frac{\alpha}{l} \int_0^l \frac{\xi^2}{D_e(\xi)} \, d\xi \qquad \text{series sites} \tag{50-1}$$

$$t_M = t_i = \frac{\alpha l^2}{3} \int_0^\infty \frac{f(\tau) \, d\tau}{D_e(\tau)} = \frac{\alpha l^2}{3} \sum_k \frac{\omega_{pk}}{D_{ek}}, \qquad \tau = \frac{\alpha l^2}{3 \, D_e(\tau)} \qquad \text{parallel sites} \tag{50-2}$$

In most situations, the distribution of D_e is not known, and one must use the uniform model to interpret t_M. Consequently, one defines an average effective diffusivity, $<D_e>$ by matching the expressions for t_M in a uniform and in a nonuniform aggregate. This ensures that the theroretical expressions for $C_p(t)$ corresponding to the various models have the same first-order moments. Equating the right-hand sides of equations (49) and (50) shows that the average diffusivity $<D_e>$ is defined by a harmonic mean whether the sorbing sites are in parallel or in series. For a broad distribution of site properties, this harmonic mean can differ greatly from other averages.

One must then wonder whether the average effective diffusivity and the uniform model are able to account for the distribution of diffusivities. Figure 6 compares the true step response of two sorbing sites in series with the best fits obtained with the uniform model and the first-order approximation (48). The two sites are of equal thickness $l/2$, but the diffusivity in the outer site, D_{eo}, is larger than the diffusivity, D_{ei}, in the inner site. Using equations (49) and (50-1), one finds

$$\frac{1}{<D_e>} = \frac{1}{8D_{ei}} + \frac{7}{8D_{eo}} \tag{51-1}$$

When $D_{eo} = 100D_{ei}$ (Fig. 6A), the uniform model fails to fit the true response which is made of a sharp front (rapid diffusion in the outer shell) followed by a long tail (slow diffusion in the inner core). Moreover, $<D_e>$ is close to $D_{ei}/8$ and not $D_{ei}/2$ as it could be expected on the basis of the respective thicknesses of the two sites. This situation is exceptional, and discriminating between the uniform and site-in-series model is in general not so easy. In most cases, the uniform model fits well the site-in-series model, as illustrated in Figure 6B, even though $D_{eo} = 10D_{ei}$. Moreover, the agreement between the first-order approximation and the uniform model is satisfactory in both cases A and B. The latter observation means that a model which assumes either some form of external mass transfer resistance or a first-order adsorption, would equally fit the experimental curve with the same t_M.

If the two sites are of equal thickness l and are now placed in parallel with $\omega_{p1} = \omega_{p2} = 0.5$ (i.e., equal fractions of sites as in the previous case of sites in series), then equations (49) and (50-2) give

$$\frac{1}{<D_e>} = \frac{0.5}{D_{e1}} + \frac{0.5}{D_{e2}} \tag{51-2}$$

Comparing (51-1) and (51-2) clearly shows the combinations of the elementary diffusivities are different and depend on the series or parallel structure of the sorbing sites. Consequently, fitted diffusivities will depend on the model chosen.

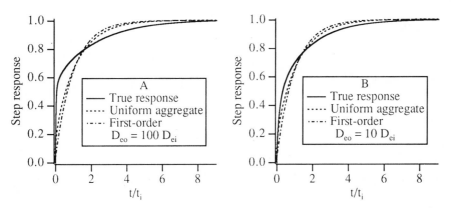

Figure 6: Step responses of a slab made of 2 sorbing slabs in series. Volumes of slabs: 50% of total volume. Uniform partition factor $<\alpha>_l = \alpha(\xi) = 1$. D_{eo} in outer slab larger than D_{ei} in inner slab.

Instead of trying to interpret the experimental t_M, one can fit the experimental curve with the model using any least-squares method. Unfortunately, the various models generally give approximately the same quality of fit (See Figure 6B). This means that choosing a model *must* rely on knowledge of the aggregate structure, and *not* on the goodness of fit.

Aggregates of distributed size are easily accounted for by the transfer time distribution concept. $L(s)$ is given by

$$L(s) = \alpha \int_0^\infty H^*(l,s) \, f(\tau) \, d\tau \tag{52}$$

where $H^*(l,s)$ is one of the functions of Table 1. Assuming that D_e, ρ_p, and α are independent of the aggregate size, the fraction *of solute sorbed at equilibrium* on a site, $f(\tau)d\tau$, is the size distribution by weight, $g_w(l)dl$, and

$$L(s) = \alpha \int_0^\infty H^*(l,s) \, g_w(l) \, dl \,, \qquad t_M = t_i = \frac{\mu}{D'_e} \int_0^\infty l^2 \, g_w(l) \, dl \tag{53}$$

It is important to note that the relevant characteristic mean size involved in t_i is the second order moment of g_w.

Pore size distribution is more difficult to take into account because the connections between narrow and wide pores must be described. A simple approach consists of assuming that the pores are not interconnected and that they are placed in parallel. If β and τ_p are independent of the pore radius R, then the *fraction of solute sorbed at equilibrium* in pores of size R is (Cui et al., 1989)

$$f(\tau) \, d\tau = \frac{\alpha(R) \, \phi(R) \, dR}{\alpha} \,, \qquad \alpha = \int_0^\infty \alpha(R) \, \phi(R) \, dR \tag{54}$$

where $\alpha(R)$ is the partition factor in an aggregate made of pores of radius R, α the average partition factor, and $\phi(R)$ the volumetric pore size distribution. When the solute adsorbs on the entire pore wall, the adsorption constant $\rho_p K_A$ depends on the pore diameter via the specific surface. If K_{AS} is the intrinsic adsorption constant (m^3 of of fluid/m^2 of pore surface area), then

$$\rho_p K_A = \frac{2\beta K_{AS}}{R} \qquad \alpha(R) = \beta \left(1 + \frac{2 K_{AS}}{R}\right) \tag{55}$$

Finally, $L(s)$ and the diffusion time are given by

$$L(s) = \int_0^\infty H^*(R,s) \, \alpha(R) \, \phi(R) \, dR \tag{56-1}$$

$$t_i = \mu \alpha l^2 \frac{\displaystyle\int_0^\infty \frac{1}{D_e(R)} (1+2K_{AS}/R)^2 \, \phi(R) \, dR}{\left[1 + 2K_{AS} \displaystyle\int_0^\infty R^{-1}\phi(R) \, dR\right]^2} = \frac{\mu \alpha l^2}{\langle D_e \rangle} \tag{56-2}$$

The equality in (56-2) defines an average effective diffusivity $<D_e>$ which eventually depends on the adsorption strength via K_{AS}. Even when $D_e(R)$ is independent of R, $<D_e>$ differs from D_e because the diffusion velocities in the pores depend on the pore radii due to the adsorption process. (56-2) reduces to (31) only when there is a single pore radius, or when D_e is independent of R and $K_{AS} = 0$.

Finally, let us show that the transfer functions of Sections 2.1 to 2.5 allow one to model arbitrarily sophisticated situations. Let us consider the sorbing aggregate of Figure 1 (right), and assume that the nonporous gravel is made of a nonsorbing material. The aggregate thus consists of two parallel sites. The first site is the globule of immobile water of characteristic size l_1. Assuming fast exchange with mobile water, the corresponding transfer function is

$$L_1(s) = \alpha_1 = \beta_1 = 1 \tag{57}$$

The second site is the porous aggregated soil of size l_2 which is fully wetted by immobile water. Assuming that the internal diffusion limitation in pore water is the limiting mass transfer process, and that adsorption on the solid is instantaneous, the associated transfer function is

$$L_2(s) = \alpha_2 H^*(s), \qquad D'_e = \frac{\beta_2 \mathcal{D}_m}{\alpha_2 \tau_p} \tag{58}$$

The transfer function for the sorption process is then

$$L(s) = w_1 L_1(s) + (1-w_1) L_2(s) \tag{59}$$

where w_1 is the volume fraction of immobile water outside the aggregated soil, with respect to the volume of the two sites. Let ψ and θ_{im} be the volume fractions of aggregated soil and of total immobile water relative to the volume of the flow system, then

$$\frac{\text{volume of aggregated soil}}{\text{immobile water outside the aggregated soil}} = \frac{1-w_1}{w_1} = \frac{\psi}{\theta_{im}-\psi\beta_2} = W_{2/1} \tag{60}$$

and finally

$$L(s) = \frac{1 + W_{2/1}\, L_2(s)}{1 + W_{2/1}}, \qquad \alpha = L(0) = \frac{1 + W_{2/1}\alpha_2}{1 + W_{2/1}}, \qquad t_M = \frac{W_{2/1}\alpha_2}{1 + W_{2/1}\alpha_2}\, \frac{\mu l^2 \alpha_2 \tau_p}{\beta_2 \mathcal{D}_m} \tag{61}$$

For a nonreactive solute, $K_{A2} = 0$ and $\alpha_2 = \beta_2$, and

$$\alpha = L(0) = \frac{\theta_{im}}{\psi}\, \frac{W_{2/1}}{1 + W_{2/1}}, \qquad t_M = \frac{W_{2/1}\beta_2}{1 + W_{2/1}\beta_2}\, \frac{\mu l^2 \tau_p}{\mathcal{D}_m} \tag{62}$$

Reactive solute transport

Solute transport can now be described on the basis of a powerful theorem stated by Villermaux (1981). Assuming that the sorbing properties (equilibrium and kinetics) and the ratio of mobile water volume to aggregate volume are uniform throughout the flow system, and provided that there is no chemical reaction, the transfer function for the solute transport process is given by

$$G(s) = G_m\{s[1 + M(s)]\}, \qquad M(s) = \frac{1-\epsilon}{\theta_m} L(s) \tag{63}$$

where θ_m and $1 - \epsilon$ are the volume fractions of mobile water and aggregates relative to the volume of the system respectively. Consequently, the theoretical developments of Sections 1 and 2 immediately give $G(s)$. From equations (63) and (6), one obtains the following general expressions for the first-order moment and variance of the residence time distribution of the solute

$$\mu_1 = t_m(1 + K'), \qquad K' = \frac{1-\epsilon}{\theta_m} \alpha \tag{64-1}$$

$$\frac{\sigma^2}{\mu_1^2} = \frac{\sigma_m^2}{t_m^2} + \frac{\sigma_{im}^2}{\mu_1^2} \tag{64-2}$$

$$\frac{\sigma_{im}^2}{\mu_1^2} = \frac{2K'}{1 + K'} \frac{t_M}{\mu_1} = 2\frac{t'_M}{t_m}, \qquad t'_M = \frac{K'}{(1 + K')^2} t_M \tag{64-3}$$

where σ_{im}^2 is the contribution to the variance of the sorption process in the aggregates, and t'_M a corrected sorption time. With the following generalized definition of the dispersion time

$$t_D = \frac{\sigma_m^2}{2 \, t_m} \tag{65}$$

one obtains

$$\frac{\sigma^2}{\mu_1^2} = \frac{2}{t_m} (t_D + t'_M) \tag{66}$$

As a result, the dynamic features of the transport process are described by K', t_m, t_D, t_M, and t'_M. Sardin et al. (1991) have shown that

 - The residence time distribution of the solute is governed by the flow pattern when $t_D \gg t'_M$. The sorption dynamics can be ignored (i.e., $L(s) = \alpha$), and the local equilibrium assumption holds. When the dispersion model describes the residence time distribution of mobile water, the residence time distribution of the solute is almost symmetrical.

 - When t'_M is of the order of t_D, flow pattern and sorption dynamics are competing.

 - When $t'_M \gg t_D$, the residence time distribution of the solute is governed by the sorption dynamics and plug flow can be assumed.

 - When $t_M \ll t_m$, the residence time distribution is almost symmetrical and the first-order approximation (48) of a detailed model is sufficient to generate the residence time distribution. Using a single residence time distribution curve, discriminating among rival models for t_M is impossible.

 - When $t_M \gg t_m$, the residence time distribution is asymmetrical. When there is no site in parallel, the residence time distribution is composed of a peak located close to $t = t_m$, followed by a long tail. When sites in parallel are involved, the leading peak (if any) can be located at $t > t_m$.

- When t_i is a minor contribution to t_M, equation (48) is sufficient to generate the residence time distribution . Using a single residence time distribution curve, discriminating among rival models for t_M is impossible.

- The exact shape of the aggregate is of little importance to the transient internal diffusion process. Any $H^*(s)$ of Table 1 will give approximately the same result.

These rules show that oversophisticated models which involve multiple sorption sites are often useless because the shape of the residence time distribution is relatively insensitive to the substructure of the sorbing aggregate.

Conclusions

The concepts presented above show that solute transport in porous media is governed by a wealth of basic phenomena. Presently, it would be useless to describe solute transport with a model accounting for all these phenomena, because fitting of the many adjustable parameters would be always succesful although meaningless. It is much more important to recognize the few predominant processes which are responsible for the shape of residence time distributions. Estimates or experimental determination of characteristic times and the sensitivity of residence time distributions to operating conditions are among the efficient tools which allow the experimentalist to progressively elucidate the governing factors of the transport process.

Sophisticated models can be used only when most parameters can be determined independently of breakthrough curve fitting. Conversely, sophisticated models *must* be used to interpret the variations of the dynamic parameters (t_M, t_D, t_i,...) with the operating conditions (flow rate, aggregate size, nature of the solute, temperature). More specifically, these possible variations suggest that a detailed model is necessary to interpret the experiments. However, care must be taken when postulating the assumptions of the detailed model, because various aggregate structures can lead to the same sensitivity with respect to an operating parameter.

The many assumptions made in models must be validated as independently as possible from each other. When an assumption is formulated, appropriate and informative test experiments must be designed, including chemical and X-ray analyses, scanning electron microscopy, etc. Frequently, a sensitivity analysis with a model helps the experimentalist to design and calibrate these experiments. Then, the experimental result must be compared with model predictions. According to the result of the comparison, the assumption is accepted or rejected. Only when the assumption is accepted, the next can be tested, and the model may grow in complexity.

Finally, note that any characteristic times involve the equilibrium parameter α. This shows that determining the dynamic properties of the system requires an accurate knowledge of the equilibrium state. Any attempt to fit a charateristic time or a mass transfer parameter in an ill-defined equilibrium context leads to meningless results.

References

CrittendenJC, Hutzler NJ, Geyer DG, Orawitz JL, Friedman G (1986) Transport of organic compounds with saturated groundwater flow, model development and parameter sensitivity. Water Resour. Res. 22: 271-284

Cui LC, Schweich D, Villermaux J (1989) Influence of pore diameter distribution on the determination of effective diffusivity in porous particles. Chem. Eng. Process. 26:121-126

Derrida B, Bouchaud JP, Georges A (1988) An introduction to random walks. In: Guyon E, Nadal JP and Pommeau Y (eds) Mixing and disorder, NATO ASI Series E, vol 152. Kluwer academic press Dordrecht.

Hughes BD, Prager S (1980) Random processes and random systems. In: Dold A and Eckmann B (eds) The mathematics of disordered media, Lecture notes in mathematics, vol 1035, Springer, Berlin

Motroll EW, Weiss GH (1965) Random walks on lattice II. J. Math. Phys. 6:176-181

Nicoud RM, Schweich D (1989) Solute transport in porous media with solid-liquid mass transfer limitations: application to ion exchange. Water Resour. Res. 25: 1071-1082

Rasmuson A (1985) The effect of particle of variable size, shape and properties on the dynamics of fixed beds. Chem. Eng. Sci. 40:621-629

Sardin M, Schweich D, Leij FJ, van Genuchten MTh (1991) Modeling the nonequilibrium transport of linearly interacting solutes in porous media : a review. Water Resour. Res. 27:2287-2307

Schweich D (1991) Determination of solid-fluid mass transfer parameters from linear chromatography. In: Meunier F and LeVan D (eds) Adsorption process for gas separation, Récents progrès en génie des procédés, Lavoisier, 17:107-112

van Genuchten MTh (1981) Analytical solutions for chemical transport with simultaneous adsorption, zero-order production and first-order decay. J. of Hydrology 49:213-233

van Genuchten MTh, Parker JC (1984) Boundary conditions for displacement experiments through short laboratory soil columns. Soil Sci. Soc. Am. J. 48:703-708

van Genuchten MTh, Wierenga PJ (1986) Solute dispersion coefficients and retardation factors. In: American Society of Agronomy (eds) Methods of soil analysis, Part I, Physical and mineralogical methods, 2nd edn, vol. 9. Soil Science Society of America, Madison, USA, p1025

Villermaux J (1981) Theory of linear chromatography. In: Rodrigues A and Tondeur D (eds) Percolation processes, theory and applications, Sijthoff and Noordhoff, Rockvill MA USA, p 83

Villermaux J (1982) Génie de la réaction chimique, conception et fonctionnement des réacteurs, 2nd edn. Lavoisier Tec et Doc, Paris

Villermaux J (1987) Chemical engineering approach to dynamic modeling of linear chromatography, a flexible method for representing complex phenomena from simple concepts. J. Chromatogr 406:11-26

Villermaux E, Schweich D (1992) Hydrodynamic dispersion on self-similar structures: a Laplace space renormalization group approach. To appear in Journal de Physique

Villermaux J, Schweich D, Sardin M (1992) Modeling of chromatographic processes: a chemical engineering approach. In: Barker M and Ganetsos G (eds) Preparative and production scale chromatographic processes and applications. To be published.

Wen CY, Fan LT (1975) Models for flow systems and chemical reactors, M Dekker, New York

List of symbols

$B(s)$, $B_k(s)$	defined by (26-2), (34-4)	l	characteristic aggregate size (Volume/external surface area)
C, C_m, C_{im}	concentration, in mobile water, immobile water		
C_p	concentration in the aggregate	L	length
		n_I	amount of solute injected
D, D_s	dispersion coefficient, statistical contribution	P, P_g, P_{gm}	Péclet number, aggregate Péclet number, molecular Péclet number at aggregate scale
D_e, D'_e	effective diffusivity, apparent effective diffusivity		
$\mathcal{D}_m$	molecular diffusivity	Q	volumetric flow rate
d_p	aggregate size	s	Laplace parameter
$E_m(t)$	residence time distribution of the mobile water phase	t	time
		t_D, t_{Ds}	dispersion time, in the statistical regime
$G(s)$, $G_j(s)$	transfer functions		
g_w	size distribution by weight	t_{des}	characteristic desorption time
$H^*(s)$	standard transfer function for the internal diffusion process	t_e	external mass transfer time
		t_i	internal diffusion time
J	number of mixing cells in series	t_M, t'_M	characteristic sorption time, corrected sorption time
K_{AS}	intrinsic adsorption constant	t_m	convection time in mobile water
K_A	adsorption constant		
k_e	external mass transfer coefficient	v	pore fluid velocity
		w_k	volumetric fraction of site
l_p	average length of a straight path	$x(t)$, $y(t)$	input, output variables
		z	abscissa

Greek symbols

α	partition factor	τ	variable in transfer time distribution
β	internal aggregate porosity		
$1 - \varepsilon$	volumetric fraction of aggregate	ω_k	fraction of solute sorbed at equilibrium on site k
$\phi(R)$	pore size distribution	ξ_D, ξ_{Ds}	characteristic dispersion length, in the statistical regime
μ_k	kth-order moment		
ψ	fraction of aggregated soil		
ρ_p	aggregate density	ξ_{Dm}	characteristic diffusion length
σ^2	variance	ξ	space variable inside aggregate
θ	volumetric fraction of water		
τ, τ_p	tortuosity factors		

Subscripts

I	inlet	im	immobile water phase
m	mobile water phase		

Other symbols

$<C>$	concentration averaged over aggregate volume	$\overline{f}(s)$	Laplace transform of f(t)

Multicomponent Wave Propagation: The Coherence Principle. An Introduction

Friedrich G. Helfferich
Department of Chemical Engineering
The Pennsylvania State University
University Park, PA 16802, U.S.A.

Convective transport phenomena can be viewed as propagation of waves. For such a purpose, a wave is defined as a variation of the values of dependent variables, usually concentrations, temperature, pressure, etc. If the species involved affect one another's behavior, as is usually the case, the waves are nonlinear (that is, obey nonlinear differential equations). Examples include multicomponent fixed-bed adsorption and ion exchange (Helfferich and Klein, 1970; Rhee *et al.*, 1970, 1986; Vermeulen *et al.*, 1984; Ruthven, 1984), preparative chromatography (Helfferich and James, 1970; Rhee and Amundson, 1982; Frenz and Horvath, 1985), enhanced oil recovery (Welge, 1952; Helfferich, 1981; Hirasaki, 1981; Orr and Taber, 1984; Lake, 1989), leach mining (Walsh *et al.*, 1984), transport with precipitation and dissolution (Bryant *et al.*, 1986, 1987; Novak *et al.*, 1988; Helfferich, 1989a), sedimentation (Kynch, 1952; Wallis G, 1969), travel of disturbances through chemical equipment that is otherwise at equilibrium or steady state (Hwang and Helfferich, 1989), and traffic flow (Whitham, 1974; Tondeur, 1987a). Migration and fate of pollutants in geological formations, especially in aquifers, also is a phenomenon of this type, if a very complex one owing to the irregular, three-dimensional geometry and the usually highly nonuniform permeability of the medium in which it occurs, to the nonuniformity and irregular distribution of soil components that interact with the pollutants, and to chemical reactions which the pollutants may undergo.

Nonlinear wave theory has been used extensively in the oil industry for the development of techniques of enhanced oil recovery (e.g., see Lake, 1989). The structure of the problems encountered is quite similar to that in pollutant migration, so that the same basic concepts are applicable in both fields. This introductory chapter reviews the principles and illustrates their application with simple examples. The emphasis is on concepts, on cause and effect, and mathematics is kept to a minimum.

NATO ASI Series, Vol. G 32
Migration and Fate of Pollutants in Soils and Subsoils
Edited by D. Petruzzelli and F. G. Helfferich
© Springer-Verlag Berlin Heidelberg 1993

Wave Equations and Velocities

The central question in any application of nonlinear wave theory is, "how fast does a given value x of a dependent variable move?" The so-called *wave equation* stating this velocity $(\partial z/\partial t)_x$ results from a mass or enthalpy balance for a differential cross-sectional layer (inflow minus outflow equals accumulation). Depending on the physics of the problem it may assume different forms. For the purpose at hand, a good place to start is with one-dimensional flow of two immiscible, incompressible fluid phases in a uniform permeable medium under ideal flow conditions. Here, the wave equation for the velocity of a "saturation" (i.e., fractional phase volume), S_j, of phase j is

$$v_{S_j} = v° \, df_j/dS_j \qquad\qquad (j = 1,2) \qquad\qquad (1)$$

(for definitions and units, see glossary of symbols at end of chapter); the velocities of S_1 and S_2 at any point are necessarily equal. In reservoir engineering, eq (1) is known as the Buckley-Leverett equation (Buckley and Leverett, 1942). It assumes ideal flow conditions and that the fractional flow is a unique function of the current saturation, that is, does not depend on prior history.

It is important to distinguish between the *wave velocity* and the *particle velocity* of a participant. The former is the velocity at which a *given value of a variable* travels, while the latter is the velocity of an *identifiable object* (e.g., see Landau and Lifshitz, 1959). The two can be quite different. For example, a hurricane may have sustained winds of over a hundred miles per hour, but move at a velocity of only, say, ten or twenty and possibly in a different direction. The wind velocity (of volume elements of air) is a particle velocity, the velocity of the storm itself (the low-pressure system) is a wave velocity. The particle velocity of a volume element of phase j in our two-phase flow is

$$v_j = v° f_j/S_j \qquad\qquad (j = 1,2) \qquad\qquad (2)$$

Types of Waves*

The essential feature of nonlinear wave propagation is that the wave velocity depends on the values of the variables. For the sake of argument, let us assume for the time being that in our two-phase flow the fractional-flow curve of phase 1 is positively curved throughout, as in Figure 1. Now consider a wave in which phase 1 gradually displaces phase 2 from the

---

* A good survey with mathematics and more detail has been given by Lake (1989).

medium. At the leading portions of this wave, where S_1 is small, the fractional-flow curve has a relatively flat slope df_1/dS_1 and, according eq (1), the wave velocity is low. The opposite is true in the trailing portions of the wave: Here, at large values of S_1, the slope df_1/dS_1 is steep and the wave velocity is high. The trailing portions, tending to travel at higher velocities, gain on the leading ones, that is, the wave sharpens (see Figure 2, left). Such a wave is called a *compressive wave* or *self-sharpening wave.*

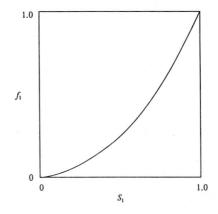

Figure 1. Hypothetical fractional-flow curve of phase 1.

Under ideal flow conditions the wave would sharpen into (or remain) a discontinuity. Unlike an ocean wave on a beach, it cannot "break," that is, higher saturations cannot overtake lower ones farther ahead because that would lead to the physical impossibility of three different saturations of the same phase existing simultaneously at the same point. Instead, the wave continues to travel as a discontinuity (the physically realized portion of the complete mathematical solution is a so-called weak solution, with two branches separated by a discontinuity.) The traveling discontinuity is called a *shock*, in analogy to shocks in physics of compressible fluids. Its velocity, obtained from a material balance across the wave, is

$$v_{\Delta S_j} = v° \Delta f_j / \Delta S_j \qquad (j = 1,2) \qquad (3)$$

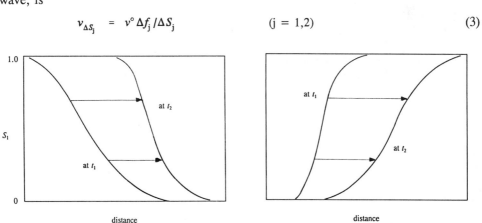

Figure 2. Saturation profiles at successive time t_1 and t_2 of self-sharpening and nonsharpening waves in displacement of phase 2 by phase 1 (left) and phase 1 by phase 2 (right) with fractional-flow curve as in Figure 1.

Under real conditions, nonidealities prevent the wave from sharpening into a discontinuity. The strengths of the dispersive effects of the nonidealities increase as the wave sharpens, so that a state is approached in which the sharpening and dispersive tendencies balance one another. The wave then travels not as a discontinuity, but with a steep, continuous saturation profile that no longer changes. If the distinction from a discontinuity is to be stressed, such a wave is called a *shock layer*, and its profile is said to be in a *constant pattern*. Its velocity, like that of the discontinuity, is given by eq (3).

Consider now a wave in which phase 2 displaces phase 1 from the medium under otherwise the same conditions. This time, the leading portions of the wave are the faster ones (steeper slopes df_1/dS_1 and thus higher velocities at higher values of S_1), so that the wave spreads (see Figure 2, right). Any nonidealities only add to that spreading and soon become negligible as the profile flattens. Such a wave is called *dispersive* or *nonsharpening*, or a *rarefaction wave*. If the wave was generated by a discontinuous variation at the starting point, it spreads in proportion to the distance or time of travel in a so-called *proportionate pattern*.

In actual permeable media, the fractional-flow curves usually are S-shaped as shown in Figure 3. Waves may then consist of self-sharpening and nonsharpening portions (so called *mixed waves* or *composite waves*). In the final pattern, the wave velocity must increase *monotonically* in the direction of flow, that is, the slope df_1/dS_1 (or $\Delta f_1/\Delta S_1$) in the f_1-versus-S_1 diagram must increase from the point corresponding to the saturation on the upstream side of the wave to that on the downstream side. In Figure 3, the wave of phase 1 displacing phase 2 consists of a leading shock followed by a nonsharpening portion.

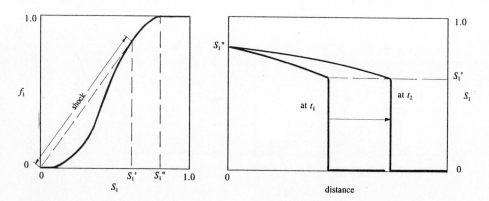

Figure 3. Typical fractional-flow curve in a permeable medium (left) and resulting composite wave in displacement of phase 2 by phase 1 (right).

Lastly, for the sake of completeness, consider the hypothetical, linear case of two phases that are immiscible but identical in their physical properties, so that they always flow at the same velocity ($f_j = S_j$, $df_j/dS_j = 1$, $v_{S_j} = v°$). Under ideal flow conditions, any wave then travels "as is," without change in its saturation profile. An initially discontinuous wave remains so, and is called a *contact discontinuity*. Any nonidealities cause spreading, but since their effect weakens as the wave loses sharpness, the spreading is in proportion not to traveled distance, but to the square root of traveled distance. In distinction from self-sharpening and nonsharpening waves, such waves are called *indifferent*.

These fundamental properties of waves are common to all fields of convective wave propagation, from flow of compressible fluids to chromatography, sedimentation, and traffic flow, even in multiphase and multicomponent systems. They do not depend on the algebraic form of the wave equation. (For waves with transport by diffusion instead of convection, see the chapter on *Diffusion, Flow, and Fast Reactions in Porous Media*).

Multiphase Systems

To appreciate the new facets that appear in systems with more than two phases or components, let us examine the behavior of three immiscible phases: water, oil, and air. So as not to be sidetracked by complications that might obscure the line of thought, we move temporarily to a simpler world in which the particle velocities of the three phases are always in constant ratios to one another, and air also is incompressible.

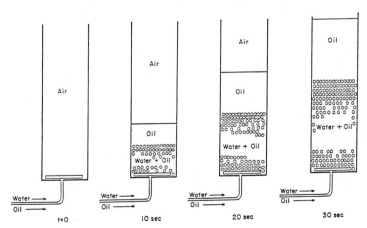

Figure 4. Injection of oil-in-water suspension into air-filled open tube (from Helfferich, 1984).

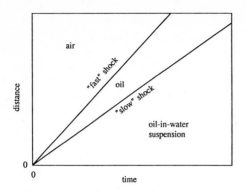

Figure 5. Distance-time diagram of water-oil injection.

Imagine a tube filled with air, into which an oil-in-water suspension is injected at a constant rate through a sparger at the bottom. The oil, being lighter, rises faster than the water in which it is suspended, so that a layer of pure oil forms on top of the suspension. As the injection is continued, the zones of suspension and pure oil widen in proportion to elapsed time (see Figure 4). There are two waves: the air/oil interface and the boundary between the oil and the suspension. A coordinate system with time as the abscissa and distance from the injection point as the ordinate can be superimposed on Figure 4. In such a *distance-time diagram* (Figure 5), the waves describe straight-line trajectories (we have chosen conditions such that both waves will be shocks).

In our thought experiment, a single, momentary variation at the inlet (the start of the injection of the suspension) has generated not one, but *two* waves that travel at different speeds. Between the two waves, a new zone (pure oil) grows whose composition could not have been generated by mixing the initial air with the injected suspension. Each wave involves a variation of the fractional volumes of *all* phases that are present. These are general facets of multicomponent nonlinear wave propagation. The number of waves arising from a variation at a single distance-time point is given by the *variance* of the system, which, in turn, is the number of dependent variables whose values can be varied independently of one another (here, two fractional phase volumes, the third being given by $\Sigma_j S_j = 1$).

The distance-time diagram tells us where the waves are at any moment. To supplement it with quantitative information about the compositions of the zones in between, we can map our experiment in the *composition space* (hodograph space) with the fractional phase volumes as coordinates. Such a so-called *composition route* (or hodograph) is shown in Figure 6 in form of arrows that point in the direction of flow: The suspension zone maps onto the point F; the "slow" wave between suspension and oil, onto the arrow F→P2; the oil zone, onto the point P2; the "fast" wave between oil and air, onto the arrow P2→I; the air zone, onto the point I.

In our dream world of in-variant particle-velocity ratios, we could repeat our experiment with different initial and injected mixtures, each time tracing the routes in the composition space. In this way a *composition-path grid* as in Figure 6 could be generated, with solid and dashed lines for the slow and fast waves, respectively. The important point here is that this grid is uniquely given by the flow equations and the parameter values and is independent of the initial and injection conditions. Therefore, once the grid has been

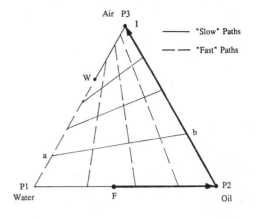

Figure 6. Composition-path grid for hypothetical water-oil-air systesm with route for injection of water-oil mixture (F) into air-filled tube (I) (for particle-velocity ratios $v_{oil}/v_{water}=2$, $v_{air}/v_{water}=4$).

constructed, the behavior for any initial and injected compositions can be immediately predicted: The route follows the "slow" path from the point of the injected mixture (the slower wave must be upstream of the faster one) and switches to the "fast" path that leads to the point of the initial composition. If the composition variation in the direction of flow along a path is such that the saturation of the fastest phase (air in our case) increases and that of slowest one (water) decreases, the wave is a shock; otherwise it is a nonsharpening wave and would give rise to a fan of composition trajectories in the distance-time diagram rather than a single shock trajectory.

When returning to the real world we find a few complications. Regimes of hydrodynamic instability cover much of our composition space. Moreover, even in stable flow and if droplet and bubble size could be taken as uniform and constant, the drift velocities rather than the particle-velocity ratios would be constant (the drift velocity is the velocity of the oil droplets or air bubbles relative to the surrounding water). This would make our composition path grid dependent on the flow rate. Fortunately, we are interested in flow in permeable media rather than open tubes, and in the former the flow is usually stable and the particle-velocity ratios depend very little on the flow rate. However, the ratios are not constant, and this results in path grids more complex than that in Figure 6, as will be seen later.

Coherence

In the infancy of enhanced oil recovery, an injected slug of surfactant, polymer, or other agent with concentrations of its constituents carefully adjusted for maximum efficiency was believed to travel through the reservoir with integrity, accumulating an oil bank in front of it. This assumes the slug to behave like an airliner flying, say, from New York to San Francisco and arriving with the same captain, crew, and passengers. Sadly, the real, multicomponent slug resembles more a Greyhound bus: Passengers board and alight at every stop, the driver is changed every six hours, and in Chicago they change the bus. Similarly, in transport phenomena in geological formations such as pollutant migration in aquifers, the wave patterns are complex and can cause unexpected effects, possibly including strong enrichment in some zones. Of course, the differential mass balances for the components can be integrated over space and time to find how a particular system will behave under exactly specified conditions - but in complex systems the computational effort might be staggering, and the results, valid only for that specific case, teach us little about cause and effect and do not provide much insight that would allow us to make predictions without extensive calculations. For that we need a better conceptual understanding. The key concept here is that of *coherence* (Helfferich and Klein 1970; Helfferich, 1989b).

Looking again at our thought experiment in Figure 4, we can say that there really were *three* waves: The initial boundary between the air and the oil-in-water suspension, although it existed for only one fleeting instant, is also a composition variation and thus a wave by our definition. It could be called "unstable," but instability in fluid mechanics has all kinds of connotations not intended here. Therefore, a different terminology has been coined. In the language of coherence theory, the initial wave is *noncoherent* and is resolved into a set of *coherent* waves. The latter, by definition, travel without further break-up; they may sharpen or spread, but do not change their routes in the composition space. In the general case of a system with variance *s*, an arbitrary initial composition variation is in general noncoherent and is resolved into a set of *s* coherent response waves:

Any arbitrary starting variation, unless further disturbed, is resolved into a set of coherent waves that become separated from one another by new plateau zones.

In our thought experiment, such behavior may seem trivial: Only a perfectionist would ask for proof that a wave or composition existing somewhere in the column at some time traveled to that point from the injection point rather than materializing at that moment from a different previous pattern. The concept comes into its own right when conditions are more complex. For example, we might vary the composition of the injected fluid gradually, thereby introducing a diffuse noncoherent wave at the inlet; a finite distance and time is then required for attainment of the coherent pattern, and the fact that it evolves at all is no longer trivial. Or we might after some time abruptly vary the composition of the fluid being injected; the faster wave from the later inlet variation then catches up at some point with the slower one from the earlier variation, generating a temporary local noncoherence that gives rise again to new coherent waves (see Figure 7). We can say, a noncoherence does not care whether it is sharp or diffuse, whether it is the result of a variation at the inlet or of wave interference at a later time within our system: It is always resolved into coherent waves by the same rules.

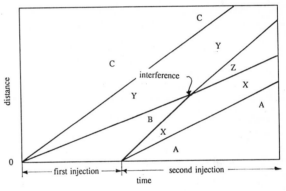

Figure 7. Shock trajectories in distance-time diagram for successive injections of two different fluids, producing interference (letters A, B, C, ... refer to example in Figure 8).

In general terms, the coherence concept resembles that of equilibrium and steady state. Each is a state a system strives to attain, and will approach if not disturbed again. That state is equilibrium if the system is closed; it is a steady state if we allow the system to be open, but impose fixed boundaries and constant boundary values; it is a state of coherence if we relax these restrictions also. Thus, coherence can be viewed as a general concept of "stability" that includes equilibrium and steady state as special cases. In the real world, coherence is only approached asymptotically, but so are equilibrium and steady state. This does not impair the utility of the concepts. In fact, they are most useful in situations far from the final, "stable" state since they tell us in which direction the system will move under complex, transient conditions. This is true for coherence as much as for equilibrium and steady state.

The notion of "stable" modes other than equilibrium and steady state is, of course, an old one in physics and mechanics. Examples that come to mind are the harmonic vibrations of an oscillator or a musical string or the flutter of an aircraft wing. It is only in physical chemistry and chemical engineering that our preoccupation with thermodynamics - which really should be called thermostatics - has let us be slow in recognizing this facet of dynamics.

A general proof for development toward coherence, independent of the exact form of the differential equations, is so far lacking - as is a proof for development toward equilibrium (the free-energy argument of thermodynamics just replaces one unproved axiom by another). Proofs for systems with hyperbolic differential equations have been given (Liu, 1977; Helfferich, 1986), but the concept has shown itself to be valid for other types of systems as well. What can be shown here, without mathematics, is that the concept is plausible. As mentioned above, coherent behavior appears trivial if there is only a single, instantaneous variation of the composition of the injected fluid, so that all waves must have originated from a single distance-time point (so-called *Riemann-type conditions*). It also appears plausible that the same wave pattern should result if, instead, the injection variation extends over a finite, but exceedingly short time (in the language of mathematics, this is so if the problem is "well-posed"). But the time over which the injection is varied is a relative quantity, and would constitute a large fraction of the total in a much shorter experiment with a much shorter column, an experiment which, in turn, is a scale-down of one with a long column and an injection variation that extends over a significant length of time. Thus, unless the minute change in conditions from discontinuous to very fast but continuous injection variation is to produce significant and long-lasting effects, development toward coherence must be expected. This is a plausible argument rather than a proof, for nature is full of systems that are not well-posed (see the "butterfly effect" of meteorology, in which a butterfly's bat of a wing could conceivably set a chain of events in motion that culminates in a typhoon thousands of miles away). However, there is absolutely no reason to suspect such behavior in the types of systems of interest here.

Mathematics

Coherence Conditions. Returning to our thought experiment in Figure 4, we can immediately identify a necessary and sufficient condition for a wave to be coherent: Any multicomponent wave can be viewed as a composite of waves of all individual phases that are present. For example, the wave between the oil and the suspension, involving variations of the fractional phase volumes of both oil and water, is at the same time an "oil wave" and a "water wave." For the wave to be coherent, the wave velocities (or shock velocities) of oil and water must be the same; if not, the wave would break up and so be noncoherent by definition. Thus, the coherence condition for multiphase flow can be written*

$$df_j / dS_j \; = \; \lambda \qquad\qquad (j = 1,...,m\text{-}1) \qquad\qquad (4)$$

for a diffuse waves, or, for a shock,

$$\Delta f_j / \Delta S_j \; = \; \Lambda \qquad\qquad (j = 1,...,m\text{-}1) \qquad\qquad (5)$$

where λ or Λ must have the same value for all phases at the same point in distance and time.

Eigenvalues, Eigenvelocities, and Eigenvectors. While seeming almost trivial in our simple example, the coherence conditions are the key to efficient mathematics of multivariant systems. This is because the wave equation by itself no longer suffices to identify the wave velocity: With more than two phases, the fractional flow f_j of a phase is a function not only of the saturation S_j of that phase, but of those of the other phases as well:

$$f_j \; = \; f_j (S_1,...,S_{m-1}) \qquad\qquad (j = 1,...,m\text{-}1) \qquad\qquad (6)$$

and the wave equation (1) becomes

$$v_{S_j} \; = \; v°(\partial f_j / \partial S_j)_z \qquad\qquad (j = 1,...,m\text{-}1) \qquad\qquad (7)$$

with a partial derivative, and thus a velocity, that can in principle assume any value. It is the coherence condition that removes this ambiguity.

In view of eq (6), the total derivative df_j can be expressed as

$$df_j \; = \; (\partial f_j / \partial S_1)_{S_i \, (i \neq 1,m)} \, dS_1 \; + \; ... \; + \; (\partial f_j / \partial S_{m-1})_{S_i \, (i \neq m-1,m)} \, dS_{m-1} \qquad (8)$$

With eq (8), the coherence condition (4) becomes, in matrix form:

$$\begin{vmatrix} \partial f_1/\partial S_1 - \lambda & \partial f_1/\partial S_2 & \partial f_1/\partial S_3 & \cdots & \partial f_1/\partial S_{m-1} \\ \partial f_2/\partial S_1 & \partial f_2/\partial S_2 - \lambda & \partial f_2/\partial S_3 & \cdots & \partial f_2/\partial S_{m-1} \\ \partial f_3/\partial S_1 & \partial f_3/\partial S_2 & \partial f_3/\partial S_3 - \lambda & \cdots & \partial f_3/\partial S_{m-1} \\ & & \cdots & & \\ \partial f_{m-1}/\partial S_1 & \partial f_{m-1}/\partial S_2 & \partial f_{m-1}/\partial S_3 & \cdots & \partial f_{m-1}/\partial S_{m-1} - \lambda \end{vmatrix} \begin{Bmatrix} dS_1 \\ dS_2 \\ dS_3 \\ \cdots \\ dS_{m-1} \end{Bmatrix} = 0 \qquad (9)$$

---

* Because of $\Sigma_j f_j = 1$ and $\Sigma_j S_j = 1$, only m-1 of the f_j and S_j of the m phases can be chosen independently; i.e., the variance of the m-phase system is m-1.

This is a typical eigenvalue problem. with m-1 eigenvalues λ that can be obtained by setting the determinant of the matrix in eq (9) equal to zero (in bivariant system as in the water-oil-air example of Figure 4, the eigenvalues are the two roots of a quadratic equation).

In an (m-1)-variant system, each composition has m-1 eigenvalues λ, each associated with an eigenvector $\{dS_1,...,dS_{m-1}\}$. The eigenvectors indicate the directions of the composition paths at the respective composition, so each composition point is at the intersection of m-1 paths; the eigenvalues indicate the "eigenvelocities"

$$v_S = \lambda v° \tag{10}$$

with which a composition $S \equiv \{S_1,...,S_m\}$ travels if it is part of a (diffuse) coherent wave. For example, in the bivariant system of Figure 6, each composition is at an intersection of a "fast" and a "slow" path, and has eigenvelocities given by its two eigenvalues, the higher one for the "fast" and the lower one for the "slow" coherent wave.

Methodology

A typical application of coherence theory involves the following steps:
- construction of the composition path grid,
- calculation of eigenvelocities,
- construction of composition routes for cases of interest,
- construction of respective distance-time diagrams, and
- construction of respective composition profiles or histories.

Composition Path Grids. In principle, path grids can be constructed by obtaining the eigenvectors at a large number of composition points. This can be done point by point by integration across the composition space from a chosen starting point in the direction of a chosen eigenvector, and repetition with other starting points and eigenvectors. This is a lengthy procedure which, however, can often be obviated. As a rule, the *topology* of the grid depends only on the form of the flow equations, not on the values of their coefficients. Once the topology is known, an approximate grid can usually be constructed by algebraic calculation of the locations of a few characteristic points. For instance, the grid in Figure 6 requires the calculation of only the location of the "watershed point" W, which is given by $S_1 = (v_2/v_1 - 1)/(v_3/v_1 - 1)$, $S_2 = 0$. All paths, which are straight lines, can then be entered with help of a rule of equal intercept ratios: e.g., for the path a-b, the ratio of the distances P1-a to P1-W equals that of P2-b to P2-P3. Other examples will be seen later.

Eigenvelocities. If the grid is constructed by point-by-point solution of the eigenvalue problem, the eigenvalues and thus eigenvelocities are obtained with the eigenvectors. If not, they must be calculated separately for a few points of interest. Again, shortcuts can often be used, as will be seen in our later examples.

Composition Routes. Composition routes for Riemann-type problems (i.e., with all waves originating from a single distance-time point) are relatively easy to establish. The procedure resembles plotting a vacation trip on a road map, the roads being the composition paths. The route runs essentially along these paths. Since faster waves must be downstream of slower ones, the route (mapped in the direction of flow) starts from the composition point of the entering fluid along the path of lowest eigenvalue, then switches successively to paths of the next higher eigenvalues. In the bivariant system of Figure 6, there is only one such switch, from a "slow" to a "fast" path. Constructions in multivariant systems call for projections onto planes or surfaces and are not as easily depicted in diagrams (e.g., see Helfferich and Klein, 1970).

If the eigenvelocity decreases in the direction of flow along a path, so that the respective wave is a shock, a route correction may be necessary. This is because two points that obey the *integral* coherence condition (5) for a shock are not in general on the same path, since the latter was calculated with the *differential* coherence condition (4). For any shock, the route must obey the simultaneous algebraic equations given by eq (5) and, unless the respective path is straight, that leads to a somewhat different route.

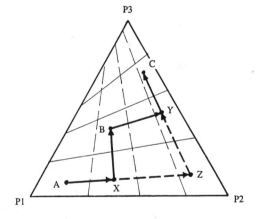

Figure 8. Composition routes for bivariant system with interference of two shocks; route before interference ———, after interference (where different) — — (for distance-time diagram, see Figure 7).

Route construction for non-Riemann problems - e.g., with successive composition variations of the entering fluid, producing wave interference - are more complex. In such cases, the route is not unique, but changes with each interference. Figure 8 shows an example, for a system with a path grid as in Figure 6 and with two successive variations leading to interference of two shocks and a distance-time diagram as in Figure 7. Each of the two variations produces at first a Riemann-type pattern of two shocks; the route before interference is A-X-B-Y-C. Soon, however, the "fast" shock (X-B) from the second variation catches up with the "slow" shock (B-Y) from the first. At the moment the two shocks merge, the composition B vanishes. The single wave that now separates X from Y is noncoherent and is resolved into two coherent shocks X-Z and Z-Y, of which the "faster" one now is downstream of the "slower." With this interference, the route has changed to A-X-Z-Y-C. (Eventually, the shocks A-X and X-Z will merge with one another, as will the shocks Z-Y and Y-C, to give a final route A-Z-C; thus, the final route is as would have arisen from a single composition variation from A to C.)

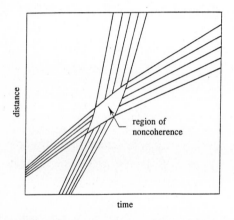

Figure 9. Noncoherent distance-time region in interference of two diffuse waves (from Helfferich, 1984).

If the interference is between two *diffuse* waves, as in Figure 9, behavior is noncoherent within a finite distance-time region. Within that region, the route does not follow any compositions paths and shifts continuously. The composition variations across the coherent waves resulting from interference and the composition of the new plateau zone between them can be obtained even in such cases by the same kind of route construction as in the previous example. However, numerical integration is usually required to establish the exact behavior within that region and the exact placement of the composition trajectories of the new waves in the distance-time plane.

In the event of multiple interferences, route construction as above may not suffice for prediction of the sequence in which they occur. In such cases, the construction of transient routes and of the distance-time diagram (see below) must proceed in tandem: The path grid shows what new waves arise from interference, the distance-time diagram shows when and where the next interferences occur.

Distance-Time Diagrams. The slope of a given composition or shock trajectory in the distance-time plane, with ordinate z and abscissa t, is given by the respective wave velocity, $(\partial z/\partial t)_s$ or $(\Delta z/\Delta t)_{\Delta s}$. This allows the distance-time diagram to be constructed once the wave or shock velocities are known. Where trajectories intersect, waves interfere and so produce noncoherence at points, along curves, or in entire regions. Route construction in the path grid then is needed to identify the new coherent waves into which such noncoherence is resolved. The construction of the distance-time diagram can then continue, the wave or shock velocities of the new waves providing the new trajectory slopes.

Profiles and Histories. Composition profiles $S(z)$ at any given time t, or histories $S(t)$ at any given location z, are easily obtained from the composition route (or routes, if wave interferences cause route changes) and the distance-time diagram. A profile is the variation of the saturations along the (vertical) line of constant, given t in the distance-time plane; the values of z at the intersections of that line with the trajectories of compositions or shocks identify where these appear in the profile; the route identifies the respective saturation values. The construction of a history, along a (horizontal) line of constant, given z, is analogous.

A convenient graphical procedure for construction of profiles and histories in flow of two immiscible phases has been worked out by Welge (1952), and has been extended to include partial miscibility (Helfferich, 1982).

Applications

Two examples will be given to illustrate the application of coherence theory to specific problems of flow in permeable media under conditions that require the rudimentary theory presented so far to be extended by inclusion of additional effects.

Example 1. *Multicomponent, Multiphase Flow:*

Injection of Surfactant Into Water-Flooded Oil Reservoir

A typical phase diagram of a pseudo-three component system with lumped components brine, oil, and surfactant at moderately high salinity is shown in Figure 10, left. The aqueous and oleic phases are partially miscible and have S-shaped fractional flow curves as in Figure 3. In the two-phase region there are six dependent variables: the volume fractions c_{ij} of the three components (i) in the two phases (j); yet, because of four constraints, the system is only bivariant, with a path grid shown in Figure 10, right.

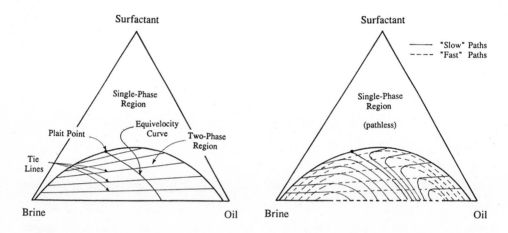

Figure 10. Phase diagram (left) and composition-path grid (right) of pseudo-three component system brine-oil-surfactant at moderately high salinity (from Helfferich, 1981).

The complication of partial miscibility of the three lumped components can be shifted from calculus to mere algebra by the use of what may be called "blind-man's variables." These are the fractional flows and saturations a man would find who can analyze for components, but cannot distinguish between liquid phases - who may not even realize that several of the latter exist:

$$F_i \equiv \sum_{j=1}^{m} f_j c_{ij}, \qquad C_i \equiv \sum_{j=1}^{m} S_j c_{ij} \qquad (i = 1,\dots,n\text{-}1) \qquad (11)$$

(for n components and m phases). Since the wave equations are obtained from the mass balances, which are in terms *components*, the wave equations for immiscible phases and the coherence conditions derived from them remain valid for fluids with partial miscibility if the f_j and S_j are replaced by the respective blind-man's fractional flows F_i and overall concentrations fractions C_i (Helfferich, 1981):

wave equations:	$v_{C_i} = v°(\partial F_i/\partial C_i)_z$	for diffuse waves	(12)
	$v_{\Delta C_i} = v° \Delta F_i/\Delta C_i$	for shocks	(13)
coherence conditions:	$dF_i/dC_i = \lambda$	for diffuse waves	(14)
	$\Delta F_i/\Delta C_i = \Lambda$	for shocks	(15)

The same is true, of course, for the formulas for the eigenvalues and eigenvectors. The rules for coherent behavior, composition routes, etc., therefore remain the same as before. The price to pay is that the (algebraic) relationships between the F_i and C_i are more complex than those between the f_j and S_j and, as is evident in Figure 10 (right), give rise to more complex composition-path grids.

Within the single-phase region, $f_1 = S_1 = 1$, no other phase being present. Accordingly, the coherence condition is met for any composition variation: the region is "pathless." Moreover, $v_S = v°$, i.e., the eigenvelocity is that of fluid flow.

- *Any composition variation within the single-phase region is a coherent, indifferent wave with velocity of fluid flow.*

The condition $S_1 = f_1 = 1$ also holds along the "binodal curve" (envelope of the two-phase region). Accordingly:

- *The binodal curve is a composition path.*

For the two-phase region, the coherence condition (14) with the definitions (11) becomes, for three components:

$$\frac{f_1(dc_{i1} - dc_{i2} + dc_{i2} + (c_{i1} - c_{i2})df_1}{S_1(dc_{i1} - dc_{i2} + dc_{i2} + (c_{i1} - c_{i2})dS_1} = \lambda \qquad (i = 1,2) \qquad (16)$$

This condition can be met in several ways. If all dc_{ij} are zero, as is true for the tie lines, condition (16) reduces to $df_1/dS_1 = \lambda$ for all components i; the (invariant) phases along a tie line thus obey the coherence condition:

- *All tie lines are composition paths.*

Also, along the equivelocity curve (the locus of compositions at which both phases flow at the same velocity, that of overall fluid flow), $f_j = S_j$ for both phases, so that $\lambda = 1$ for both phases, satisfying condition (16):

- *The equivelocity curve is a composition path.*

It can also be shown with condition (16) that the only paths leading into the interior of the two-phase region from the binodal curve are the tie lines or, at the plait point, the equivelocity curve. For systems with S-shaped fractional-flow curves as in Figure 3, the wave velocities for a tie-line path and a non-tie line path are found to be equal at two points along each tie line, one on each side of the equivelocity curve; at these singular points, whose locations can be found by equating the two eigenvalues, the two paths must be tangential. With this knowledge, an approximate path grid as in Figure 10 (right) can be constructed without point-by-point solution of the eigenvalue problem.

One additional rule is needed, for waves with routes that enter the two-phase region (downstream) from the single-phase region (upstream). With fractional-flow curves as in Figure 3, such waves are shocks or composite waves because of the low wave velocities in the two-phase region at low saturation of one phase (flat initial portion of the f_j-vs-S_j curve). The shock portion must meet the coherence condition (15), which can be shown to require the single-phase composition on the upstream side to be a linear combination the compositions of two phases on the downstream side. Accordingly:

- *Routes can enter the two-phase region only along a linear extension of a tie line. The respective wave or wave portion is a shock with a velocity lower than that of overall fluid flow* (for an exception, see below).

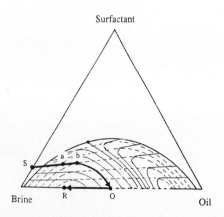

If the point of the single-phase composition on the upstream side is outside the region of tie-line extensions, the point of the two-phase on the downstream side must be on the equivelocity curve, and the wave is indifferent and moves at the velocity of overall fluid flow (piston-type displacement).

Figure 11. Composition route for injection of surfactant solution (S) into a water-flooded oil reservoir (residual saturation R); O = oil bank; portions A→a and O→R are shocks, a→b is spreading, b→O is indifferent.

The route for injection of a surfactant solution (S) into a reservoir with residual oil saturation (R) can now be constructed in the grid (see Figure 11). Point S is in the region of tie-line extensions, so the route enters the two-phase region along such an extension. The only way it can reach the downstream point R without ever switching from a "fast" to a "slow" path is by following the tie line to the point of tangency, then continuing on the "slow" path as it curves to the brine-oil axis, and finally following that axis on a "fast" path to the point R. The last route portion, O-R, is a shock, and O indicates the composition of the "oil bank" that the injection throws up. A distance-time diagram and surfactant and oil profiles are shown in Figure 12. For more details and applications to more complex conditions, see Helfferich (1981) and Hirasaki (1981).

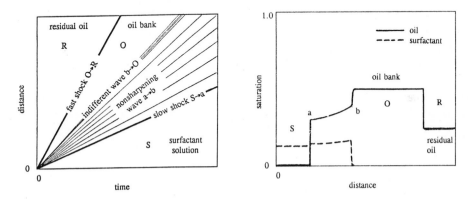

Figure 12. Distance-time diagram (left) and oil and surfactant profiles (right) for injection of surfactant solution into oil-bearing reservoir.

Example 2.   *Precipitation-Dissolution Waves:*

Erosion of a Precipitate AX by Water Containing Ions A, B, and X.

An interesting and important phenomenon of flow through permeable media is convective transport of ions that are formed by dissolution of minerals and that can in turn precipitate in form of other minerals. Here, no functional relationship exists between the amount of precipitate and the liquid-phase concentrations of its ionic constituents. Rather, the latter concentrations are constrained by the solubility product; e.g., for a precipitate AX:

$$c_A c_X \leq K_{AX} \qquad \text{if AX is absent}$$
$$c_A c_X = K_{AX} \qquad \text{if AX is present} \tag{17}$$

Nevertheless, the coherence concept is found to remain valid. Moreover, the methods of route construction in path grids can be applied here, too, if the dependent variables are taken to be the liquid-phase and "overall" concentrations, c_i and C_i, respectively, per unit volume of liquid phase. The latter concentrations are defined as the amounts of i present in both the dissolved and precipitated states:

$$C_i \equiv c_i + \bar{c}_i \tag{18}$$

The overall concentrations, C_i, are used as the coordinates of the composition space, as is standard practice for phase diagrams. A phase diagram (with path grid) for a system with ions A, B, and X and precipitates AX and BX is shown in Figure 13 farther below.

In terms of the liquid-phase and overall concentrations, the wave equations and coherence conditions are found to be:

wave equations:	$v_{C_i} = v°(\partial c_i/\partial C_i)_z$	for diffuse waves	(19)
	$v_{\Delta C_i} = v° \Delta c_i/\Delta C_i$	for shocks	(20)
coherence conditions:	$dc_i/dC_i = \lambda$	for diffuse waves	(21)
	$\Delta c_i/\Delta C_i = \Lambda$	for shocks	(22)

Where no precipitate is present, the overall concentrations equal the liquid-phase concentrations, so that $dc_i = dC_i$ for all species under all conditions. Thus, any composition variation in within the no-precipitate region is coherent and is propagated at the velocity of liquid-phase flow; that region is "pathless."

Whether or not a precipitate is present, $c_i = C_i$ and thus $dc_i = dC_i$ for any species that remains exclusively in the liquid phase; for such a species, the wave equation (19) shows that the wave velocity is that of liquid-phase flow, $v°$, and is independent of composition. On the other hand, any variation of the amount of a precipitate results in $dC_i > dc_i$, for which eq (19) shows the wave velocity to be lower than that of liquid-phase flow. Thus, coherent waves, being subject to condition (21) or (22), can *either* involve concentration variations of non-precipitating species (and then travel at the velocity of liquid flow), *or* involve precipitation or dissolution (and then have a lower velocity), but not both. This knowledge suffices to identify the composition paths in the AX and BX regions of our system A,B,X,AX,BX: "fast" path along which the liquid-phase composition varies but the amount of precipitate remains constant; and "slow" paths along which the amount of precipitate varies but the liquid-phase composition remains constant (see Figure 13). The wave equation (19) shows the eigenvelocities to be zero along the "slow" paths.*

---

* For the AX region: For $dc_B=0$, electroneutrality requires $dc_A=dc_X$. Because of the constraint $c_A c_X=K_{AX}=$const., this is possible only if $dc_A=dc_X=0$. With $dc_A=0$ and $dC_A \neq 0$, eq (19) shows the wave velocity to be zero.

In the AX+BX region, the liquid-phase composition is invariant (that of the triple point, T). Thus, any composition variation within that region involves only the precipitates, is coherent, and is a zero-velocity wave ("standing wave").

Lastly, coherent waves with precipitate on the downstream side and none on the upstream side must be considered. Because the amount of precipitate varies across the wave, the velocity is lower than that of liquid flow. This requires the con-

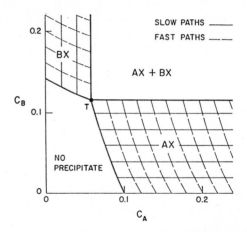

Figure 13. Phase diagram and composition path grid of system A,B,X,AX,BX (with K_{AX} =0.01, K_{BX}=0.02; T=triple point) (from Helfferich, 1989).

centrations of any species remaining exclusively in the liquid to remain constant. With, say, AX downstream and unsaturated solution upstream, B is such a species, and $dc_B = 0$ requires the route to be parallel to the C_A axis. Along such a route, the eigenvelocity is $v°$ in the upstream (no-precipitate) portion, and zero along the downstream portion ("slow" path). Accordingly, the wave is a shock. Its velocity is given by eq (20), with ΔC_A and Δc_A, respectively, being the total length of the route of the shock and the length of the portion in the no-precipitate region (note that c_A does not vary along the portion in the AX region).

These considerations show that there can only be three types of coherent waves in systems of this kind:

- *indifferent waves that involve a variation of only the liquid-phase composition and travel at the velocity of liquid-phase flow* (so-called tracer waves),

- *standing waves across which only of the amounts of precipitates vary,* and

- *shocks that involve dissolution of a precipitate and travel at a velocity lower than that of liquid-phase flow.*

As an example illustrating the application of coherence theory, the composition routes and the distance-time diagram of erosion of a bank of precipitate AX (composition P) by a solution rich in B (composition U) are shown in Figure 14. The concentration of B in the

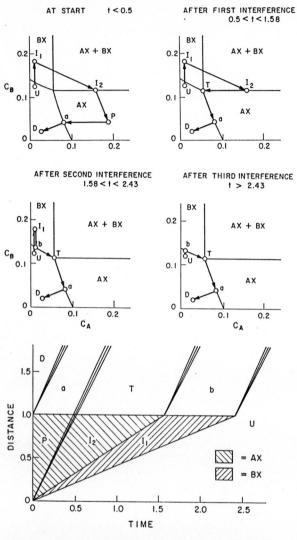

Figure 14. Transient composition routes (top) and distance-time diagram (bottom) for erosion of AX bank (P) initially between unsaturated solutions rich in B upstream (U) and lean in A and B downstream (D) (from Helfferich, 1989).

solution on the upstream side (U) is higher than at the triple point (T). Therefore, the route of the wave pattern originating from the upstream end of the AX bank (U-P at start) cannot enter the AX region directly from U. Rather, it enters the BX region with a slow shock U-I_1 and involves a second, faster shock I_1-I_2 and a tracer wave I_2-P. The development of the zone I_1 between the two shocks shows that the high concentration of B in the upstream liquid causes BX to precipitate where AX dissolves. However, BX is redissolved (although more slowly) at the upstream end of its new bank. Downstream of the AX bank, saturated solution (a) displaces the initially present solution (D) at velocity $v°$; the wave P-a is a standing wave, so the downstream end of the AX bank neither advances nor recedes. Before any interferences, the route is U-I_1-I_2-P-a-D.

With successive interferences the route changes as shown in Figure 14. The first interference (of I_2-P with P-a) marks the breakthrough of solution (T) saturated with both AX and BX from the downstream end of the AX bank. With the second interference (of I_1-I_2 with I_2-T), dissolution of the original AX bank is completed; the solution (b) exiting from the remaining BX bank is no longer saturated with AX. That bank, which had been growing in length, now begins to shrink. It disappears with the third interference (of U-I$_1$ with I$_1$-b). With no precipitate left, the route is U-b-T-a-D, with all waves being indifferent and traveling at liquid-phase velocity. Dispersion eventually smears out the individual waves give a single, diffuse wave U-D (not shown). At no time does precipitation or dissolution occur at any point downstream of the original AX bank.

More details are given in the chapter, on *Propagation of Dissolution/Precipitation Waves in Porous Media*. That presentation is based on an earlier approach (Walsh *et al.*, 1984; Bryant *et al.*, 1986, 1987; Novak *et al.*, 1988) that uses the coherence concept, but not the method of graphical or numerical construction of routes in the composition space as demonstrated here. For systems with large numbers of species and precipitates, the latter method (Helfferich, 1989) still awaits the development of algorithms, but promises to be able to obviate the heuristics of the earlier approach.

Other Types of Systems

Various other types of wave phenomena are of interest for migration and fate of pollutants in soils. Foremost among these are "chromatographic" behavior, transport with chemical reaction, and transport by diffusion rather than convection.

Chromatographic behavior involves convection of a liquid phase combined with adsorption or ion exchange with stationary solids. This theory is well-developed and documented (e.g., Helfferich and Klein, 1970; Rhee *et al.*, 1986). Adsorption can also be accounted for within the theory of multicomponent, multiphase flow shown here by treating the adsorbate as a separate phase with zero mobility (Helfferich, 1981).

For chromatographic transport combined with chemical reactions *at equilibrium*, established theories exist (Schweich *et al.*, 1980; Hwang *et al.*, 1988). In this context, a useful concept is that of "moieties," that is, of atoms or groups of atoms that remain intact, the reactions only combining them in different ways. The moieties can be viewed as a kind of blind-man's variables, those a man would see who can analyze only for them and might not even know that reactions take place. Their use preserves the mass balances and wave equations, again shifting the added complication from calculus to algebra, which now must include the reaction equilibria. The combination of transport with *slow* reactions (including slow dissolution) is more complex, as the next chapter will show.

Systems with diffusive instead of convective transport show a different behavior and have their own chapter. Although the differential equations now are parabolic, the coherence concept remains applicable, but the composition paths are no longer independent of the initial condition. However, paths that *are* independent of the initial condition can again be obtained in a coordinate space with fluxes instead of concentrations as axes (Tondeur, 1987b).

Coherence *versus* Method of Characteristics

The coherence concept and the method of characteristics are related, but differ in essential respects. The method of characteristics (e.g., Acrivos, 1956; Rhee *et al.*, 1986) is a superb, exact, and popular *mathematical tool* for solving simultaneous partial differential equations. In contrast, coherence is a *concept*. It states explicitly a behavior inherent in the differential equations of wave propagation, making it possible to predict without much calculation. Moreover, while the method of characteristics is restricted to hyperbolic equations, the coherence principle has proved to be of much broader validity in wave propagation, e.g., to the parabolic equations of systems with diffusive transport. The similarities are in that both coherence and the method of characteristics deal with the same phenomena. The composition paths of coherence theory are characteristics in the hodograph space. Coherent waves, if diffuse, are "simple" waves. Translated into the language of the method of characteristics, the coherence principle states: *An arbitrary starting variation, if embedded between sufficiently large regions of constant state, is resolved into simple waves or shocks which become separated by new regions of constant states.*

What Wave Theory Offers

In practical problems as complex as migration and fate of pollutants in soils, nonlinear wave theory - of which coherence theory is a part - can be useful in two quite different respects. First, because it is primarily a conceptual rather than mathematical theory, it can help in obtaining a better *qualitative* insight into cause and effect, and so in acquiring better judgment. Second, It can provide simpler, if approximate, mathematics for *quantitative* numerical modeling. For instance, numerical step-by-step integration can be obviated in space-time regions in which coherent waves travel without interference. Few of the opportunities offered by wave theory have as yet been fully tapped.

Glossary of Symbols

c_i	normalized concentration of component i in single-phase liquid	-
$\bar{c}_i$	normalized amount of component i present as precipitate, per unit volume of liquid	-
c_{ij}	concentration (volume fraction) of component i in phase j	-
C_i	overall concentration of component i [eq (11) or (18)]	-
f_j	fractional flow of phase j	-
F_i	overall fractional flow of component i [eq (11)]	-
K_{ij}	solubility product of precipitate ij, in terms of normalized concentrations [eq (17)]	-
m	number of phases	-
n	number of components	-
s	variance	-
S	$\equiv \{S_1,...,S_{m-1}\}$, composition of multiphase liquid	-
S_j	fractional phase volume (saturation)	-
t	time	t
v°	velocity of liquid-phase flow	$\ell\, t^1$
v_j	particle velocity of phase j	$\ell\, t^1$
v_S	eigenvelocity of composition S	$\ell\, t^1$
v_{S_j}	wave velocity of phase j	$\ell\, t^1$
$v_{\Delta S}, v_{\Delta C}$,	eigenvelocity of composition shock ΔS or ΔC	$\ell\, t^1$
$v_{\Delta S_j}, v_{\Delta C_i}$	shock velocity of phase j or component i	$\ell\, t^1$
z	distance coordinate	ℓ
Δ	finite difference across shock	operator
λ	eigenvalue of diffuse wave	-
Λ	eigenvalue of shock	-

Indices i refer to components, j to fluid phases.

References

Acrivos A (1956) Method of characteristics technique. Application to heat and mass transfer problems. *Ind Eng Chem* **48**:703-710

Bryant SL, Schechter RS, Lake LW (1986) Interaction of precipitation/dissolution waves and ion exchange in flow through permeable media. *AIChE J* **32**:751-764

Bryant SL, Schechter RS, Lake LW (1987) Mineral sequences in precipitation/dissolution waves. *AIChE J* **33**:1271-1287

Buckley SE, Leverett MC (1942) Mechanism of fluid displacement in sands. *Trans AIME* **146**:107-116

Frenz J, Horvath C (1985) High performance displacement chromatography: Calculation and experimental verification of zone development. *AIChE J* **31**:400-409

Helfferich F, James DB (1970) An equilibrium theory for rare-earth separation by displacement development. *J Chromatogr* **46**:1-28

Helfferich F, Klein G (1970) *Multicomponent chromatography*. Dekker, New York (out of print, available from University Microfilms, Ann Arbor MI, USA)

Helfferich FG (1981) Theory of multicomponent, multiphase displacement in porous media. *SPE J* **21**:51-62

Helfferich FG (1982) Generalized Welge construction for two-phase flow in porous media with limited miscibility. SPE 9730, presented at SPE Meeting, New Orleans, LA

Helfferich FG (1984) Conceptual view of column behavior in multicomponent adsorption and ion-exchange systems. *AIChE Symp Ser* **80** No 233 1-13

Helfferich FG (1986) Multicomponent wave propagation: Attainment of coherence from arbitrary starting conditions. *J Chromatogr* **44**:275-285

Helfferich FG (1989a) The theory of precipitation/dissolution waves. *AIChE J* **35**:75-87

Helfferich FG (1989b) Coherence: power and challenge of a new concept, in Keller GE II, Yang RT (eds) *New directions in sorption technology*. Butterworths, Boston, Chap 1

Hirasaki GJ (1981) Application of the theory of multicomponent, multiphase displacement to three-component, two-phase surfactant flooding. *SPE J* **21**:191-204

Hwang Y-L, Helfferich FG, Leu W-J (1988) Multicomponent equilibrium theory for ion-exchange columns involving reactions. *AIChE J* **34**: 1615-1626

Hwang Y-L, Helfferich FG (1989) Dynamics of continuous countercurrent mass-transfer processes. III. Multicomponent systems. *Chem Eng Sci* **44**:1547-1568

Kynch GJ (1952) A theory of sedimentation. *Trans Faraday Soc* **48**:166-176

Landau LD, Lifshitz EM (1959) *Fluid mechanics*. Pergamon, London

Lake LW (1989) *Enhanced oil recovery*. Prentice Hall, Englewood Cliffs NJ, Chap 5

Novak CF, Schechter RS, Lake LW (1988) Rule-based mineral sequences in geochemical flow processes. *AIChE J* **34**:1607-1614

Orr FM Jr, Taber JJ (1984) Use of carbon dioxide in enhanced oil recovery. *Science* **244**:536-539

Rhee H-K, Aris R, Amundson NR (1970) On the theory of multicomponent chromatography. *Phil Trans Roy Soc Lond* **267A**:419-455

Rhee H-K, Amundson NR (1982) Analysis of multicomponent separation by displacement development. *AIChE J* **28**:423-433

Rhee H-K, Aris R, Amundson NR (1986) *First-order partial differential equations*, vol 1. Prentice Hall, Englewood Cliffs NJ

Ruthven DM (1984) *Principles of adsorption & adsorption processes*. Wiley, New York, Chap 9

Schweich D, Villermaux J, Sardin M (1980) An introduction to the nonlinear theory of adsorptive reactors. *AIChE J* **26**:477-486

Tondeur D (1987a) Unifying concepts in non-linear unsteady processes. Part I: Solitary travelling waves. *Chem Eng Process* **21**:167-178

Tondeur D (1987b) Unifying concepts in non-linear unsteady processes. Part II: Multi-component waves, competition and diffusion. *Chem Eng Process* **22**:91-105

Vermulen T, LeVan MD, Hiester NK, Klein G (1984) Adsorption and Ion Exchange. In: Perry RH, Green DW, Maloney JO (eds) *Chemical Engineers' Handbook*, 5th edn., Sec 16. McGraw-Hill, New York

Wallis G (1969) *One-dimensional two-phase flow*. McGraw-Hill, New York

Walsh MP, Bryant SL, Schechter RS, Lake LW (1984) Precipitation and dissolution of solids attending flow through porous media. *AIChE J* **30**:317-328

Welge HJ (1952) A simplified method for computing oil recovery by gas or water drive, *Trans AIME* **195**:91-98

Whitham GB (1974) *Linear and nonlinear waves*. Wiley, New York

Propagation of Dissolution/Precipitation Waves in Porous Media[*]

Craig F. Novak

Fluid Flow and Transport Department 6119

Sandia National Laboratories

P.O. Box 5800

Albuquerque, New Mexico 87185-5800 USA

S. David Sevougian

Department of Petroleum Engineering

University of Texas at Austin

Austin, Texas 78712 USA

The transport of a chemically reactive fluid through a permeable medium is governed by many classes of chemical interactions. Dissolution/precipitation reactions are among the interactions of primary importance because of their significant influence on the mobility of aqueous ions. In general, advective or diffusive transport coupled with dissolution/precipitation reactions leads to the propagation of coherent concentration waves. These waves can cause local buildup of concentrations in the aqueous and mineral phases, but can also cause local reductions in concentrations. A decrease in concentration may be desirable if it leads to contaminant concentrations below recognized hazard levels. An increase in concentration might aid removal of a contaminant by concentrating the contaminant in a relatively small region.

This chapter provides an overview of the types of wave phenomena observed in one-dimensional (1D) and two-dimensional (2D) porous media for systems in which mineral dissolution and precipitation, either in local equilibrium or kinetically controlled, are the dominant types of chemical reaction. For local equilibrium, it is demonstrated that minerals dissolve in sharp waves in 1D advection-dominated transport, and that these waves separate zones of constant concentrations in the aqueous and mineral phases. Analytical solutions based on coherence methods are presented for solving 1D advection-dominated local equilibrium transport problems with constant and variable boundary conditions. Numerical solutions of diffusion-dominated local equilibrium transport in porous media show that sharp dissolution/precipitation fronts occur in this system as well. A final local equilibrium example presents a simple dual-porosity system with advection in an idealized fracture and solute diffusion into an adjacent porous matrix. This example illustrates delay of contaminant release from the 2D domain due to a combination of physical and chemical retardation. For partial local equilibrium, in which one or more reactions are controlled by kinetic expressions, an example

[*] This work was prepared in part by Sandia National Laboratories, Albuquerque, New Mexico 87185 USA and Sandia National Laboratories, Livermore, California 94550 USA for the United States Department of Energy under Contract DE-AC04-76DP00789.

NATO ASI Series, Vol. G 32
Migration and Fate of Pollutants in Soils and Subsoils
Edited by D. Petruzzelli and F. G. Helfferich
© Springer-Verlag Berlin Heidelberg 1993

demonstrates that coherent waves can still be generated. Waves influenced by kinetic reactions have a finite width rather than being step changes in concentration, but still propagate as a unit once developed.

Transport through porous media has applications in environmental systems such as contaminant transport away from waste disposal sites and radionuclide migration from radioactive waste repositories, as well as in other areas such as ore body formation and oil reservoir stimulation. Multiple variations on these and other systems are given by, to list a few, Cussler (1982), Cussler et al. (1983), Gruber (1990), Helgeson (1979), Helgeson et al., (1984), Lichtner (1985; 1988; 1991), Lichtner et al. (1986), Ortoleva et al. (1987a; 1987b), and Weare et al. (1976). This chapter interprets transport from the perspective of propagating waves and is based in part on the work of Novak et al. (1988; 1989), Schechter et al. (1987), Sevougian (1992), Walsh and Lake (1989), and Walsh et al. (1984).

PARTIAL DIFFERENTIAL EQUATIONS FOR TRANSPORT

The term "advection" signifies chemical transport by bulk flow of a fluid phase. The equations that describe transport in porous media are referred to as *advection-diffusion-reaction* equations, which are material balances on chemical species, and *advection-diffusion* equations which are material balances on chemical elements. The term "dispersion" is sometimes used in place of the term "diffusion" in these names; they will be used interchangeably here. The advection-diffusion-reaction and advection-diffusion equations are related through stoichiometry; the advection-diffusion equations are more convenient to use with a local equilibrium description of reactions while the advection-diffusion-reaction equations are more appropriate when one or more reactions are kinetically controlled. This section presents both sets and shows their interrelationship. In addition, the advection-diffusion equations are transformed into dimensionless variables for later application. The chemical coupling between the element or species transport equations is also demonstrated, with more detail given in the next section of this chapter.

The advection-diffusion-reaction equations are differential material balances on a set of S chemical species that can undergo R independent reactions. These equations are constrained by a set of I conservation equations for the chemical elements, where $S \geq I$; these element balances are the advection-diffusion equations. In general, there are $R = S - I$ independent chemical reactions. In a local equilibrium description, all the reactions are described by R nonlinear algebraic equations among the local species concentrations. In a partial local equilibrium description, some of the R nonlinear algebraic equations from the local equilibrium system are replaced with nonlinear differential equations that provide reaction rate as a function of local species concentrations. The assumption of local equilibrium may more appropriate for homogeneous (intraaqueous) reactions, while dissolution or precipitation rate equations may be more appropriate for describing heterogeneous (aqueous-mineral) reactions. The wave behavior discussed in this chapter is due to nonlinear coupling of the advection-diffusion or advection-diffusion-reaction equations through chemical reactions, both equilibrium and kinetic.

The advection-diffusion-reaction equations in two dimensions for J aqueous species and K mineral species, where $J + K = S$, and for a single flowing phase are

$$\frac{\partial}{\partial t}[\phi_1 C_j^a] + \frac{\partial(C_j^a u)}{\partial x} + \frac{\partial(C_j^a v)}{\partial y} - \frac{\partial}{\partial x}\left[D\phi_1\frac{\partial C_j^a}{\partial x}\right] - \frac{\partial}{\partial y}\left[D\phi_1\frac{\partial C_j^a}{\partial y}\right] = \sum_{r=1}^{R} v_{jr} r_r$$

$$j=1,...,J \qquad (1a)$$

$$\frac{d}{dt}[\phi_k^m C_k^m] = \sum_{r=1}^{R} v_{kr} r_r \qquad\qquad k=1,...,K \qquad (1b)$$

where

C_j^a $\quad$ = concentration of species j in the aqueous phase, micromoles per dm^3 aqueous volume

C_k^m $\quad$ = molar density of mineral k, micromoles per dm^3 mineral volume

ϕ_1 $\quad$ = aqueous phase volume fraction, aqueous volume per bulk volume

ϕ_k^m $\quad$ = mineral k phase volume fraction, mineral k volume per bulk volume

r_r $\quad$ = bulk reaction rate of the rth independent reaction, micromoles per dm^3 bulk volume per second

v_{jr} (v_{kr}) = stoichiometric coefficient of the jth (k^{th}) species in the rth reaction

D $\quad$ = effective diffusion coefficient, m^2/s, assumed the same for all species

u $\quad$ = fluid velocity in x-direction, m/s

v $\quad$ = fluid velocity in y-direction, m/s

t $\quad$ = time, s.

The total element concentrations are given by

$$C_{i1} = \sum_{j=1}^{J} a_{ij} C_j^a \qquad\qquad \phi_2 C_{i2} = \sum_{k=1}^{K} a_{ik} \phi_k^m C_k^m \qquad\qquad i=1,...,I \qquad (2)$$

where a_{ij} (a_{ik}) are members of the formula matrix which indicates the number of atoms of element i in species j (k), C_{i1} is the concentration of element i in the aqueous phase, C_{i2} is the concentration of element i in all mineral phases, and ϕ_2 is the combined volume fraction of all mineral phases,

$$\sum_{k=1}^{K} \phi_k^m = \phi_2 = 1 - \phi_1 \qquad (3)$$

The advection-diffusion equations for the I elements are obtained by linear combination of the advection-diffusion-reaction equations multiplied by the stoichiometric coefficients of the elements in each of the species,

$$\frac{\partial}{\partial t}[\phi_1 C_{i1} + \phi_2 C_{i2}] + \frac{\partial(C_{i1}u)}{\partial x} + \frac{\partial(C_{i1}v)}{\partial y} - \frac{\partial}{\partial x}[D\phi_1 \frac{\partial C_{i1}}{\partial x}] - \frac{\partial}{\partial y}[D\phi_1 \frac{\partial C_{i1}}{\partial y}] = 0$$

$$i = 1,...,I \qquad (4)$$

All reactions terms have canceled with this linear combination because of conservation of mass, i.e.,

$$\sum_{j=1}^{J} a_{ij} \sum_{r=1}^{R} v_{jr}r_r + \sum_{k=1}^{K} a_{ik} \sum_{r=1}^{R} v_{kr}r_r = 0 \qquad i = 1,...,I$$

Although the units listed with the above equations will be used through this chapter, any consistent set can be used. Equations 1 and 4 allow for spatial and temporal variation in flow velocity, but all calculations and discussions in this chapter use constant velocity.

1D Advection-Diffusion Equation in Dimensionless Form

For a one-dimensional domain of length L_x, with constant porosity and constant velocity, Equation 1 can be simplified to a form that is dimensionless in all but concentration,

$$\frac{\partial(C_{i1} + \bar{C}_i)}{\partial T_D} + \frac{\partial C_{i1}}{\partial X_D} - \frac{1}{N_{Pe}} \frac{\partial^2 C_{i1}}{\partial(X_D)^2} = 0 \qquad i = 1,...,I \qquad (5)$$

where

$$X_D = \frac{x}{L_x} \qquad \text{dimensionless distance}$$

$$T_D = \frac{ut}{L_x \phi_1} \qquad \text{dimensionless time}$$

$$N_{Pe} = \frac{uL_x}{D\phi_1} \qquad \text{Peclet number}$$

$$\bar{C}_i = \frac{\phi_2}{\phi_1} C_{i2} \qquad \text{moles i in the mineral phases per dm}^3 \text{ aqueous volume.}$$

When micromolar, μM, units are used for mineral concentrations, these refer to the symbol $\bar{C}_i$. The Peclet number indicates the importance of advection relative to diffusion/dispersion; a large Peclet number signifies an advection-dominated system, while a small Peclet number signifies a diffusion/dispersion dominated system. Dimensionless time is often referred to as "pore volumes" because a dimensionless time of $T_D = 1$ is required for complete replacement of all the fluid in the domain. The transformation to dimensionless variables defines the fluid velocity as unity, or $V_D = 1$. Concentration waves in the systems discussed herein cannot travel faster than the fluid and thus have dimensionless velocities between 0 and 1. Mineral dissolution/precipitation waves similar to those discussed in this chapter can travel faster than the fluid velocity during two-phase gas/aqueous flow. These fast waves are caused when chemical reactions among the phases provide a source or sink for the gas phase, such as production of $CO_2(g)$ by acid dissolution of carbonate minerals (Novak et al., 1991).

The boundary conditions used in this chapter to solve Equation 5 are the initial concentrations or conditions, I.C., at $T_D = 0$,

$$C_{i1}(X_D, T_D = 0) = C_{i1}^{I.C.}(X_D) \qquad \bar{C}_{i1}(X_D, T_D = 0) = \bar{C}_{i1}^{I.C.}(X_D)$$

and the boundary concentrations or conditions, B.C., at $X_D = 0$,

$$C_{i1}(X_D = 0, T_D) = C_{i1}^{B.C.}(T_D) \qquad \bar{C}_{i1}(X_D = 0, T_D) = \bar{C}_{i1}^{B.C.}(T_D)$$

which can also be called injected or entry conditions for advection-dominated problems.

For transport of elements in the aqueous phase that do not interact with the mineral phases, Equation 5 reduces to a linear PDE,

$$\frac{\partial C_{i1}}{\partial T_D} + \frac{\partial C_{i1}}{\partial X_D} - \frac{1}{N_{Pe}} \frac{\partial^2 C_{i1}}{\partial (X_D)^2} = 0 \qquad\qquad i = 1, \ldots, I \qquad (6)$$

For a semi-infinite domain ($0 < x < \infty$) with uniform initial conditions and constant boundary conditions, Equation 6 has an exact solution (Marle, 1981) which shows that the initially abrupt concentration change becomes less abrupt as it propagates. Such waves are called *tracer waves* because they propagate with the fluid, and thus trace the fluid velocity. (In the language of coherence these are *indifferent* waves.) A tracer wave initiated in response to a step change gradually loses its sharpness with time, and as a function of dispersion. Thus, for a constant value of dispersion, the wave will become more diffuse as it propagates, and a tracer wave in a system with smaller dispersion will be sharper than a tracer wave in an equivalent system with more dispersion.

COUPLING THE ADVECTION-DISPERSION AND ADVECTION-DISPERSION-REACTION EQUATIONS THROUGH CHEMICAL REACTIONS

The relationships among concentrations in the aqueous and mineral phases can be given by either a kinetic or a local equilibrium description of chemical reactions. The local equilibrium description assumes that the time required for chemical reactions to reach equilibrium is much shorter than the time required for transport. The kinetic description is more general, and allows reactions to occur on any time scale. The local equilibrium description requires basic thermodynamic data (generally Gibbs free energies of formation or equilibrium constants, and data for activity coefficient models), and the kinetic description requires information on reaction rates in addition to the basic thermodynamic data. This section presents the local equilibrium approach first, followed by a kinetic description for mineral dissolution and precipitation that are rate limited.

Coupling through Chemical Reactions Described by Local Equilibrium

Nonlinear chemical coupling, as noted earlier, greatly increases the complexity of solving the advection-diffusion equations. For adsorption phenomena, the advection-diffusion equations are often linearized by postulating a constant ratio of element concentrations in the aqueous and mineral phases, i.e.,

$$C_{i2} \approx K_d \, C_{i1} \tag{7}$$

This equation is a linear isotherm using a proportionality "constant" K_d called a distribution coefficient. Approximating C_{i2} in this way transforms the nonlinear Equation 5 into the linear Equation 6 with an additional constant term, a linear PDE for which an analytical solution exists. However, this assumption is only applicable under very restrictive conditions; even for adsorption the K_d model is frequently inadequate because the value of K_d is not constant, but rather is a function of aqueous and mineral compositions. The model is clearly inapplicable to dissolution/precipitation phenomena because any arbitrary amount of precipitate C_{i2} can be present in equilibrium with a saturated solution. Unfortunately, the K_d concept has been incorrectly applied as encompass all aqueous and mineral interactions occurring during transport. The linear isotherm model, while appealing in its simplicity in transport calculations, does not adequately describe the chemical aspects of transport in many systems (Brown et al., 1991; Gruber, 1990; Novak, 1992; Payne and Waite, 1991).

Thermodynamic models for chemical equilibrium are based on minimization of the total Gibbs free energy of the system, resulting in sets of nonlinear algebraic equations describing the equilibrium concentrations. Detailed discussions of the principles of chemical equilibrium can be found in books such as Garrels and Christ (1990) and Denbigh (1981). Numerical algorithms for solving chemical equilibrium problems are given in Smith (1980) and Smith and Missen (1982).

Solubility product expressions are commonly used to represent mineral equilibria. A reaction such as

$$AX(s) \leftrightarrow A^+ + X^- \tag{8a}$$

has the equilibrium expression

$$[A^+] \, [X^-] \leq K^{sp}_{AX(s)} \tag{8b}$$

where $[A^+]$ and $[X^-]$ represent the activities of aqueous species A^+ and X^-, respectively, and $K^{sp}_{AX(s)}$ is a constant called the solubility product of mineral $AX(s)$. For dilute solutions, activity can be approximated by molality (moles per kg H_2O), an assumption used throughout this chapter. Thus, one can consider Equation 8b as a constraint on species concentrations rather than activities. If no mineral $AX(s)$ is present, the aqueous concentrations of A^+ and X^- can assume any values within the restriction imposed by Equation 8b. If $AX(s)$ is present, the product of A^+ and X^- concentrations must equal the solubility product; the actual amount $AX(s)$ is unconstrained by equilibrium but is controlled by the mass balance equations. For dilute solutions, molar and molal concentrations are effectively the same (micromolar quantities are used in the transport equations), so these will be taken as identical in this chapter. However, this assumption is not valid for concentrated (brine) solutions, where conversion between these units is required; approximation of activity by concentrations is also not valid in brines.

The mathematical complexity introduced by chemical nonideality and by reactions within the aqueous phase do not qualitatively change the wave phenomena observed during propagation of dissolution/precipitation waves, at least for local equilibrium systems (Novak,

1990). It is likely that chemical nonideality will not qualitatively change the wave phenomena in partial local equilibrium systems. In any case, ideal chemical reactions are assumed throughout this chapter, and none of the examples include intraaqueous reactions. However, phenomena such as ion exchange and adsorption, because of equilibrium expressions that explicitly depend on immobile phase concentrations, do cause different wave propagation character; these are considered in papers such as Brown et al. (1991) and Bryant et al. (1986).

Coupling through Kinetically Limited Chemical Reactions

The most convenient form of the transport equations for a partial local equilibrium system is a hybrid between Equations 1 and 4. These equations are combined to eliminate rate terms for all local equilibrium reactions, while keeping the rate terms for kinetic reactions. Assume there are K^k minerals with kinetically controlled dissolution and K^e minerals that undergo equilibrium dissolution and precipitation, where $K^k + K^e = K$. In general, all the chemical reactions involving kinetic minerals can be written such that each kinetic mineral is involved in only one reaction (Smith and Missen, 1982). With reactions written this way, and with the stipulation that minerals with kinetically controlled dissolution appear as the first K^k reactions, Equation 1b becomes

$$\frac{d}{dt}[\phi_k^m C_k^m] = v_{kk} r_k \qquad\qquad k=1,...,K^k \qquad (9a)$$

$$\frac{d}{dt}[\phi_k^m C_k^m] = \sum_{r=1}^{R} v_{kr} r_r \qquad\qquad k=K^k+1,...,K^k+K^e \qquad (9b)$$

where the equilibrium and kinetic parts have been separated. Next, Equation 9a, containing all the kinetic mineral contributions, is summed over kinetic species (cf. Equation 2) and the result is subtracted from Equation 4, to achieve a form of the transport equations useful for partial local equilibrium systems,

$$\frac{\partial}{\partial t}\left(\phi_1 C_{i1} + \phi_2 C_{i2} - \sum_{k=1}^{K^k} \phi_k^m C_k^m \right)$$

$$+ \frac{\partial(C_{i1} u)}{\partial x} + \frac{\partial(C_{i1} v)}{\partial y} - \frac{\partial}{\partial x}[D\phi_1 \frac{\partial C_{i1}}{\partial x}] - \frac{\partial}{\partial y}[D\phi_1 \frac{\partial C_{i1}}{\partial y}]$$

$$= - \sum_{k=1}^{K^k} a_{ik} v_{kk} r_k \qquad\qquad i=1,...,I \qquad (10a)$$

$$\frac{d}{dt}[\phi_k^m C_k^m] = \nu_{kk} r_k \qquad\qquad k=1,...,K^k \quad (10b)$$

Or, cast into 1D dimensionless form similar to Equation 5,

$$\frac{\partial}{\partial T_D}\left(C_{i1} + \bar{C}_i - \sum_{k=1}^{K^k} \bar{C}_k^m\right) + \frac{\partial C_{i1}}{\partial X_D} - \frac{1}{N_{Pe}}\frac{\partial^2 C_{i1}}{\partial(X_D)^2}$$

$$= -\sum_{k=1}^{K^k} a_{ik}\nu_{kk} R_k \qquad\qquad i=1,...,I \quad (11a)$$

$$\frac{d\bar{C}_k^m}{dT_D} = \nu_{kk} R_k \qquad\qquad k=1,...,K^k \quad (11b)$$

where

$$R_r = \frac{r_r L}{u} \qquad \text{effective rate constant}$$

$$\bar{C}_k^m = \frac{\phi_k^m}{\phi_1} C_{i2} \quad \text{moles mineral k per aqueous volume.}$$

The remaining reaction terms are then described by equations giving the rate of reaction as a function of an intrinsic rate constant at the temperature of interest, the concentrations of participating species, and the surface area of the mineral involved. Although precipitation reactions and intraaqueous reactions such as electron transfer may be kinetically limited, this chapter focuses only on kinetically limited heterogeneous reactions describing mineral dissolution. All intraaqueous and precipitation reactions will be treated with local equilibrium.

Assuming that reaction rates are not limited by boundary layer mass transfer (Bird et al., 1960), the reaction rate for mineral dissolution is a function of the reacting mineral surface area per bulk volume, S_k, and the surface reaction rate, σ_k, in micromoles per surface area per time. In the absence of empirically measured surface rate equations, rate equations may be based on transition state theory (Helgeson et al., 1984). For kinetic dissolution reactions, one form of the transition state theory rate equation is

$$r_k = S_k \sigma_k = S_k k_k q_k \left(1 - \frac{Q_k}{K_k^{sp}}\right) \qquad\qquad (12)$$

or

$$R_k = \frac{S_k k_k L}{u} q_k \left(1 - \frac{Q_k}{K_k^{sp}}\right) \qquad\qquad (13)$$

where k_k is the intrinsic forward rate constant, K_k^{sp} is the equilibrium constant, Q_k is the activity product, and q_k is the forward activity product. For the dissolution reaction of Equation 8a, these expressions are

$$q_k = [AX(s)] = 1 \qquad Q_k = [A^+][X^-]$$

In the wave propagation examples discussed in this chapter, we assume a constant reaction surface area for the dissolving mineral, such as may exist when the dissolving mineral is present as a coating on a substrate.

ONE-DIMENSIONAL SOLUTIONS: DISSOLUTION/PRECIPITATION WAVE PROPAGATION IN LOCAL EQUILIBRIUM SYSTEMS

Walsh et al. (1984) performed numerical simulations of reactive transport systems with uniform initial conditions and constant boundary conditions, called Riemann problems, and noted that mineral dissolution/precipitation waves propagate with constant velocity. These observations led to an analytical solution in the absence of diffusion/dispersion based on the concept of coherence (Helfferich and Klein, 1970). A noncoherent disturbance is introduced into the domain at the inlet as a step change from initial to boundary conditions. The noncoherence resolves into a series of coherent waves that are step changes in aqueous and mineral phase concentrations. These abrupt concentration changes separate the domain into regions of constant aqueous and mineral composition.

The dimensionless velocity of element i in a coherent step or shock wave is given by a material balance across the step wave,

$$V_{Di} = \left[1 + \frac{\Delta \bar{C}_i}{\Delta C_{i1}} \right]^{-1} \tag{14}$$

where the delta indicates the difference in concentration across the wave. The *coherence condition* requires that the velocity of each element in the wave be the same; thus $V_{D1} = V_{D2} = \ldots = V_{DI}$.

Along with the coherence condition and local chemical equilibrium, the general solution to these problems often requires a condition relating concentrations upstream of a wave to concentrations downstream of the wave. The "Downstream Equilibrium Condition," DEC, requires the aqueous compositions downstream of a wave to be in equilibrium with the minerals upstream of the wave, whether or not those minerals are present in the downstream region

(Walsh et al., 1984). The DEC can be derived from a detailed analysis of the structure of dissolution/precipitation waves including diffusion/dispersion (Schechter et al., 1987) where it arises from the matching condition at a wave requiring continuity of aqueous concentrations. The description of the DEC as given here is sufficient for most situations but is not complete. A more thorough discussion is given by Bryant et al. (1987), although that interpretation must be modified by considerations given by Dria (1988) and Novak (1990).

Analytical solutions to 1D advection-dominated transport problems with dissolution/precipitation reactions can be calculated using coherence, local equilibrium, and downstream equilibrium, provided the order in which the minerals dissolve and precipitate is known (Bryant et al., 1987, Lichtner, 1991). When this order is unknown, candidate mineral sequences can be tested until the answer is found. A rule-based method for generating candidate sequences is given by Novak et al. (1988), along with heuristics that speed calculation through candidate sequences; simulators to perform these calculations are discussed in Novak (1990). For systems with few precipitates, composition route constructions give direct answers without the need to search through mineral sequences, as discussed in the chapter on *Multicomponent Wave Propagation*. Unfortunately, a general recursion algorithm necessary to extend this to real-world problems with large numbers of precipitates is still lacking.

Example: Dissolution/Precipitation Waves in 1D Local Equilibrium Advective Transport with Constant Boundary Conditions

Consider advection-dominated transport in a 1D permeable medium of length L_x, with the four ionic species A^+, B^+, X^-, and Y^-, where B^+ and Y^- are considered contaminants, and the mineral AX(s), BX(s), and AY(s) can form. The initial concentrations (conditions), I.C., are uniform throughout the domain and are given along with the boundary concentrations (conditions), B.C., in Table 1; solubility product data is given as well. The boundary conditions will also be referred to as the injected conditions. As indicated in Table 1, the initial aqueous phase is saturated with respect to AX(s) and undersaturated with respect to AY(s) and BX(s), while the boundary composition is undersaturated with respect to all minerals. Mineral AX(s) would be expected to dissolve during transport because the boundary condition is undersaturated with respect to this solid. This does indeed happen (see solution below), and this dissolution causes AY(s) and BX(s) to precipitate. Because the boundary conditions are undersaturated with respect to AY(s) and BX(s), these minerals eventually dissolve as well. Also shown in Table 1 is the inert tracer T(aq), which provides a means for tracking the bulk

Table 1.   Chemical species, equilibrium data, and initial and boundary compositions for all examples

Species	Solubility Product $(\mu M)^2$	I.C.: Initial Concentration (μM)	B.C.: Boundary Concentration (μM)
A^+	-	1.0	0
B^+	-	0	1.0
X^-	-	1.0	0
Y^-	-	0	1.0
$T(aq)$	-	0	1.0
AX(s)	1.0	4.0	0
AY(s)	0.60	0	0
BX(s)	0.10	0	0

water flow and distinguishing between the initial water and the water that is introduced into the domain.

The analytical solution using the coherence methods outlined above is given in Figure 1 for $T_D = 1$; the compositions in each of the regions are given in Table 2. The tracer wave, moving with $V_D = 1$, is located at $X_D = 1$ at $T_D = 1$ and is thus not visible in the figure. The fastest dissolution/precipitation wave shows AX(s) dissolution with precipitation of AY(s) and BX(s); AY(s) dissolves at the next wave, moving more slowly, followed by BX(s) dissolution at the slowest wave. The waves partition the domain into regions of constant concentrations containing different minerals. The resulting mineral sequence from the initial to boundary compositions, regions I.C.-R1-R2-R3-B.C. in Table 2, is

$$\text{I.C.} - \{AX(s)\} - \left\{ \begin{array}{c} AY(s) \\ BX(s) \end{array} \right\} - \{BX(s)\} - \text{B.C.}$$

where minerals in a region are enclosed in curly brackets and the waves between regions are indicated by hyphens. The mineral sequence is shown in a history rather than a profile because this is the causal direction, e.g., AX(s) dissolution causes AY(s) and BX(s) precipitation, not the other way around as a profile would seem to indicate.

Contaminants B^+ and Y^- undergo *chemical retardation* through incorporation into the mineral phase at the R1-R2 wave. In R1, downstream of this wave, B^+ and Y^- are at about 12% and 50% of the boundary concentrations, respectively; in R2, upstream of this wave, the total (aqueous plus mineral) concentrations of B^+ and Y^- are about 340% and 360% of the boundary concentrations. Thus, the precipitation of AY(s) and BX(s) caused by AX(s)

Table 2.   Analytical solution to local equilibrium 1D advection example through $T_D = 1$, based on coherence methods

	Initial Condition, I.C.	R1: Region 1	R2: Region 2	R3: Region 3	Boundary Condition, B.C.
Species					
A^+	1.000	1.206	0.4220	0	0
B^+	0	0.1206	1.092	1.092	1.000
X^-	1.000	0.8292	0.09161	0.09161	0
Y^-	0	0.4975	1.422	1.000	1.000
$T(aq)$	0	1.000	1.000	1.000	1.000
$AX(s)$	4.000	4.000	0	0	0
$AY(s)$	0	0	2.164	0	0
$BX(s)$	0	0	2.273	2.273	0

dissolution has reduced B^+ and Y^- concentrations in one region while increasing concentrations in another.

The wave structure of these problems is more easily recognized on a distance-time, or X-T, diagram, as shown in Figure 2; however, only the portion of Figure 2 from $0 \leq T_D \leq 1$ applies to this example as discussed thus far. In X-T coordinates, the slope of the line separating two regions of constant composition represents the dimensionless velocity of thewave between the regions. A vertical slice on an distance-time diagram results in a profile of

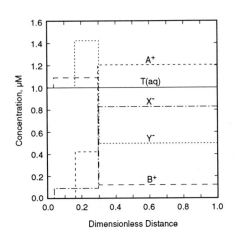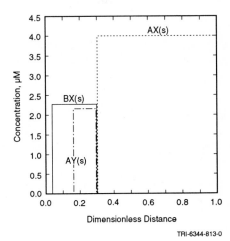

TRI-6344-813-0

Figure 1.   Concentration profiles at $T_D = 1$ for analytical solution to 1D local equilibrium advection example.

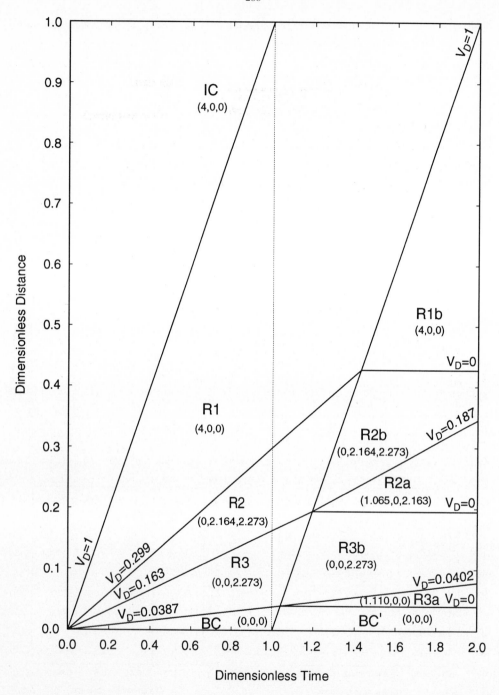

Figure 2.   Distance-time diagram for local equilibrium advection example.

TRI-6344-815-0

the domain (i.e., at a constant time) such as in Figure 1, while a horizontal slice yields a history (i.e., the time-evolution of concentrations at a constant position). An effluent history, the concentrations leaving the domain versus time, is obtained by a slice at $X_D = 1$ on an X-T diagram. The point $(X_D = 0, T_D = 0)$ can be considered as an intersection between waves carrying information on the initial and boundary conditions. The distance-time diagram illustrates that this point of noncoherence is resolved into the four coherent waves shown in Figure 2. The ordered triples on Figure 2 give the concentrations of AX(s), AY(s), and BX(s), respectively, in each region. The dimensionless velocity of any wave can be calculated from Equation 13; an example using Y^- concentrations across the R2-R3 wave is

$$V_{Di} = \left[1 + \frac{(2.164 - 0.0)}{(1.422 - 1.0)}\right]^{-1} = 0.163$$

Example: Dissolution/Precipitation Waves in 1D Local Equilibrium Advective Transport with a Step Change in Boundary Conditions at $T_D = 1$

The example thus far simulates continuous injection of one pore volume of aqueous phase contaminated with B^+ and Y^-. Often, contaminants are introduced for a finite period of time in a pulse, after which an uncontaminated aqueous phase flows into the medium. We continue the example by imposing for $T_D > 1$ a new boundary condition B.C.' with the same aqueous composition as the initial condition. Such a problem results in waves intersections which can each be considered as an independent Riemann problem. Wave interference is discussed in the chapter on *Multicomponent Wave Propagation* in general, and in Dria et al. (1987) for specific examples of the analytical approach used in this chapter. The distance-time diagram for this example extending through $T_D = 2$ is shown in Figure 2. Compositions for each region appearing after $T_D = 1$ are in given in Table 3.

Figure 2 illustrates the intersections resulting from a faster wave overtaking slower waves. When the tracer wave from the second injection intersects waves caused by the first injection, the waves interfere to produce new waves. In addition to the dissolution/precipitation and tracer waves already seen, the intersections cause standing (stationary) waves at region boundaries B.C.'-R3a, R3b-R2a, and R2b-R1b. These waves do not propagate and occur when there is a change in mineral identities and/or concentrations with no change in aqueous concentrations. The stationary wave closest to the inlet forms because BX(s) dissolution causes AX(s) precipitation, yet the boundary condition B.C.' is saturated with respect to mineral AX(s) so it cannot dissolve. Standing waves complete the set of three wave types observed in these systems as discussed in the chapter on *Multicomponent Wave Propagation*.

Table 3.   Analytical solution to 1D local equilibrium advection example for regions created between $T_D = 1$ and $T_D = 2$, based on coherence methods

Species	R1b	R2b	R2a	R3b	R3a	B.C.'
A^+	1.206	1.206	0.9535	0.9535	1.000	1.000
B^+	0.1206	0.1206	0.09535	0.09535	0	0
X^-	0.8292	0.8292	1.049	1.049	1.000	1.000
Y^-	0.4975	0.4975	0	0	0	0
T(aq)	0	0	0	0	0	0
AX(s)	4.000	0	1.065	0	1.110	0
AY(s)	0	2.164	0	0	0	0
BX(s)	0	2.273	2.163	2.273	0	0

The complexity associated with calculating concentration profiles using Riemann invariants and coherent wave theory often renders a numerical solution helpful if not essential. It is not our purpose here to discuss the details of the numerical solution, which can be found in such places as Gambolati et al., Neretnieks, (both this publication) and Novak and Gelbard (1991). Figure 3 shows concentration profiles at $T_D = 2$, as calculated with the finite-difference simulator FMT (Novak and Gelbard, 1991). Because the numerical solution gives nonphysical oscillations in solid concentrations when dispersion/diffusion is identically zero, there is a small amount of numerical dispersion in the simulation, resulting in the smoothing of waves most noticeable in the T(aq) profile. These numerical solutions were generated with 50 nodes, $\Delta X_D = 0.02$, $\Delta T_D = 0.01$, and with numerical dispersion resulting in an equivalent Peclet number of 200 (Lantz, 1971).

The profile in Figure 3 corresponds to a vertical slice of the distance-time diagram in Figure 2 at $T_D = 2$, and the concentrations in each region from Table 3. The aqueous concentrations in Figure 3 change in three places, X_D ~0.05, ~0.3, and 1.0, corresponding to the waves between R3a-R3b, R2a-R2b, and R1b-R1, respectively. While it is a little more difficult to discern the zones of constant mineral concentration in the numerical solution at $T_D = 2$, Figure 3, the numerical and the analytical solution, Figure 4, compare well.

Contaminants undergo chemical retardation through reactions that incorporate them into the mineral phases (e.g., by dissolution/precipitation, adsorption, ion exchange, etc.). Effluent history and cumulative release diagrams (Figure 5) illustrate chemical retardation for the above example. The pulse of contaminant was injected for $0 \leq T_D \leq 1$, after which water with the initial composition was injected. The effluent history reflects the tracer pulse moving with the fluid velocity, and thus advecting out of the domain between $1 \leq T_D \leq 2$, with slight spreading due to numerical dispersion. The peak concentration of contaminant Y^- is reduced to ~50% of the injected value because of AY(s) precipitation, and is advected out of the domain (i.e., crosses the $X_D = 1$ boundary) between $1 \leq T_D \leq 3$. Contaminant B^+ undergoes more chemical retardation in this example, with peak concentration reduced to ~10% of the boundary concentration; it is retarded to such an extent that ~9% of the mass that entered the domain remains in it after a dimensionless time of ten, as shown in the cumulative release diagram.

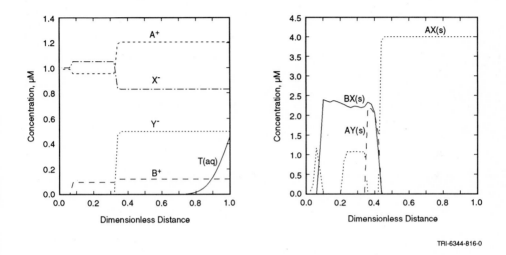

TRI-6344-816-0

Figure 3.   Concentration profiles at $T_D = 2$ for numerical solution to 1D local equilibrium advection example.

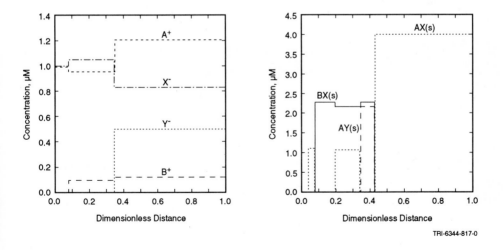

TRI-6344-817-0

Figure 4.   Concentration profiles at $T_D = 2$ for analytical solution to 1D local equilibrium advection example.

Effluent histories are frequently the only data practically obtainable from laboratory column experiments, it being very difficult to determine what is actually happening in the core during the course of the transport experiment. Also, this type of information is often the basis for regulations, and thus can be very important from that standpoint.

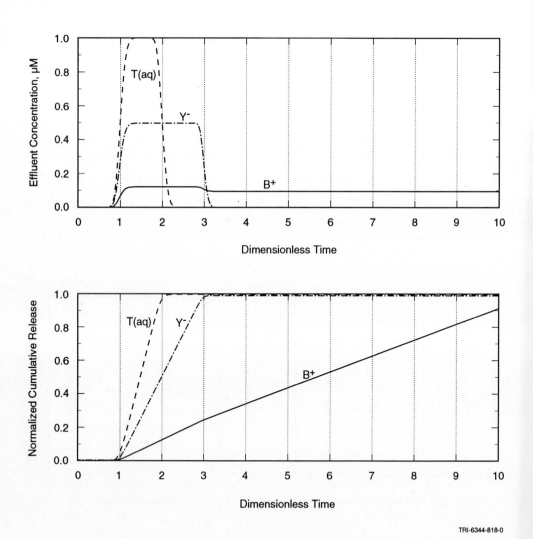

TRI-6344-818-0

Figure 5.   Effluent concentration histories and cumulative releases ($X_D = 1$) through $T_D = 10$ from numerical solution to 1D local equilibrium advection example.

Example: Dissolution/Precipitation Waves in 1D Local Equilibrium Diffusive Transport with Constant Boundary Conditions

Shock waves are commonly found in solutions to advection and convection equations such as the advection-diffusion equation, but are less common in solutions to the diffusion equation (Equation 1 with u = v = 0). Thus, it is noteworthy that similar abrupt changes in mineral concentration are observed during diffusion-only transport in porous media with mineral dissolution/precipitation (Kim and Cussler, 1987; Kopinsky et al., 1988; Lichtner et al., 1986; Novak et al., 1989). The sharp waves are again caused by the nonlinear coupling of the transport equations through dissolution/precipitation reactions. Because this issue is discussed in detail in the chapter on *Diffusion, Flow and Fast Reactions in Porous Media*, it is only briefly given here. In diffusion systems, transport partitions the domain into regions with distinct mineral assemblages, but the zones of constant concentration seen in the advection-dominated system are not present. This system is amenable to an analytical treatment, but the solution appears to be much less generalizable than the advection-dominated case.

Concentration profiles calculated using FMT are shown in Figure 6 for transport by diffusion with the initial and boundary conditions of Table 1 and a no-flux boundary condition at $y = L_y$. The simulation used a 15-node variable grid given by $\Delta y_1 = 5.014 \times 10^{-4}$ m, $\Delta y_i = 1.3209 \Delta y_{i-1}$, shown in Figure 7, a time step $\Delta t = 50$ seconds, and the diffusion coefficient $D = 10^{-9}$ m^2/s. The results of integration through 2.5×10^4 seconds (equivalent to

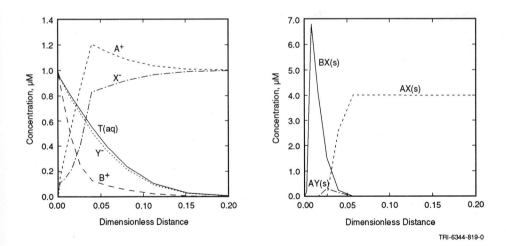

TRI-6344-819-0

Figure 6.   Concentration profiles at equivalent $T_D = 1$ for numerical solution to 1D local equilibrium diffusion example.

$T_D = 1$ in the advection example) are given. Only 20% of the domain is shown in Figure 6 because beyond that there are no changes by this time. There are again sharp fronts where minerals dissolve and precipitate (the sharpness is limited by the coarseness of the grid in this example). At each sharp change in mineral concentration, there is a discontinuity in aqueous concentrations for species that comprise the affected minerals, as required by a material balance about the wave (Novak et al., 1989).

Example: Dissolution/Precipitation Waves in 1D Advective Transport with Constant Boundary Conditions and One Kinetically Dissolving Mineral

The wave interpretation of kinetically controlled concentration fronts in porous media is in early stages. Therefore, this chapter will focus on only one kinetic example. The example again uses the same initial and boundary conditions of Table 1; however, AX(s) dissolution is not an equilibrium reaction in this case. Rather, AX(s) dissolution is kinetically controlled with

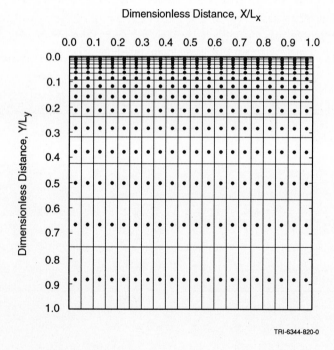

TRI-6344-820-0

Figure 7.   Domain and discretization used for 1D diffusion example (Y-direction only) and for 2D example.

the rate expression (cf. Equations 10b, 11, and 12)

$$\frac{d[AX(s)]}{dT_D} = (-1)\,(7.884)\left(1 - \frac{[A^+][X^-]}{1.0}\right) \tag{15}$$

It appears that a smaller rate constant merely scales the wave behavior, so that profiles for a larger rate constant at, say $T_D = 0.1$, would be similar to profiles for a smaller rate constant at one pore volume. In the limit of no AX(s) dissolution, i.e., a rate constant of zero, the system reverts to the trivial case with only a tracer wave and no mineral dissolution or precipitation because there is no source of A^+ or X^- to initiate BX(s) or AY(s) precipitation.

The example is solved with KGEOFLOW, a one dimensional finite difference simulator that includes kinetic reactions (Sevougian, 1992). KGEOFLOW and FMT are descendants of the same code, but have been developed differently. At this point an option for a 2D domain is not included in KGEOFLOW, and a chemical kinetics option is not included in FMT.

Concentration profiles for dimensionless times of 0.1, 0.2, and 0.5 PV are given in Figure 8, and for 1.0, 1.5, and 2.0 pore volumes in Figure 9. Curves were generated with 200 nodes $\Delta X_D = 0.005$, $\Delta T_D = 0.0025$, and an equivalent Peclet number of 800. However, no difference was noted between these profiles and those generated with 50 nodes, $\Delta X_D = 0.02$, $\Delta T_D = 0.01$, and an equivalent Peclet number of 200, other than in the degree of spreading of the tracer wave due to numerical diffusion. This indicates that, excluding the tracer wave, the aqueous and mineral concentration profiles are not significantly affected by numerical diffusion, and that the numerical solution has converged.

The early time profiles, plotted in Figure 8 for half the domain, show very different wave behavior than the corresponding local equilibrium example, Figure 1. Indeed, the profiles are similar to those seen in the diffusion dominated local equilibrium example, Figure 6. Ignoring the part of the domain downstream of the tracer wave ($X_D > T_D$), for which no change from the initial condition is expected, there are as yet no zones of constant concentration. Rather there is gradual dissolution of AX(s) with gradual precipitation of BX(s) and AY(s). Although difficult to discern on Figure 8 for 0.1 pore volumes, there is some AY(s) precipitate located at $X_D \sim 0.1$, the leading edge of the BX(s) precipitation wave. The profiles for A^+ and Y^- at $T_D = 0.1$ show that it is only near $X_D = 0.1$ that the aqueous phase is saturated with respect to AY(s), $K^{sp}_{AY(s)} = 0.6$. This is notable because the AY(s) precipitation and dissolution waves do not begin propagating from $X_D = 0$ at $T_D = 0$ as seen in local equilibrium systems, but rather appear downstream of the inlet. This is more obvious at the later times in Figure 8. By $T_D = 0.5$, the peak of BX(s) concentration just inside the domain has reached a value significantly larger than that seen in the local equilibrium advection example in Figure 1. The

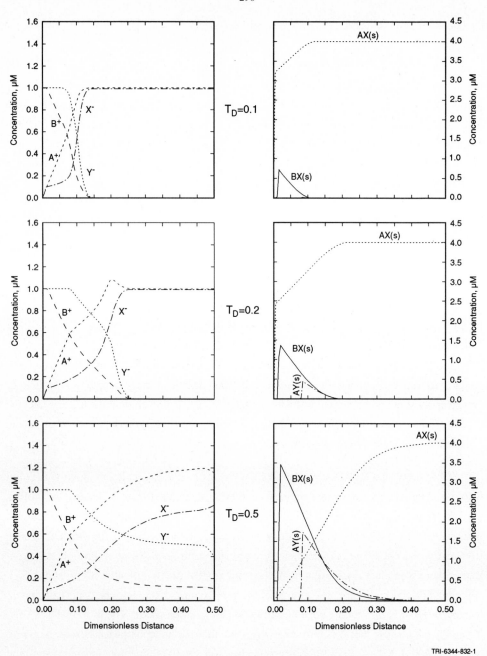

Figure 8.   Early time profiles for 1D advection example with kinetically controlled AX(s) dissolution.

297

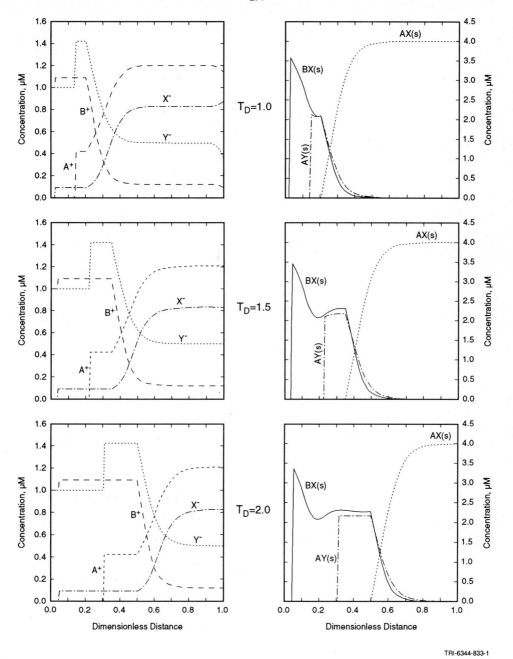

TRI-6344-833-1

Figure 9. Late time profiles for 1D advection example with kinetically controlled AX(s) dissolution.

BX(s) dissolution wave does not appear to move initially or moves only very slowly. Also, the AY(s) dissolution wave does not appear to move for the first 0.5 pore volumes, suggesting that this is a wave stationary at early times.

The later time profiles, for $T_D = 1.0$, 1.5, and 2.0, plotted in Figure 9 over the entire domain, show behavior more similar to the local equilibrium analog. By $T_D = 1$, AX(s) has dissolved in approximately two tenths of the domain. Upstream of the last appearance of AX(s), the system is once again in local equilibrium, and the aqueous concentrations are identical to those given in Figure 1 and Table 2. The mineral concentrations, however, are not the same. The influence of kinetic AX(s) dissolution at early time has caused the BX(s) concentration to be dramatically larger, and the AY(s) concentration to be slightly larger, than in the complete equilibrium case. The aqueous concentrations are the same as those in the local equilibrium case because it is the presence of the minerals, not the amount of the minerals, that determines the aqueous concentrations. By one pore volume, the BX(s) profile has started to develop a shoulder on the downstream end. As the later profiles show, this shoulder has a concentration less than the local equilibrium concentration. The profile undergoes an oscillation, most obvious in the $T_D = 2.0$ profile, until the local equilibrium concentration is attained. The reason for this oscillation is unclear, but it does not appear to be an artifact of the numerics. The AY(s) profile behaves similarly, although it is less dramatic. The profiles in Figure 9 show that the wave of equilibrium AY(s) and BX(s) precipitation with kinetic AX(s) dissolution has attained a constant width of about 30% of the domain by about one pore volume, and thereafter propagates as a unit in a coherent wave.

TWO-DIMENSIONAL SOLUTIONS: DISSOLUTION/PRECIPITATION WAVE PROPAGATION IN LOCAL EQUILIBRIUM SYSTEMS

One practical application for transport in porous media involves geologic disposal of nuclear waste. The concept of geologic disposal involves emplacing nuclear waste underground within a series of manmade and natural barriers designed to prevent or delay radionuclide migration away from the repository. One of the more likely mechanisms for radionuclide migration away from a repository involves radionuclide dissolution and subsequent transport with the aqueous phase by advection and/or diffusion.

The Waste Isolation Pilot Plant (WIPP) is a U.S. Department of Energy facility intended to demonstrate safe containment of transuranic nuclear waste. The WIPP is located in Southeastern New Mexico within thick halite, NaCl(s), beds chosen for their geologic stability and low porosity and permeability. Overlying the repository is a stratigraphic layer called the Culebra Dolomite Member of the Rustler Formation. Hydrologic testing has indicated that the Culebra is the most transmissive unit above the repository, yet the flow velocities are less than one meter per year. Calculations assessing repository performance with respect to environmental regulations for radionuclide containment indicate that radionuclides might reach the accessible environment through transport in the Culebra should a repository breach occur (Bertram-Howery et al., 1990).

Both hydrologic and geologic observations have demonstrated that the Culebra is appreciably fractured and suggest that transport in the Culebra may occur by a dual porosity mechanism. Dual porosity systems are so named because advection occurs primarily through fracture networks, with diffusion of solutes from the fracture into the porous matrix. The fracture porosity allows flow to occur while the matrix porosity provides the bulk of solute storage capacity. A contaminant introduced into a dual-porosity system will undergo *physical retardation*, that is, the contaminant rate of transport will be slowed relative to the bulk water flow, because the contaminant will diffuse out of the fractures where it is present in high concentration and into the matrix (Neretnieks, 1980).

The idealized domain for this 2D example consists of one planar fracture coupled to a porous matrix. One boundary is along the centerline of the fracture where a no flux boundary condition is imposed. The half-width of the fracture is represented by the Δy_1. The domain and the discretization used in the numerical solution are given in Figure 7. A no diffusive flux boundary condition is imposed on all boundaries except the fracture inlet and outlet where there is advection into and out of the domain. The fluid velocity is constant in the fracture, and there is no advection in the matrix. The parameters for this example are: $u = 4 \times 10^{-6}$ m/s; $D = 1 \times 10^{-9}$ m^2/s; $L_x = 0.1$ m; $L_y = 0.1$ m; $\Delta x = 0.005$ m; $\Delta y_1 = 5.014 \times 10^{-4}$ m; $\Delta y_i = 1.321 \Delta Y_{i-1}$; and $\Delta t = 50$ seconds. Dimensionless parameters can be calculated based on the fracture, giving: $\Delta T_D = 0.002$; $\Delta X_D = 0.05$; $\Delta Y_{D1} = 5.014 \times 10^{-3}$; $\Delta Y_{Di} = 1.321 \Delta Y_{Di-1}$; and a fracture Peclet number of 40.

This example uses the initial and boundary compositions of Table 1. Analogous to the 1D local equilibrium advection example, the contaminant boundary condition is imposed for one pore volume (based on fracture volume), after which the boundary condition is set equal to the initial condition.

Consider first the mineral concentration profiles at $T_D = 1$, plotted in Figures 10 through 12. Only a subset of the domain is shown because mineral concentrations are constant outside the range plotted. The fracture is along the back of the base plane, and fracture inlet is at the

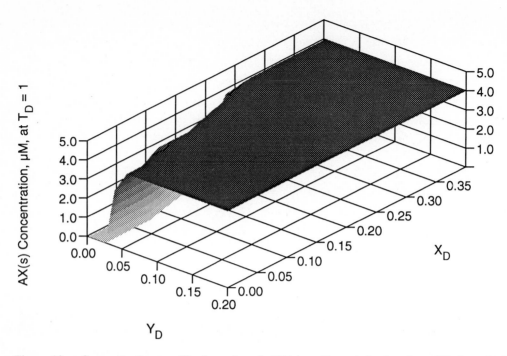

Figure 10. Concentration profile for mineral AX(s) at $T_D = 1$ for local equilibrium dual porosity example.

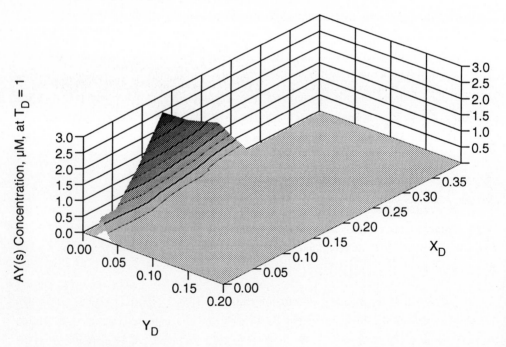

Figure 11. Concentration profile for mineral AY(s) at $T_D = 1$ for local equilibrium dual porosity example.

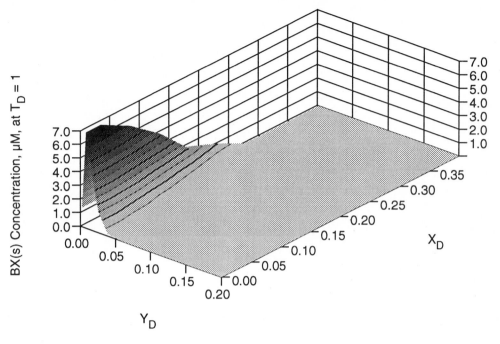

Figure 12. Concentration profile for mineral BX(s) at $T_D = 1$ for local equilibrium dual porosity example.

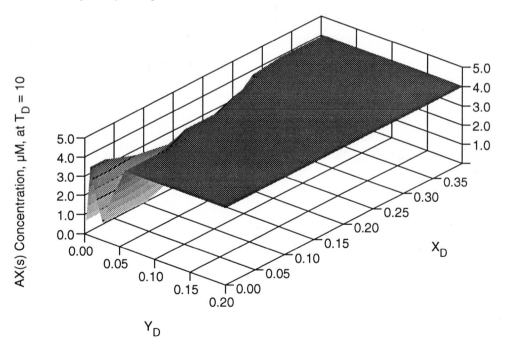

Figure 13. Concentration profile for mineral AX(s) at $T_D = 10$ for local equilibrium dual porosity example.

left-most corner of these figures. Because after $T_D = 1$ no additional contaminants B^+ or Y^- can enter the domain, these mineral profiles represent the approximate maximum mineral concentrations in this example. There might be slight additional increases in these mineral concentrations until the waves propagating the boundary compositions for $T_D \geq 1$ have reached the regions of precipitation. Because B^+ and Y^- were incorporated into precipitates, we expect to observe chemical retardation of the elements, and more for B^+ than Y^- because the amount of B^+ incorporated into the minerals is greater.

The mineral concentration profile for AX(s) at $T_D = 10$, given in Figure 13, shows the long-time response of the minerals in the domain to the new boundary condition. All AY(s) and BX(s) have dissolved by this time. There has also been precipitation of AX(s) near the inlet to the domain, just as observed in the 1D multiple-injection example. The aqueous concentration profiles (not shown) are within ~6% of their steady-state (i.e. long time) values by $T_D = 10$.

Figure 14 shows the incremental and normalized cumulative releases, $C_j^a(X_D = 1, T_D)$ and

$$\frac{C_j^a(X_D = 1, T_D)}{C_j^{a,B.C.}(0 \leq T_D \leq 1)}$$

respectively, for the tracer and the two contaminants that undergo chemical retardation in this example. Also shown for reference on these figures are the incremental and normalized cumulative releases for tracer flow in a system with the same geometry but in which there is no diffusion out of the fracture into the matrix (i.e., for transport through the fracture only, the 1D analogous case). These effluent curves represent waves of chemical concentrations passing through the domain. Comparison of the curves for T(aq)-fracture and T(aq) reveals the effect that physical retardation can have on species composition. Recall that species T(aq) does not react chemically in this system. The curve T(aq)-fracture represents the two step changes in concentration that entered the domain at $T_D = 0$ and $T_D = 1$, and exited the domain at $T_D = 1$ and $T_D = 2$, respectively, with spreading because of dispersion. The tracer wave is more diffuse for T(aq)-fracture, Figure 14, than for T(aq) in the 1D example, Figure 5, because there is more dispersion; the Peclet numbers are 40 and 200, respectively. The peak concentration for T(aq) (Figure 14) occurs at a slightly later time than for T(aq)-fracture and at a significantly reduced concentration, both of which are common observation for dual porosity systems (Grisak and Pickens, 1980). Also, although in the 1D case most of the T(aq) has left the system after $T_D \sim 2.5$ PV, about 19% of the tracer remains in the domain after $T_D = 10$ in the 2D case. Thus, the physical retardation due to matrix diffusion is significant.

303

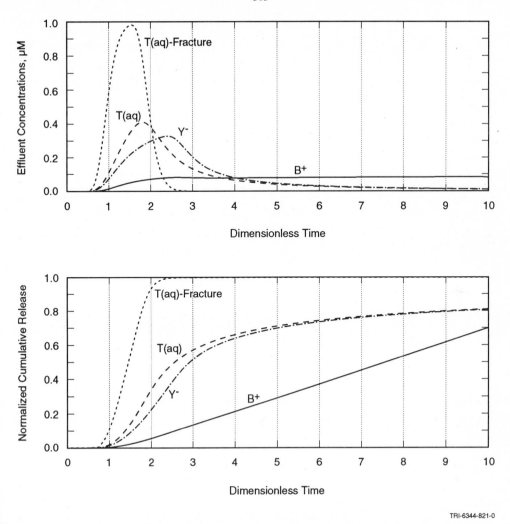

Figure 14. Effluent concentration histories and cumulative releases ($X_D = 1$) through $T_D = 10$ from numerical solution to local equilibrium dual porosity example.

Consider the Y^- effluent profiles relative to the T(aq) profiles in Figure 14. Here there is a slight delay of maximum Y^- concentration relative to T(aq) concentration because Y^- undergoes both chemical and physical retardation in this system. The amount of chemical retardation of species Y^- in this example is slight, as demonstrated by comparison of its cumulative release curve with that for T(aq) cumulative release. However, there is significant retardation compared to the 1D case in Figure 5.

Element B undergoes a significant amount of chemical retardation due to incorporation into the mineral phase. Figure 14 shows that the release concentration of B^+ never exceeds ~10% of its inlet concentration, just as observed in the 1D case in Figure 5, and consequently takes a much longer time to leave the domain than T(aq) or Y^-. Combined chemical and physical retardation cause ~30% of the injected B^+ to remain in the domain even after $T_D = 10$, which is appreciably more than the ~9% remaining in the 1D domain after $T_D = 10$, Figure 5.

CONCLUDING REMARKS

The analytical and numerical solutions presented here demonstrate that chemical retardation caused by mineral precipitation can delay and/or reduce peak concentrations and thus limit the release concentration of contaminants from simple one-dimensional and two-dimensional systems. These results are generalizable to systems with more complicated representations of the important chemical reactions. Acting separately, mineral precipitation and diffusion from fractures into the surrounding matrix can both delay contaminant release and reduce release concentrations. Acting together, chemical retardation (e.g., precipitation in this chapter) and physical retardation can further reduce and delay contaminant release. The combination suggests significant retardation of radionuclides may occur in dual-porosity transport systems if radionuclide precipitation occurs.

REFERENCES

Bertram-Howery SG, Marietta MG, Rechard RP, Swift PN, Anderson DR, Baker BL, Bean JE, Beyeler W, Brinster KF, Guzowski RV, Helton JC, McCurley RD, Rudeen DK, Schreiber JD, Vaughn P (1990) Preliminary comparison with 40 CFR part 191, subpart B for the Waste Isolation Pilot Plant, December 1990. SAND90-2347. Sandia National Laboratories, Albuquerque, NM

Bird RB, Stewart WE, Lightfoot EN (1960) Transport phenomena. John Wiley and Sons, New York

Brown PJ, Haworth A, Sharland SM, Tweed CJ (1991) Modeling studies of the sorption of radionuclides in the far field of a nuclear waste repository. Radiochim Acta 52-53:439-443

Bryant SL, Schechter RS, Lake LW (1986) Interactions of precipitation/ dissolution waves and ion exchange in flow through permeable media. Am Inst of Chem Eng J 32:751-764

Bryant SL, Schechter RS, Lake LW (1987) Mineral sequences in precipitation/dissolution waves. Am Inst Chem Eng J 33:1271-1287

Cussler EL (1982) Dissolution and reprecipitation in porous solids. Am Inst Chem Eng J 28:500-508

Cussler EL, Kopinsky J, Weimer JA (1983) The effect of pore diffusion on the dissolution of porous mixtures. Chem Eng Sci 38:2027-2033

Denbigh K (1981) The principles of chemical equilibrium. Cambridge University Press, Cambridge

Dria MA, Bryant SL, Schechter RS, Lake LW (1987) Interacting precipitation/dissolution waves: the movement of inorganic contaminants in groundwater. Water Resour Res 23:2076-2090.

Dria MA (1988) Chemical and thermochemical wave behavior in multiphase fluid flow through permeable media: wave-wave interactions. Ph.D. Thesis, University of Texas at Austin. University Microfilms, Ann Arbor, Michigan

Garrels RM, Christ CL (1990) Solutions, minerals and equilibria. Harper and Row, New York

Grisak GE, Pickens JF (1980) Solute transport through fractured media. 1. The Effect of Matrix Diffusion. Water Resour Res 16:719-730

Gruber J (1990) Containment accumulation during transport through porous media. Water Resour Res 26:99-107

Helfferich FG (1989) The theory of precipitation/dissolution waves. Am Inst Chem Eng J 35:75-87

Helfferich F, Klein G (1970) Multicomponent chromatography. Marcel Dekker, New York

Helgeson HC (1979) Mass transfer among minerals and hydrothermal solutions. Geochemistry of hydrothermal ore deposits. H.L. Barnes, ed. John Wiley and Sons New York

Helgeson HC, Murphy WM, Aagaard P (1984) Thermodynamic and kinetic constraints on reaction rates among minerals and aqueous solutions. II. rate constants, effective surface area, and the hydrolysis of feldspar. Geochim Cosmochim Acta 48: 2405-2432.

Kim JL, Cussler EL (1987) Dissolution and reprecipitation in model systems of porous hydroxyapatite. Am Inst Chem Eng J 33:705-710

Kopinsky J, Aris R, Cussler EL (1988) Theories of precipitation induced by dissolution. Am Inst Chem Eng J 34:2005-2010

Lantz RB (1971) Quantitative Evaluation of Numerical Diffusion (Truncation Error). Soc Pet Eng J Sept: 315-320.

Lichtner PC (1985) Continuum model for simultaneous chemical reactions and mass transport in hydrothermal systems. Geochim Cosmochim Acta 49:779-800

Lichtner PC (1988) The quasi-stationary state approximation to coupled mass transport and fluid-rock interaction in a porous medium. Geochim Cosmochim Acta 52:143-165

Lichtner PC (1991) The quasi-stationary state approximation to fluid/rock reactions: local equilibrium revisited. In: Ganguly J (ed) Diffusion, atomic ordering, and mass transport. Springer Verlag, Berlin Heidelberg New York, p. 452-560

Lichtner PC, Oelkers EH, Helgeson HC (1986) Exact and numerical solutions to the moving boundary problem resulting from reversible heterogeneous reactions and aqueous diffusion in a porous medium. J Geophy Res B 91:7531-7544

Marle C (1981) Multiphase flow in porous media. Gulf Publishing Company, Houston, Texas, p. 206-215

Neretnieks I (1980) Diffusion in the rock matrix: an important factor in radionuclide retardation? J Geophy Res B 85:4379-4397

Novak CF (1990) Metasomatic patterns produced by infiltration or diffusion in permeable media. Doctoral Dissertation, The University of Texas at Austin. University Microfilms, Ann Arbor, Michigan

Novak CF (1991) Geochemical modeling of two-phase flow with interphase mass transfer. Am Inst Chem Eng J 37:1625-1633

Novak CF (1992) An evaluation of radionuclide batch sorption data on Culebra Dolomite for aqueous compositions relevant to the human intrusion scenario for the Waste Isolation Pilot Plant (WIPP). SAND91-1299. Sandia National Laboratories, Albuquerque, NM

Novak CF, Gelbard FM (1991) Modeling mineral dissolution and precipitation in dual-porosity systems. SAND91-0430. Sandia National Laboratories, Albuquerque, NM

Novak CF, Schechter RS, Lake LW (1988) Rule-based mineral sequences in geochemical flow processes. Am Inst Chem Eng J 34:1607-1614

Novak CF, Schechter RS, Lake LW (1989) Diffusion and solid dissolution/precipitation in permeable media. Am Inst Chem Eng J 35:1057-1072

Ortoleva P, Merino E, Moore C, Chadam J (1987a) Geochemical self-organization I: reaction-transport feedbacks and modeling approach. Am J Sci 287:979-1007

Ortoleva P, Chadam J, Merino E, Sen A (1987b) Geochemical self-organization II: the reactive-infiltration instability. Am J Sci 287:1008-1040

Payne TE, Waite TD (1991) Surface complexation modelling of uranium sorption data obtained by isotope exchange techniques. Radiochim Acta 52/53:487-493

Schechter RS, Bryant SL, Lake LW (1987) Isotherm-free chromatography: propagation of precipitation/dissolution waves. Chem Eng Commun 58:353-376

Sevougian SD (1992) Partial local equilibrium and the propagation of mineral alteration zones. Doctoral Dissertation, The University of Texas at Austin. University Microfilms, Ann Arbor, Michigan

Smith WR (1980) Computational aspects of chemical equilibrium in complex systems. Theor Chem: Adv Perspect 5:185-259

Smith WR, Missen RW (1982) Chemical reaction equilibrium analysis: theory and algorithms, John Wiley and Sons, New York

Walsh MP, Bryant SL, Schechter RS, Lake LW (1984) Precipitation and dissolution of solids attending flow through porous media. Am Inst Chem Eng J 30:317-328

Walsh MP, Lake LW (1989) Applying fractional flow theory to solvent flooding and chase fluids. J Pet Sci Eng 2:281-303

Weare JH, Stephens JR, Eugster HP (1976) Diffusion metasomatism and mineral reaction zones: general principles and application to feldspar alteration. Am J Sci 276:767-816

Fate of Non-Aqueous Phase Liquids: Modeling of Surfactant Effects

Jeffrey H. Harwell[1], David A. Sabatini[2], and Thomas S. Soerens[2]

[1]School of Chemical Engineering and Materials Science and

[2]School of Civil Engineering and Environmental Science

The University of Oklahoma

Norman, Oklahoma 73019, U.S.A.

When a sufficient volume of an organic liquid is released into the environment, it may persist in the environment in the form of a residual saturation in either the unsaturated or the saturated zones of the subsurface. In the case of organic liquids less dense than water (referred to as LNAPLs, Light Non-Aqueous Phase Liquids), if the spill is of sufficient volume to reach the water table (Figure 1), some fraction of the liquid will remain as a residual saturation in the pores of the soil in the unsaturated zone. The liquid that reaches the water table will form a liquid lens, with some depression of the water table. As the volume of the LNAPL lens decreases, the water table may return to its original height, resulting in the formation of a residual saturation of the LNAPL in the aquifer medium (Weber, et

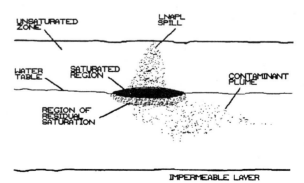

Figure 1. Formation of a residual saturation of an LNAPL in an aquifer.

NATO ASI Series, Vol. G 32
Migration and Fate of Pollutants in Soils and Subsoils
Edited by D. Petruzzelli and F. G. Helfferich
© Springer-Verlag Berlin Heidelberg 1993

310

al., 1991; Johnson, et al., 1989). If the liquid is a Dense Non-Aqueous Phase Liquid (DNAPL)--a chlorinated hydrocarbon solvent such as trichloroethylene is an example--its fate in the soil above the water table is essentially the same as that of the LNAPL. Upon contact with the ground water, however, the chlorocarbon, being up to 60% more dense than water, can continue to move down through the aquifer, displacing the ground water (Figure 2). Within the aquifer, the chlorocarbon may also become

Figure 2. Formation of a residual saturation of a DNAPL in the subsurface.

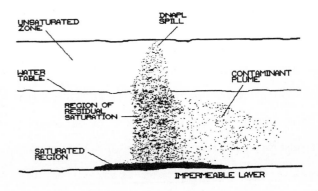

trapped in pores of the aquifer medium by capillary forces. If the size of the spill is sufficient, pools of the chlorocarbon may collect in the aquifer on an impermeable layer. Though the chlorocarbon is immiscible with the ground water, it has sufficient solubility to contaminate the ground water to an unacceptable level. Chlorocarbon dissolved in the ground water will not only move away from the site of the residual saturation, but will also partition between the ground water and the solid phase of the aquifer as it moves away from the spill (Huling and Weaver, 1991). A residual saturation is particularly persistent in the environment because only a limited interfacial area may be exposed to the ground water; this has the effect of reducing the overall mass transfer rate away from the site of the residual saturation which in turn increases the persistence of the contamination. In case of a residual saturation in a ground water environment, traditional pump-and-treat methods may require that many hundreds of pore volumes of water be circulated through the contaminated region before the residual saturation has been removed. This situation has led to an interest in the use of surfactants for enhancing the rate of removal of DNAPLs from a subsurface environment (Palmer and Fish, 1992).

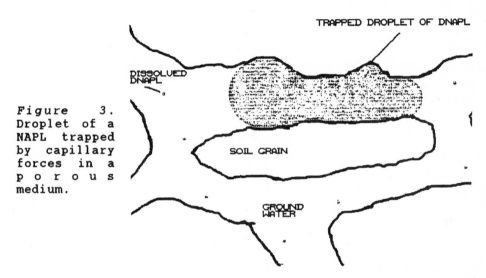

Figure 3. Droplet of a NAPL trapped by capillary forces in a porous medium.

This paper presents a simple model of surfactant enhanced removal of a residual saturation of a DNAPL along with the solution of the model in terms of shock wave propagation. The shock wave solution is also compared to a numerical solution of the model equations.

Surfactant Behavior in Ground Water

The molecular architecture of a surfactant molecule causes it to concentrate very efficiently at interfaces. Figure 4 presents a

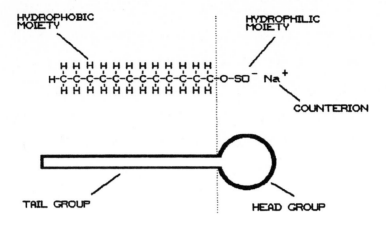

Figure 4. Molecular architecture of a typical surfactant molecule.

schematic representation of a typical surfactant molecule, sodium dodecylsulfate (SDS), a compound which has United States Food and Drug Administration status as a direct food additive. It is found in many consumer products such as toothpaste and shampoo. The key to the design of an effective surfactant is the creation of well defined hydrophobic and hydrophilic regions together with sufficient water solubility and a large enough hydrophobic group to confer a high degree of surface activity. SDS satisfies all these criteria. Surfactants have been studied in the literature both as ground water contaminants and as potential aids to remediation efforts (West and Harwell, 1992; Jafvert, et al., 1991; Abdul, et al., 1991).

One potential mode of applying surfactants in ground water remediation is analogous to that in which they have been used in enhanced oil recovery operations (Hirasaki, 1981; Lake, 1989). In this enhanced oil recovery mode the surfactant system is designed to reduce the capillary forces which resulted in the formation of the residual saturation by lowering the interfacial tension between the non-aqueous liquid and the ground water. If the interfacial tension reaches a sufficiently low value, the droplets will become mobilized and may be pumped directly to the surface. Because of concern about the possible downward migration of released droplets of DNAPLS, however, the current emphasis in the United States on research regarding surfactants for remediation applications is based on the phenomenon of solubilization.

Solubilization occurs when surfactants form micelles in solution (Rosen, 1989). Micelle formation is a common characteristic of good surfactants. A micelle is a transient aggregate of 50-200 surfactant molecules in solution (Figure 5). Surfactants can form micelles at very low concentrations in water, generally on the order of millimoles per liter. The concentration at which a surfactant begins to form micelles is called its Critical Micelle Concentration (CMC). The surfactants in the aqueous micelle are organized so that the hydrophilic moieties are in contact with the aqueous solution and the hydrophobic moieties are on the interior of the micelle. Though a number of different shapes of micelles are known, many surfactants form spherical micelles over a wide concentration range. The interior of the micelle is very much like a droplet of an organic liquid. Since like dissolves like, any hydrophobic compounds in the solution will tend to partition into the interior of the micelle. This phenomenon is called solubilization. Micellar solubilization of a hydrophobic compound can increase its concentration in water several orders of magnitude above its solubility limit (Mittal, 1977; Scamehorn, 1986). At concentrations at or below the CMC, a surfactant will have little if any effect on the solubility of a compound.

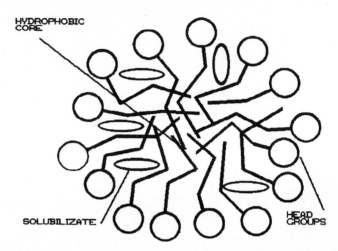

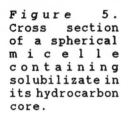

Figure 5. Cross section of a spherical m i c e l l e c o n t a i n i n g solubilize in its hydrocarbon core.

Theory

In an application of surfactants to enhance the remediation of an aquifer through solubilization of a residual saturation of a DNAPL, an aqueous solution of surfactant at a concentration above the CMC of the surfactant would be injected into the contaminated region where the surfactant would greatly increase the concentration of the DNAPL in the flowing aqueous phase by the mechanism of solubilization (West and Harwell, 1992). With a properly selected surfactant the interfacial tension between the DNAPL and the ground water would not be reduced sufficiently to result in the mobilization of the DNAPL. A one dimensional analog of this process would be the removal of a residual saturation of DNAPL from a soil column in the laboratory. Material balance equations to describe the one dimensional analog may be obtained by performing an integral material balance for the surfactant and the DNAPL on the volume element between x_n and $x_n + \Delta x$ and between the time t_n and $t_n + \Delta t$. The material balance for the surfactant yields:

$$
\int_{t_n}^{t_n+\Delta t} \left[(V_x A_c \epsilon (1-S_D) C_s)_{x_n} - (V_x A_c \epsilon (1-S_D) C_s)_{x_n+\Delta x} \right] dt =
$$

$$
\int_{x_n}^{x_n+\Delta x} \left[(A_c \epsilon (1-S_D) C_s)_{t_n+\Delta t} - (A_c \epsilon (1-S_D) C_s)_{t_n} \right] dx
$$

(1)

The first term on the left hand side of the material balance represents the molar flow rate of surfactant into the volume element; the remaining term on the left hand side is the molar flow rate of surfactant out of the volume element. The first term on the right hand side of the equation is the amount of surfactant in the element at the end of the time interval; the remaining term is the amount of surfactant in the volume element at the beginning of the time interval. Note that the area accessible to flow of the aqueous phase and the volume occupied by the surfactant solution are reduced by the volume occupied by the non-flowing non-aqueous phase, which is accounted for by the term $(1-S_D)$, where S_D is the residual saturation of the DNAPL. The equation does not yet require assumption of local equilibrium between the surfactant solution and the residual DNAPL. By the Mean Value Theorems, in the limit as Δx and Δt go to zero, we obtain:

$$
\frac{\partial}{\partial t}(A_c \epsilon (1-S_D) C_s) + \frac{\partial}{\partial x}(V_x A_c \epsilon (1-S_D) C_s) = 0
$$

(2)

Since the cross sectional area, A_c, and the porosity, ϵ, of the column are constant, but the residual saturation, S_D, is not, eq

(2) becomes:

$$(1-S_D)\frac{\partial C_s}{\partial t} - C_s\frac{\partial S_D}{\partial t} +$$
$$v_x(1-S_D)\frac{\partial C_s}{\partial x} + C_s\frac{\partial}{\partial x}(v_x(1-S_D)) = 0 \tag{3}$$

The first term in the equation represents the accumulation of surfactant in the ground water, the second term represents the filling with surfactant solution of the space left by the DNAPL after it has been solubilized, and the third term represents the flux of the aqueous surfactant solution. The last term in the equation represents the flux from an increase in the upstream area available to the flowing surfactant solution; however, under the assumption of a constant volumetric flow rate (the volume of aqueous solution increases as the DNAPL is solubilized into it and so fills the space previously occupied by the DNAPL) this last term is equal to zero.

Similarly, the integral material balance equation for the DNAPL in the column is:

$$\int_{t_n+\Delta t}\int_{x_n}^{x_n+\Delta x}\left[(\rho_D\varepsilon A_c S_D)_{t_n} - (\rho_D\varepsilon A_c S_D)_{t_n+\Delta t}\right]dx =$$
$$\int_{t_n}\left[(v_x A_c\varepsilon(1-S_D)C_D)_{x_n+\Delta x} - (v_x A_c\varepsilon(1-S_D)C_D)_{x_n}\right]dt \tag{4}$$

$$\therefore \qquad \frac{\partial S_D}{\partial t} = -\frac{v_x(1-S_D)}{\rho_D}\frac{\partial C_D}{\partial x} -$$
$$\frac{C_D}{\rho_D}\frac{\partial}{\partial x}(v_x(1-S_D)); \qquad \frac{\partial C_D}{\partial x} > 0$$
$$\frac{\partial S_D}{\partial t} = 0; \qquad \frac{\partial C_D}{\partial x} \leq 0 \tag{5}$$

The restriction of the applicability of the balance equation to a positive gradient in C_D is necessary because the residual saturation does not increase when the concentration of DNAPL in solution increases because of micellar solubilization; without this restriction the numerical solution will generate a non-physical increase in the residual saturation at the wave front of the propagating surfactant solution when it is saturated with solubilized DNAPL. In the same way as with eq (3), the last term on the left hand side of eq (5) is equal to zero under the assumption of a constant volumetric flow rate. Eq (5), however, does make the assumption of local equilibrium between the surfactant solution and the residual DNAPL. A third balance equation is not required because the concentration of DNAPL in the solution is a function of the surfactant concentration:

$$
\begin{aligned}
C_D &= C_D^*(1+K_s\,(C_s-CMC\,)); & S_D &> 0, & C_s &> CMC \\
C_D &= C_D^*; & S_D &> 0, & C_s &\leq 0 \\
C_D &= 0; & S_D &= 0
\end{aligned}
\tag{6}
$$

In these expressions K_s is the partition coefficient for the DNAPL between the micelles and the concentration of DNAPL outside of the micelles. Under the assumption of local equilibrium, the aqueous concentration of DNAPL, C_D, reduces to the solubility limit of the DNAPL in water, C_D^t, when no micelles are present and must reduce to zero when the residual saturation is reduced to zero.

Together with appropriate initial and boundary conditions, eqs (3), (5), and (6) can be solved simultaneously to obtain a description of the surfactant enhanced remediation process under the assumption of local equilibrium. The numerical solution is obtained more easily by first making the equations dimensionless by introducing the following dimensionless variables:

$$\tau = \frac{V_{x0}}{l} \qquad \xi = \frac{x}{l}$$

$$y_1 = \frac{C_s}{C_{s,inj}} \qquad y_2 = \frac{S_D}{S} - D, 0 \qquad y_3 = \frac{C_D}{C_{D,0}} \qquad (7)$$

We obtain, therefore, the following dimensionless equations:

$$\frac{\partial y_1}{\partial \tau} - \frac{y_1 S_{D,0}}{1 - y_2 SD, 0} \frac{\partial y_2}{\partial \tau} + \frac{1}{1 - y_2 SD, 0} \frac{\partial y_1}{\partial \xi} = 0$$

$$\frac{\partial y_2}{\partial \tau} = -\frac{C_{D,0}}{\rho_D S_{D,0}} \frac{\partial y_3}{\partial \xi} \qquad (8)$$

These equations were solved for this paper using the Method of Lines (Harwell, et al., 1985), along with the appropriate dimensionless restrictions and initial and boundary conditions. The numerical solution is presented in the Results section of the paper.

The system described by the equations can, however, be expected to give rise to coherent waves which will propagate through the column (Helfferich and Klein, 1970; Helfferich, 1986, 1989). Further, the equations describing the variation of y_1 and y_2 with τ and ξ--though subject to some unusual constraints--are first order hyperbolic equations whose solution can be expected to give rise to propagating wave fronts (Rhee and Amundson, 1986). Making the appropriate substitutions we can obtain expressions (eqs (9)) for the characteristic directions along which boundary values will propagate into the solution spaces, $y_1(\tau,\xi)$ and $y_2(\tau,\xi)$. From these expressions we make the observation that

$$\frac{dy_1}{d\tau} = 0, \qquad \frac{d\xi}{d\tau} = \frac{1 + y_1 \left(K_s C_{s,INJ} \dfrac{C_{D,0}}{\rho_D} \right)}{1 - y_2 S_{D,0}}$$

$$\frac{dy_2}{d\tau} = -\frac{C_{D,0} K_s C_{s,INJ}}{\rho_D S_{D,0}} \frac{dy_1}{d\xi}, \qquad \frac{d\xi}{d\tau} = 0 \qquad (9)$$

since y_1 increases in an upstream direction and y_2 decreases in an upstream direction, upstream variations in y_1 will propagate at a greater velocity than downstream variations and upstream values of y_2 will decrease at a greater rate than downstream values of y_1. These are conditions under which shock waves form. From coherence theory, for a system of three components (water and the two dissolved substances) we anticipate that two shock waves will be set up by a disturbance at the boundary of the system. The propagation of these waves and the variation of the dependent variables in the system across the waves can be represented schematically, as in Figure 6, or with a distance-time diagram, as in Figure 7 (Helfferich and Kline, 1970; Helfferich, 1986, 1989).

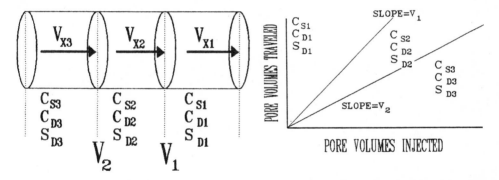

Figure 6. Schematic of shock waves moving through a column.

Figure 7. Distance-time diagram showing shock trajectories.

All of the concentrations and saturations downstream of wave 1 and upstream of wave 2 are known from the initial and boundary conditions on the system. Combining these diagrams with the integral-material balance expressions (eq (1) and eq (4)), we can obtain mass balances for the surfactant and DNAPL across the waves. For the surfactant balance across the fast wave, the wave of velocity V_1, we obtain:

$$\Delta x \, (\varepsilon A_C \, (1-S_{D2})C_{S2} - \varepsilon A_C \, (1-S_{D1})C_{S1})=$$
$$\Delta t \, (V_{X2} \varepsilon A_C \, (1-S_{D2})C_{S2} - V_{X2} \varepsilon A_C(1-S_{D2})C_{S1}) \qquad (10)$$

We know that the surfactant concentration downstream of the fastest wave, C_{S1}, is zero and that the DNAPL concentration downstream of the fastest wave, C_{D1}, and the residual saturation downstream of the fastest wave, S_{D1}, correspond to the initial conditions in the column. Additionally, our assumption of a constant volumetric flow rate requires that $V_{X2}(1-S_{D2})=V_{X1}(1-S_{D1})$. Introducing a new variable, the reduced shock velocity, $\Lambda_1=(\Delta x/\Delta t)/(V_{X1}(1-S_{D1}))$, we obtain the following simple expression for the velocity of the fastest wave:

$$\Lambda_1 = \frac{1}{(1-S_{D2})} \qquad (11)$$

Similarly, the integral-material balance equation for the DNAPL, eq (4), can be used to obtain a material balance equation for the DNAPL across the slow shock wave and the wave velocity, V_2:

$$\Delta x \, (\varepsilon A_C \, S_{D2} - \varepsilon A_C \, S_{D3})\rho_D=$$
$$\Delta t \, (V_{X2} \varepsilon A_C \, (1-S_{D2})C_{D2} - V_{X3} \varepsilon A_C \, (1-S_{D3})C_{D3}) \qquad (12)$$

$$\Lambda_2 = \frac{C_{D2}}{\rho_D S_{D2}} \qquad (13)$$

Solving for the surfactant balance across the slow shock, we obtain the following expression:

$$\Lambda_2 = \frac{C_{S3} - C_{S2}}{C_{S3}-(1-S_{D2})C_{S2}} \qquad (14)$$

We encounter a difficulty with the DNAPL balance across the fast shock wave, however; Eq (15) cannot be satisfied because $C_{D1}<C_{D2}$ but we do not allow S_{D2} to be greater than S_{D1} (remember the

$$\Lambda_1 = \frac{C_{D1}-C_{D2}}{\rho_D\,(S_{D1}-S_{D2})} \tag{15}$$

constraint that had to be placed on the differential balance equation for the DNAPL, eq (5)); we conclude, therefore, that $S_{D2}=S_{D1}$. We can now re-write eq (11) as:

$$\Lambda_1 = \frac{1}{1-S_{D1}} \tag{16}$$

Eliminating Λ_2 from eq (14), replacing S_{D2} with S_{D1}, and recalling the relationships between C_D and C_S (eq (6)), we can obtain an expression which can be solved for C_{S2}:

$$\frac{C_D^*(1+K_S\,(C_{S2}-CMC))}{\rho_D S_{D1}} = \frac{C_{S3}-C_{S2}}{C_{S3}-(1-S_{D1})C_{S2}} \tag{17}$$

Results

The model equations were solved for four cases. The parameter values associated with each case are summarized in Table 1.

	C_D^* mols/l	$C_{S,inj}$ mols/l	CMC mols/l	K_S 1/mol	ρ_D mols/l	$S_{D,0}$
Case 1	10^{-3}	0	0	0	12.3	0.01
Case 2	10^{-3}	0	0	0	12.3	0.1
Case 3	10^{-3}	1	10^{-3}	218	12.3	0.01
Case 4	10^{-3}	1	10^{-3}	218	12.3	0.1

Table 1. Parameter values used in case studies.

Case 1 and Case 2 correspond to traditional pump-and-treat scenarios, where no surfactant is added to enhance the solubility of the DNAPL, with residual saturations of 1% and 10%, respectively. A distance-time diagram for Case 1 is shown in Figure 8. With no added surfactant only one wave is generated and the concentration of the DNAPL in the flowing aqueous phase is not enhanced.   The DNAPL is finally completely eluted from the column

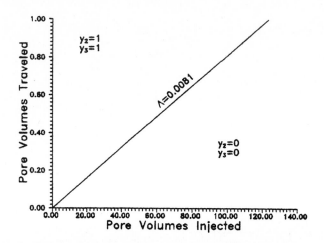

Figure 8.   Elution of 1% residual saturation of DNAPL with ground water.   Breakthrough occurs at 123 pore volumes.

after 123 pore volumes.   A distance-time diagram for Case 2 is shown in Figure 9.   When the residual saturation is increased to 10%, with no added surfactant the breakthrough of the shock wave

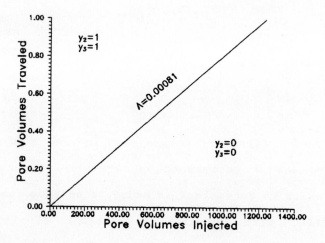

Figure 9.   Elution of 10% residual saturation of DNAPL with ground water.   Breakthrough occurs at 1230 pore volumes.

does not occur until 1230 pore volumes of water have been circulated through the system.

Case 3 corresponds to Case 1 but with a 1M solution of surfactant injected instead of ground water. The distance-time diagram for Case 3 is shown in Figure 10. The striking difference between Figures 8 and 10 is the velocity of the waves in the case of the surfactant injection. The residual DNAPL is now completely eluted from the column in only 1.01 pore volumes. In fact, the velocities of the slow wave and the fast wave are nearly identical. An interesting additional observation is that the concentration of the surfactant in the plateau region between the waves is only 56% of the injected concentration. This low

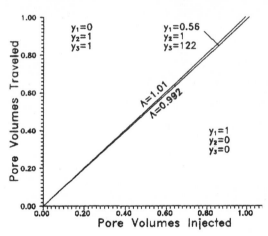

Figure 10. Distance-time diagram for Case 3. A low residual saturation of DNAPL is rapidly eluted by injected surfactant solution.

concentration arises because the solubilized DNAPL is all concentrated into the narrow zone between the two waves while all of the surfactant solution which has filled the void left by the DNAPL been taken from this same narrow zone.

Figure 11 is a distance-time diagram for Case 4, which corresponds to Case 2 with addition of surfactant to the injected solution. While 1230 pore volumes were required to elute a DNAPL

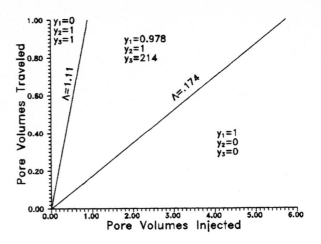

Figure 11. Distance-time diagram for Case 4. A residual DNAPL saturation of 10% is eluted in 5.7 pore volumes by a concentrated surfactant solution.

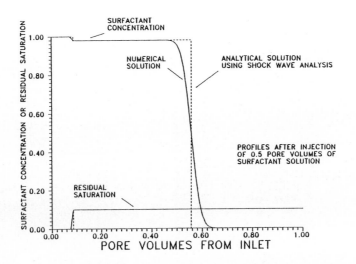

Figure 12. Numerical and analytical solutions of the variation of the surfactant concentration and the residual saturation after injection of 0.5 pore volumes of surfactant solution in Case 4.

residual saturation of 10% with only ground water injection, the surfactant solution now elutes the DNAPL in only 5.7 pores volumes. In Figures 12 and 13 we have concentration profiles of the column after the injection of 0.5 pore volumes of the surfactant solution. For comparison the numerical solution and

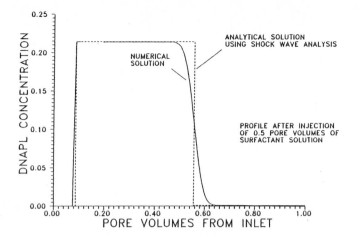

Figure 13. DNAPL concentration profile after injection of 0.5 pore volumes of surfactant solution for the system in Case 4.

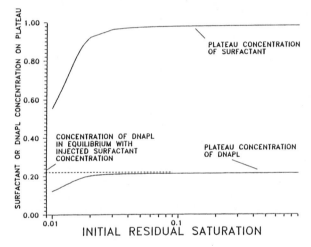

Figure 14. Effect of the initial residual saturation on the plateau concentrations of surfactant and DNAPL for Case 4.

the analytical solution are plotted on the same figures. The numerical solution was obtained by the Method of Lines. The dimensionless forms of the governing equations for the system were discretized in the ξ-direction into 640 nodes ($\Delta\xi$=0.0015625) and the resulting set of 1280 ordinary differential equations was

integrated simultaneously using a 2nd order Runge Kutta algorithm
($\Delta\tau$=0.00078125).   The distortion of the concentration profile
obtained in the numerical solution is caused by numerical
dispersion which arises from the large truncation error associated
with estimating the first derivatives in the ξ-direction by a
first backward difference.   The numerical dispersion in the
numerical solution is analogous to the dispersion that would be
observed in a laboratory column experiment. It is also important
to note, however, that the essential features of the column

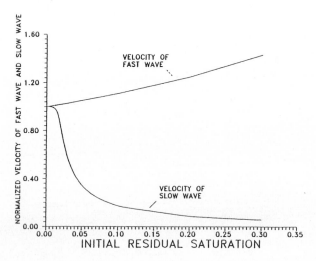

Figure 15. Effect of the initial residual saturation on the
shock wave velocities in Case 4.

profile are well described by the shock wave analysis though the
actual expressions for the velocities of the shock waves are
extremely simple.

Figures 14 and 15 summarize, respectively, the analytical
solutions for the plateau concentrations of the surfactant and the
DNAPL and the velocities of the waves as a function of the initial
residual saturation in the column.

Glossary of Symbols

A_C	cross sectional area of column	m^2
C_{D_*}	aqueous phase concentration of DNAPL	mol/l
C_D^*	solubility limit of DNAPL in ground water	mol/l
C_{Di}	aqueous phase concentration of DNAPL in zone i	mol/l
C_S	aqueous phase concentration of surfactant	mol/l
C_{Si}	aqueous phase concentration of surfactant in zone i	mol/l
C_{Sinj}	injected aqueous phase concentration of surfactant	mol/l
l	column length	m
S_D	DNAPL residual saturation	–
S_{Di}	DNAPL residual saturation in zone i	–
S_{D0}	initial DNAPL residual saturation	–
t	time	s
v_x	superficial fluid velocity	m/s
v_{xi}	superficial fluid velocity in zone i	m/s
x	distance	m
y_1	dimensionless surfactant concentration, $C_S/C_{S,inj}$	–
y_2	dimensionless residual saturation, S_D/S_{D0}	–
y_3	dimensionless DNAPL concentration, C_D/C_D^*	–
ϵ	porosity	–
Λ_i	normalized velocity of shock wave i	–
ξ	dimensionless distance, x/l	–
τ	dimensionless time, tv_{x3}/l	–
ρ_D	DNAPL phase molar density	mol/l

References

Abdul A S, Gibson, T L (1991) Laboratory studies of surfactant-enhanced washing of polychlorinated biphenyl from sandy material. *Environmental Sci Technol* 25: 665

Harwell J H, Schechter, R S, Wade W H (1985) Surfactant Chromatographic Movement: An Experimental Study. *AICHE Journal* 31: 415

Helfferich F, Klein G (1970) *Multicomponent chromatography.* Dekker, New York.

Helfferich F G (1986) Multicompenent wave propagation: Attainment of coherence from arbitrary starting conditions. *J. Chromatogr* 44:275-285

Helfferich F G (1989) Coherence: power and challenge of a new concept, in Keller G E II, Yang R T (eds) *New directions in sorption technology.* Butterworths, Boston, Chap 1

Huling S G, Weaver J W (1991) *Dense nonaqueous phase liquids.* United States Environmental Protection Agency Publication EPA/540/4-91-002, EPA Center for Environmental Research Information, Cincinnati, OH

Jafvert C T, Heath J K (1991) Sediment- and saturated-soil-associated reactions involning and anionic surfactant (dodecylsulfate). 1. Precipitation and micelle formation. *Environ Sci Technol* 25: 1031

Johnson R L, Keely J F, Palmer C D, Suflita, J M, Fish, W (1989) *Transport and fate of contaminants in the subsurface.* United States Environmental Protection Agency Seminar

Publication EPA/625/4-89/019, EPA Center for Environmental Research Information, Cincinnati, OH

Mittal K L (ed) (1977) *Micellization, solubilization, and microemulsions.* Plenum, New York, NY

Palmer C D, Fish W (1992) *Chemical enhancements to pump-and-treat remediation.* United States Environmental Protection Agency Publication EPA/540/S-92/001, EPA Center for Environmental Research Information, Cincinnati, OH

Rhee H-K, Aris R, Amundson N R (1986) *First-order partial differential equations, vol 1.* Prentice Hall, Englewood Cliffs NJ

Rosen M J (1989) *Surfactants and interfacial phenomena.* Wiley-Interscience, New York, NY

Scamehorn J F (ed) (1986) *Phenomena in mixed surfactant systems.* ACS Symposium Series 311. American Chemical Society, Washington D.C.

Weber W J, McGinley P M, Katz L E (1991) Sorption phenomena is subsurface systems: concepts, models and effects on contaminant fate and transport. Wat. Res. 25: 499

West C C , Harwell J H (1992) Surfactants in the environment. Our current state of knowledge and the key issues relevant to purposeful use. *Environ Sci Technol* in press

Diffusion, Flow and Fast Reactions in Porous Media

Ivars Neretnieks
Department of Chemical Engineering
Royal Institute of Technology
S-100 44 Stockholm, Sweden

When fluid flows through a porous medium, the dissolved constituents can react with the solid material. Solid phases may dissolve, precipitate, or react with some constituents. Rapid reactions produce a sharp reaction front propagating through the medium. Low fluid velocity makes diffusion important, as in slow degradation of concrete by diffusing acid, carbonate, sulphate and magnesium ions. Redox fronts in the ground form where pyrite reacts with oxygenated rainwater, forming sulphuric acid, which, in turn, reacts with other minerals. Sharp fronts in liquid and mineral composition, pH and Eh (oxidation potential) develop and move, as in the leaching of sulphidic copper ores. In these and numerous other cases the reactions are fast in comparison with the rates of transport.

Ion exchange and adsorption are used industrially to recover solutes or to purify water. In hydrometallurgy, sulphide ores are leached by reaction with oxygen dissolved in water. Ores are formed in nature by similar processes, where a metal is dissolved upstream and precipitated when entering a zone with reducing minerals or organics. (Walsh et al., 1984; Ortoleva et al., 1986). Mine tailings and solid waste deposits are leached by intruding water transporting the dissolved constituents. During transport, dissolved species react by precipitating to form new minerals, or by ion exchange or sorption on the surfaces of the minerals encountered downstream. Where reactions are fast, the system is practically at local equilibrium everywhere and can be modelled by mass balances and equilibrium relations, whereas in systems where reactions are slow kinetics must be taken into account. Often diffusion and dispersion must also be accounted for.

For ion exchange and sorption there is a well-developed theoretical basis for the propagation of concentration waves through such systems (Rhee et al., 1970; Rhee and

NATO ASI Series, Vol. G 32
Migration and Fate of Pollutants in Soils and Subsoils
Edited by D. Petruzzelli and F. G. Helfferich
© Springer-Verlag Berlin Heidelberg 1993

Amundson, 1970). There is a strong similarity between the waves in chromatography and systems with precipitation or dissolution fronts, as reviewed by Schechter et al. (1987) and Helfferich (1989). Wave propagation in general and the propagation of dissolution and precipitation waves are treated in other chapters in this book.

When sharp fronts dominate, certain simplifying assumptions are possible. If the total concentration of a component in the fluid phase is much less than its concentration in the solid phases, the mass balance is dominated by the solids. The accumulation in the liquid can then be neglected, so that the system of partial differential equations, which are used to describe the system, becomes a system of ordinary differential equations. This assumption has long been used with the shrinking core model to describe poisoning of catalyst pellets and the solidification of melts. In those cases only one reaction takes place but the principle can be extended to simultaneous multicomponent transport.

This chapter presents the basic theory of coupled chemistry and transport problems, the underlying principles of some common computer programs, including both the one-step and two-step programs, and the basis for the pseudo-stationary state approach. Application of these programs to real problems is also shown.

Recent common program packages for equilibrium calculations and modelling are reviewed. The presence of a large amount of solids dominating the dynamics of a chemical system cause difficulties when applying the usual codes. A method which facilitates the computation of such systems, based on the pseudo-stationary state approximation is discussed.

BASIC PROCESSES UNDERLYING GEOCHEMICAL TRANSPORT MODELS

Equilibrium, reaction rate, and mass balance relations are used to describe the present problem. For illustrative purposes the system can be thought to consist of a sequence of connected reactors. In the reaction step, fluid and solids are processed in each reactor to equilibrium, or for a certain time. Thereafter, a fraction of the liquid is transported from one reactor to the next. Diffusive (and dispersive) transport can also be directly included in the transport step. Many computer codes are based on this two-step way of solving the equations; examples include TRANQL (Cederberg, 1985), CHEQMATE (Haworth et al., 1988), and others (Walsh et al., 1984; Bryant et al., 1986) which can readily be extended to two or three dimensions (DYNAMIX, Liu and Narasimhan, 1989a,b; HYDROGEOCHEM, Yeh and

Tripathi, 1991). In this approach, the composition in one "reactor" is solved independently of the composition in the others, at each time step. With rapid reactions the reaction step can be treated as an equilibrium problem, otherwise kinetics are included. This is the two-step method where the reaction step is solved independently of the transport step. In one-step methods, transport equations and reaction equations are solved simultaneously.

Equilibrium problem

In a system with I dissolved *species* it is possible to choose J "primary species" or *components* which are linearly independent, from which all other species can be derived by mass action relationships. Note that $J \leq I$. When concentrations of components are known, simple algebraic equations (mass action relations) determine the concentrations of all other species. By the use of mass action relations, a system of I unknown species can be reduced to a much smaller system of J unknowns. For every one of the J unknown components, a mass balance can be formulated.

Solid species can be defined by the components via mass action and solubility product relations. However, concentrations of solid phases do not depend on the concentrations of the components. The unknown mineral concentrations must be determined by keeping track of how much of a mineral dissolves or precipitates during an equilibration step. The main difficulty is to determine which minerals actually form or dissolve.

Redox reactions and acid-base reactions are expressed through similar basic equations. For the mass balance of protons, the concentration of water would have to be accounted for. As this is impractical, because of the large amount of water, a charge balance is used instead. For redox reactions, the "electron activity" is not needed if, e.g., the oxygen fugacity is used instead.

Adsorption by surface complexation and ion exchange has been included in transport programs. Miller (1983) uses a known and constant number of sites, whereas Yeh and Tripathi (1991) allow the number of sites to vary with the amount of minerals present. This increases the number of unknown concentrations by including those species bound to the sites, but they can be expressed, as for dissolved species, using the algebraic mass action equations.

Kinetics of reactions with solids

In most programs dealing with solid-solute interactions, reactions forming complexes (i.e. species formed by a combination of components) in the liquid phase are assumed to be instantaneous, as are reactions forming surface complexes and species bound to ion exchange sites. Radioactive decay is readily included as a first-order reaction (Yeh and Tripathi, 1991). Dissolution and precipitation can also be included, provided rate expressions are known. Rate equations are then used instead of mass action-equilibrium relations for solids (Steefel and Cappellen, 1990). Noorishad et al. (1987) include kinetics in the code CHMTRNS. Lichtner (1988) makes this a central issue in the code MPATH based on the pseudo-stationary state approximation. To use rate equations, the reaction rates as well as the specific surface area must be known. The reaction rates depend on concentrations of solute species and pH, while the specific surface area depends on both the amount and physical structure of the solid. The driving force for the dissolution-precipitation is often assumed to be proportional to the difference between the ion activity product and the solubility product (Lichtner, 1988). A more general expression is used by Steefel and Cappellen (1990).

Transport

The transport equations are based on the principle of conservation of mass. The difference between inflow and outflow by any combination of advective flow, diffusion, dispersion, injection or withdrawal, equals the accumulation (depletion if negative) in the reactor. This principle is applied to every component by stating that what leaves one "reactor" must enter its neighbour(s). If transport involves diffusion or dispersion in addition to advection, net transfer may occur not only in, but also against the direction of advective flow.

Continuous tube reactors or a continuous porous medium such as an aquifer do not consist of a series of reactors. In most programs the continuum is discretized by finite difference or integrated finite difference approximations which act as the reactors. Yeh and Tripathi (1991) use a finite element formulation.

The problem can be formulated more easily and in a more compact way if the system is described as a continuum, but when the equations are to be solved, we shall return to discretization.

FORMULATION OF GOVERNING EQUATIONS

Here only reactions between dissolved species and minerals will be accounted for, leaving out ion exchange and surface complexation. Formulations including these processes can be found in Miller (1983) and Yeh and Tripathi (1991).

The basic mass balance equation for component j which moves by advection and diffusion or dispersion and dissolves or precipitates in the porous medium is

$$\theta\frac{\partial Y_j}{\partial t} + \sum_{r=1}^{R} v_{jr}\frac{\partial X_r}{\partial t} = -V\frac{\partial Y_j}{\partial x} + \theta D_L\frac{\partial^2 Y_j}{\partial x^2} \tag{1}$$

θ is the porosity, Y_j is the total concentration of component j in solution, v_{jr} is the stoichiometric coefficient for component j in mineral r, X_r is the concentration of mineral r, V the water flux, D_L the dispersion coefficient, t is time and x the length coordinate.

The equation is written for one dimension, assuming that all species have the same diffusion or dispersion coefficient. Y_j is the concentration of component j both as a free component and bound in all the complexes.

$$Y_j = C_j + \sum_{i=1}^{N_{complexes}} v_{ji} C_i \tag{2}$$

The concentrations C_i of the complexes are obtained from the mass action relation

$$C_i = \frac{K_i}{\gamma_i} \prod_{j=1}^{J} (C_j\gamma_j)^{v_{ji}} \tag{3}$$

The mass action equations for the minerals demand that at equilibrium

$$K_r \prod_{j=1}^{J} (C_j\gamma_j)^{v_{jr}} - 1 = 0 \tag{4}$$

If the reaction is not at equilibrium, the rate of approach to equilibrium is assumed to be proportional to the actual value of the expression on the left hand side of Equation (4). The reaction rate becomes

$$\frac{\partial X_r(x,t)}{\partial t} = \frac{k_r a_r}{K_r} \left(K_r \prod_{j=1}^{J} (C_j \gamma_j)^{\nu_{jr}} - 1 \right)$$ (5)

where k_r is the reaction rate constant and a_r the specific surface of the mineral. There is the obvious condition that a mineral which is not present cannot dissolve. With a fast reaction rate it may be assumed that there is local equilibrium between the minerals and the dissolved species. Equation (5) does not apply and the rate of reaction is directly determined by the difference in solute transport to and from every location according to Equation (1). The difficulty is in determining which of all possible minerals are forming or dissolving.

Solution of equilibrium problem.

The equilibrium problem is solved based on the assumption that the total concentration of every component in a reactor is known. This is the natural starting point for the problem in this context, because after the previous time step, the concentration of the component in solution as well as the amount (concentration) of minerals in a reactor has been determined. In preparation for the next time step, the species are transported by diffusion, dispersion and advection in and out of the reactor so that a net accumulation or depletion of the species results. The total concentration of every component T_j is thus known. The components must distribute themselves between the different species i and minerals r. A system of J equations stating this can be set up.

$$T_j = C_j + \sum_{i=1}^{N_{complexes}} \nu_{ji} C_i + \sum_{r=1}^{R} \nu_{jr} X_r$$ (6)

The above equations are solved together with Equations (3) and (4). This method is used in HYDROGEOCHEM (Yeh and Tripathi, 1991). Other equilibrium programs, e.g., EQ3/6 (Wolery, 1983) and PHREEQE (Parkhurst, 1985) use similar approaches. PHREEQE is used for the equilibrium calculations in the transport codes CHEQMATE and DYNAMIX. One of several numerical difficulties is that the amount of a component present in the mineral is sometimes many orders of magnitude larger than the dissolved component, so accuracy may be lost when T_j is determined. The concentration of oxygen in some common redox reactions typically varies from about 10^{-60} to 1. In the equilibrium case, Gibbs' phase rule must be obeyed, which states that when the temperature and pressure are set, the number of minerals in equilibrium cannot exceed the number of components.

MICROQL (Westall, 1979), used in the transport code TRANQL (Cederberg, 1985), adopts a different approach in the equilibration step. The number of components is reduced by one for every solid present, so that the system of Equations (2) is solved with a reduced number of components $J\text{-}R$. The resulting C_j´s are then used to determine the C_i´s together with the precipitated or dissolved mineral amounts which are obtained by mass balance.

To avoid having to account for the mass of water in the system, the mass balance of hydrogen is replaced by a charge balance. The electron is sometimes used as a component when redox reactions are involved. The activity coefficients γ_i are near unity for dilute aqueous systems and are often estimated by Davies' equation (Stumm and Morgan, 1981).

Determination of mineral sequences

The interaction of dissolved species with solids leads to the development of precipitation-dissolution waves similar to those in multicomponent chromatography (Rhee at al., 1970; Rhee and Amundson, 1970). Similar waves exist for precipitation-dissolution reactions (Helfferich, 1989). Bryant et al. (1987) and Novak et al. (1988) have demonstrated that there are many possible solids which could form or dissolve in the system but that a unique solution can be assumed to exist. This is governed by the initial and boundary conditions. They demonstrate a method to establish a unique sequence of minerals. Liu and Narasimhan (1989) in their equilibrium based code have developed a sequential search procedure which minimizes Gibbs' free energy at each moving front. Yeh and Tripathi (1991) use a similar scheme, checking the over- and undersaturation of all possible minerals and precipitating or dissolving them accordingly. Lichtner (1988) uses a kinetic formulation for the formation and dissolution of minerals at the fronts, which gives the correct mineral sequences without having to use other criteria.

Solution of transport problem

After the reaction step when changes in mineral amounts have been determined in every reactor, species are transported between reactors during a certain time, the so-called time step, and new equilibria are determined. This is repeated until the problem is solved. If the changes in concentration between time steps in a reactor are small, this two-step procedure is sufficient (TRANQL: Cederberg, 1985; CHEQMATE: Haworth et al., 1988; DYNAMIX: Liu and Narasimhan, 1989). If the concentration changes are larger, iterations between the transport and equilibrium steps are necessary (HYDROGEOCHEM: Yeh and Tripathi, 1991).

Lichtner et al. (1986) have also studied systems in which there is only diffusion and find that moving fronts which develop in these can be computed by finite difference techniques and that the solutions converge to the expected steady states.

COMPARISON OF SOLUTION PROCEDURES

The two-step approach is easy to conceptualize and there are several codes that have implemented the procedure successfully. One disadvantage is that it uses an explicit time stepping procedure. Then small time steps must be used to obtain sufficient accuracy. Another way of solving the system of parabolic partial differential equations, Equation (1), is to use an implicit solution technique such as the Crank-Nicolson scheme or a backward scheme (Lapidus and Pinder, 1982). Discretization of the system can be performed by finite difference, integrated finite difference or finite element methods. The difference between the explicit two-step method and the implicit one-step method can be described in the following way.

In the two-step method there are N reactors and one needs to solve for the J unknown component concentrations N times for every time step. The $N*J$ unknowns for a given time are obtained after the two steps.

In the implicit formulation the equations are solved simultaneously for the $N*J$ unknowns in every time step. This has been successfully done in the CHEMTRN family of codes, CHEMTRN (Miller, 1983), CHMTRNS (Noorishad et al., 1987) and THCC (Carnahan, 1986). The advantage of the two-step approach is that it takes much less time to solve a system of J equations N times than $N*J$ equations once. This advantage must be weighed against the possible need to iterate between equilibrium and transport steps and the fact that shorter time steps must be used in the two-step method. Two-step methods have been successfully applied to two-dimensional transport problems (Liu and Narasimhan, 1989; Yeh and Tripathi, 1991). In both codes the two-step method was chosen because of advantages in computing times. Yeh and Tripathi, (1989) discuss the different approaches for solving transport, reaction rate and equilibrium equations, comparing the differences in computer storage and time needed when different primary dependent variables are chosen. They conclude that for two- and three-dimensional problems, in practice, only the two-step method can be used.

Some properties of the programs

In the explicit two-step methods the largest time step which can be used is determined by the stability criterion which is:

$$\frac{D_L \Delta t}{\Delta x^2} \leq \frac{1}{2} \tag{7}$$

Δt is the time step and Δx is the length (volume) of the element. To illustrate the number of time steps needed the following example is used. A medium of length x_0 is discretized into 20 cells. For D_L unity Δt must be smaller than $x_0{}^2/800$. For a step concentration increase, the time for a component to diffuse from the inlet to the outlet, so that it reaches a concentration at $x=x_0$ about 90% of that at the inlet, takes on the order of $t=x_0{}^2/D_L$ (Bird et al., 1960). To simulate the evolution of the concentration up to this time, would take 800 time steps.

This demand can be rewritten to show the number of time steps needed to reach time $t=x_0{}^2/D_L$. From Equation (7) the number of time steps N_t becomes

$$N_t = 2*N_c^2 \tag{8}$$

N_c is the number of elements. For a discretization into 20 elements, something in the order of thousands of time steps, N_t, would be needed if there were no mineral reactions. This is no problem even for desktop computers. When there are mineral reactions the time to exhaust (or fill up) a mineral in an element depends on its amount and solubility. If the mineral concentration is 10 000 times larger than its solubility the element must be "flushed" 10 000 times to exhaust the mineral. The same applies if an incoming reactant is present in small concentration and the mineral in large concentration. The time step is still, however, determined by the previous criterion and the number of time steps has increased by 10 000 times to, on the order of $10\ 000*1000 = 10^7$. This takes significant time even on Megaflop computers. In several instances we have encountered still much larger ratios. That makes it impossible to simulate the problem directly.

Sometimes it is possible to use scaling of mineral concentrations to circumvent this problem. The chemistry of the system is not influenced by the amounts of mineral. Their concentration can be scaled down by, say a factor of 100. This reduces the number of time steps needed by the same ratio. This was done for a system with pyrite oxidation involving also secondary neutralization reactions (Cross et al., 1991). Time scaling can also be achieved by increasing the water flux and the diffusion or dispersion coefficient in the simulations.

An alternative approach, valid for the same conditions when components are predominantly present in minerals rather than in solution, has been used for catalyst decay, solidification of melts and in adsorption with extremely nonlinear isotherms. Its extension to multicomponent systems was proposed by Lichtner (1988).

Pseudo-stationary state approach

The idea is that if the rate of change in the system is dominated by mineral reactions, the local concentration change in the water contributes little to the total mass in the system. The mass will be dominated by the rate of change in mineral abundance. The principle can be illustrated by the following example. Consider a stream tube which contains a mineral with a very low solubility. Undersaturated water flows in and dissolves more and more mineral. The mineral dissolves, generating a dissolution front which slowly moves downstream. If dissolution is fast the front will be sharp. A lot of water must pass to move the front a small distance. In that very narrow region a great deal of mineral dissolves while the local concentration change in the water, at most, can be from zero to the solubility concentration. For an observer who moves with the front, the concentration profile does not change with time. A pseudo-stationary state is attained. If conditions around the front could be determined, only mass balances on both sides of the front would be needed to determine its rate and movement. In chromatography this is referred to as a constant pattern front. See the chapter on "Multicomponent Wave Propagation" in this book for further details.

The mass balances are dominated by the rate of change in the mineral composition and the first term can be neglected compared with the second in Equation (1). This implies that only the net change in mass due to diffusion and advection determines the mass balance for minerals. The errors introduced are at most the mass of the components that can be present in water, compared with the mass present in solids. For solids of low solubility the errors obviously are small.

The system of partial differential equations reduces to a system of ordinary differential equations in addition to the algebraic equations of mass action relations, thus becoming especially simple when diffusion and dispersion can be neglected. Then the equations reduce to first-order ordinary differential equations in Y_j, given by Equation (9).

$$\sum_{r=1}^{R} v_{jr}\frac{\partial X_r}{\partial t} = -V\frac{dY_j}{dx} \tag{9}$$

The two systems of ordinary differential equations (5) and (9) with the algebraic equations, (2) and (3) are then solved in tandem. This scheme was proposed by Lichtner (1988). If reaction rates become large, the system approaches the equilibrium case. This happens when the reaction zone becomes so narrow that it occupies only a fraction of the distance between two different mineral fronts. For large reaction rates, the rates do not influence the behaviour of the system any more. Lichtner (1991) has implemented the idea in the code MPATH for one-dimensional flow. In this method the correct mineral sequences are found directly. If the problem is solved by assuming equilibrium, other methods must be used to determine the correct mineral sequences as dicussed earlier (Novak et al., 1988).

Some related problems

A class of problems which have attracted much attention lately is the migration of radionuclides in the subsurface by advection and diffusion. In such cases sorption may significantly retard many of the nuclides relative to other species which are not sorbed. This is because most of the nuclides of interest are metals and have positive charges or are neutral in typical groundwaters. Their sorption properties are however very much influenced by their redox state, by the pH and by the presence of complexing agents such as carbonate. These entities are themselves influenced by the rock-water reactions as water moves through the system. A further complication is present in porous fractured rocks where advective flow takes place in the fractures and the dissolved constituents, by diffusion, move in and out of the porous rock matrix where they sorb. Two-step, two-dimensional codes like DYNAMIX and HYDROGEOCHEM could address these problems.

The radionuclide migration problems have been treated mainly with the simplifying assumption that sorption is instantaneous, with a linear isotherm, and is not affected by the changes in water chemistry along the pathway. Neretnieks et al. (1985) give an overview of some analytical solutions for such cases, as well as some numerical solution procedures.

In systems with strong changes in redox potential and pH, sorption is strongly influenced by the water chemistry. This has recently been demonstrated by Yeh and Tripathi (1991). Nevertheless these problems are still being simulated with the linear assumption because codes are lacking which can efficiently handle the problem for the radionuclide migration cases. These are more complex, including chain decay and dual porosity media.

SOME ANALYTICAL SOLUTIONS

There is a class of simple problems for which analytical solutions can be found when the pseudo-stationary state assumption is valid. One such case is when a component is transported by simultaneous advection and diffusion in a streamtube and irreversibly reacts with a solid. Another is the case when there is flow in the pores of a medium and all the reactions take place in the solids into which the reactant diffuses and reacts irreversibly and instantaneously with a stationary component. Cooper and Liberman (1970) give a solution for spherical particles. Similar solutions have been derived for transport in flat fractures and cylindrical conduits with diffusion into the surrounding solid in which the reaction takes place. These solutions have been used to describe transport in fractured rocks where the flow takes place in fracture planes or in networks of channels, (Romero et al., 1991).

The solution to Equation (1) is given by Equation (10) for simultaneous flow and diffusion (Zhu, 1988). X_o is the concentration of the stationary reacting mineral, C_o the concentration of the dissolved species at the inlet boundary, and f is the stoichiometric coefficient of the reaction. The expression is implicit in x_f. At short times the terms on the left hand side dominate the rate of front movement. At long times and distances the second term becomes negligible and the rate of front movement is constant and equal to $\dfrac{C_o}{f\,X_o}$ V. This is the reason why it is reasonable to neglect the diffusion term in Equation (1) to give Equation (9).

$$x_f + \frac{\theta D_L}{V}\left(e^{\left(\frac{Vx_f}{\theta D_L}\right)} - 1\right) = \frac{C_o}{f\,X_o}\,V\,t \qquad (10)$$

For transport by diffusion only, the rate of front movement is

$$v_f = \frac{dx_f}{dt} = \left(\frac{C_o D_L \theta}{X_o\ 2f}\ \frac{1}{t}\right)^{1/2} \qquad (11)$$

SOME SAMPLE CASES

The first case is the slow degradation of concrete by diffusing reactive species such as Mg^{2+}, SO_4^{2-}, Ca^{2+} and CO_3^{2-}. The interest in this problem stems from the need to estimate the lifetime of a silo repository for intermediate-level radioactive waste. A lifetime of thousands of years is needed and even slow diffusive processes are important. Detailed thermodynamic data were not available for many of the important reactions, but the problem had to be addressed. A combination of analytical solutions and numerical calculations were performed to gain insight into the problem.

The second case is also concerned with processes around a nuclear waste repository, this time for high level waste, where alpha radiolysis may generate oxidizing species which will change the naturally reducing waters in the near field of the repository to oxidizing conditions. This could enhance the dissolution and mobility of some radionuclides. A redox front might develop and slowly move away from the repository (Neretnieks, 1983). Time scales for these processes range up to hundreds of thousands to millions of years. It is clearly impossible to design experiments for such long time scales. However, mechanistically analogous natural phenomena that have existed for millennia can be found. One such natural analog is at an open pit uranium mine at Poços de Caldas in Brazil, where oxygen saturated intruding waters react with pyrite, causing a redox front to develop. Upstream of the front, the uranium is mobilized and precipitates once the water enters the reduced zone. At the slowly moving front uranium has accumulated over long time to form uranium ore.

Both cases have been studied with coupled codes and the numerical results have been verified at least partly by comparison with analytical solutions.

The concrete degradation problem was solved using the one-step CHEMTRN code. The redox front problem has been studied using the two-step code CHEQMATE. Codes based on the pseudo-stationary state method (Lichtner, 1991 and Nyman et al., 1991) have also been used.

Degradation of concrete

The concrete silo for the intermediate-level nuclear waste is 30 m in diameter and 50 m high, with 80 cm thick steel-reinforced concrete walls. The silo is located in a cavern in crystalline rock, 50 m below the bottom of the Baltic. The silo is surrounded by bentonite clay which contains large amounts of sodium in ion exchange positions and some calcium sulphate. Transport is diffusion-dominated because the hydraulic gradient over the silo and the permeabilities of the bentonite and concrete are very low. Concrete can in this context be considered to be a homogeneous porous medium. In the underground location below the water table the pores are water filled. Concrete consists of a mixture of ballast, assumed inert (stone) and a cement paste. The latter consists mainly of calcium-, aluminate- silicate- and iron-containing compounds. There is also a minor amount of KOH and NaOH. The pH of the pore water is 13.4 while the latter two constituents dissolve in pore water. When these are depleted the pH will be buffered by the solid $Ca(OH)_2$ and kept at about 12.4. When this compound is depleted the concrete becomes much weaker and the structures may collapse. Other reactions also influence the strength of the concrete.

The following reactions were identified as the most important for concrete degradation in the silo.

1   KOH and NaOH neutralization in the concrete by diffusive loss of OH^-

2   Loss of $Ca(OH)_2$ by dissolution and diffusion of Ca^{2+} and OH^-

3   Intrusion of CO_3^{2-}, which may react with Ca^{2+} to precipitate $CaCO_3^{2-}$

4   Intrusion of SO_4^{2-}, which may react with tri-calcium aluminate to precipitate ettringite, releasing OH^- in the process. Ettringite is a swelling compound which could potentially induce cracks

5   The SO_4^{2-} may also precipitate as $CaSO_4$

6   Mg^{2+} intrusion and precipitation as $Mg(OH)_2$, which also is a swelling compound.

In reaction 1 the hydroxides are dissolved in the pore water of the concrete and diffuse out rapidly. The pH of the system is then determined by the $Ca(OH)_2$ dissolution. No accurate thermodynamic data on ettringite are available. However, since the process is known to occur, a value of the solubility product was chosen so that ettringite precipitates in the model.

Fifteen aqueous species and eight minerals which possibly could form were considered at first. Preliminary equilibrium calculations and the low Mg content of the bentonite reduced the compounds to be accounted for to the following: Ca^{2+}, CO_3^{2-}, SO_4^{2-}, H^+, OH^- and X^{2+}, where X stands for the tri-calcium aluminate. The following minerals were included: $Ca(OH)_2$, $CaCO_3$, $CaSO_4$, XSO_4 (the ettringite), and $X(OH)_2$. In addition six aqueous complexes were considered: $CaOH^+$, $CaHCO_3^+$, $CaCO_3$, $CaSO_4$, HCO_3^-, and H_2CO_3. Other complexes were excluded based on preliminary equilibrium calculations. Thermodynamic data were taken from Stumm and Morgan (1981) except for XSO_4 and $X(OH)_2$. The solubility products for these were chosen such that sulphate intrusion ensures release of OH^- and precipitation of XSO_4. The rather arbitrary choice of constants does not influence the results of the rate of migration of the ettringite front. Other silica and alumina minerals were excluded because they were found not to influence the system appreciably during the time period when there is still $Ca(OH)_2$ present. The porosity of the concrete was taken to be 0.15 and the pore diffusivity to be $2 \cdot 10^{-10}$ m^2/s. The concrete contains 24.9 mol portlandite ($Ca(OH)_2$) per liter pore water and 0.9 mol/l

$X(OH)_2$. The boundary conditions are based on the composition of the bentonite in contact with the concrete and are; pH=8.5, $[SO_4^{2-}]$=10.4 mmol/l, $[Ca^{2+}]$=0.45 mmol/l, $[CO_3^{2-}]$=0.01 mmol/l. There is no X at the boundary.

CHEMTRN was used for the computations. The results are shown in Figures 1, 2 and 3. Figure 1 shows the concentration profile of the components into the concrete after 500 years. Figure 2 shows the concentration of the complexes and Figure 3 the amount of solid per volume of pore water.

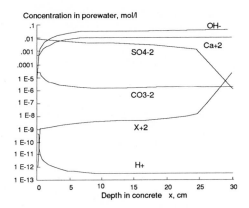

Figure 1. Concentration distribution of components in concrete after 500 years.

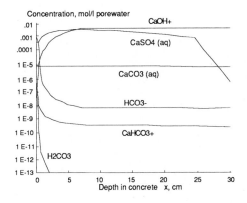

Figure 2. Concentration distribution of complexes in concrete after 500 years.

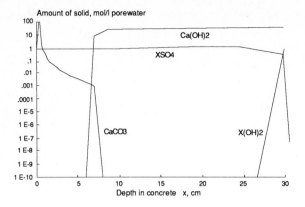

Figure 3. Concentration distribution of minerals in concrete after 500 years.

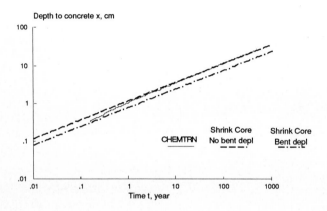

Figure 4. Penetration depth of ettringite front into the concrete over time.

Figure 3 shows the location of the fronts. The "ettringite" front has progressed to about 30 cm in 500 years. The $CaCO_3$ front has progressed to 7 cm and has a sharp peak very near the surface. Figure 1 shows the very sharp concentration changes near the surface and the sharp drop in SO_4^{2-} at the ettringite front. No other sulphate minerals are formed in the system.

The diffusion of sulphate and its reaction with $X(OH)_2$ can also be calculated independently using the simple shrinking core models, e.g., Equation (11). The results are shown in Figure 4. The agreement between the CHEMTRN results and the analytical solution is acceptable. The results for a case which includes the simultaneous depletion of the sulphate mineral gypsum ($CaSO_4$) in the bentonite in contact with the concrete are also shown in the figure. This could not be simulated with the CHEMTRN and an analytic solution was used (Zhu, 1988).

There were some cases which we would have liked to study in more detail but were unable to with the available techniques. These include the simultaneous diffusions of several components between the concrete and the bentonite.

Movement of redox and other fronts at Poços de Caldas

The open pit uranium mine at Poços de Caldas in Brazil was intensively studied in an international program (Chemical Geology, special issue in preparation). One of the aspects was the simulation of the development of the sharp front and the associated enrichment of uranium.

The front is estimated to have formed and moved over 70-90 million years with a rate of movement of some tens of meters per million years.

The bedrock is crystalline and consists mainly of phonolites and nepheline-syenites. The deeper portions of the rock contain about 2% by weight of pyrite (FeS_2) and are strongly reducing. The upper portions of the rock have become oxidized by infiltrating rainwater. Pyrite has become oxidized to ferric-oxy-hydroxides of varying degrees of crystallinity. There is a very sharp front separating the upper oxidized rock from the lower reduced rock. The boundary is not horizontal. There are many "fingers" of oxidized rock extending much further downward than the average depth of the front. These are associated with fractures and fracture zones which have higher permeability than the rock matrix itself. They slope at various angles, so that the front and fingers are not moving vertically but at some angle to the vertical. The front is rather ragged when looked at from a scale of 10-100 m.

The rock matrix has a porosity of 4-20%. The oxidized region has the higher porosity and the reduced region the lower. The hydraulic conductivity of the matrix is an order of magnitude lower than that of the overall rock including the fractures.

Uraninite nodules (UO_2) with typical sizes between 0.5-1 cm and nearly spherical in shape are found in many places just below the front in the reduced rock. It is thought that the uranium which is present as U(IV) oxides in the reduced rock is oxidized to U(VI), which is much more soluble. It dissolves and moves with flowing water or by diffusion back into reducing sections of the rock, where it is reduced to U(IV) by reactions with pyrite and subsequently precipitates.

Two modes of movement of the front and the associated uranium migration can be distinguished as shown in Figure 5. In the advective mode, the rock behaves like a porous medium with flow in the downward direction. The incoming water contains about 10 mg/l dissolved oxygen. This reacts with pyrite (2% by weight) and with the uranium (30 mg/kg rock). Dissolved uranium moves with water into the reduced part of the rock, where its solubility drops to very low values and it precipitates a little distance past the front. In the context of uranium ore formation and this called roll front movement (Walsh et al., 1984).

In the diffusive mode the following processes are thought to be active, as shown on the right side of Figure 5. Water moves along a fracture and oxygen diffuses from the fracture, through the rock matrix, to the redox front where it reacts with the pyrite. In the portion of the rock where the pyrite is oxidized the uranium is also oxidized and can be dissolved in the water. If uranium moves into the reduced rock by diffusion it would precipitate just beyond the boundary. There it would be reoxidized and redissolved as the front overruns it. In this case the uranium may diffuse towards the water in the fracture as well as further into the rock. A gradient can build up with high uranium concentration at the front and a lower concentration in the water in the fracture. This will permit the uranium to diffuse out to the moving water. The water flows downward along the fracture picking up uranium until the water passes the tip of the redox wedge where the uranium precipitates, producing large uranium oxide lumps.

One would expect to find most of the uranium at the lower tips of redox saw-teeth if a purely diffusive mode prevails. This is not the case generally. Often much uranium is found outside the sides of the wedges and also at the locations where the wedges meet higher up in the rock. One possible explanation is that the two modes are not as clear-cut as in the above description. If there is some advection through the rock matrix in the diffusion mode, flow may be superimposed on diffusion and uranium may be transported past the "diffusive" redox front by advection.

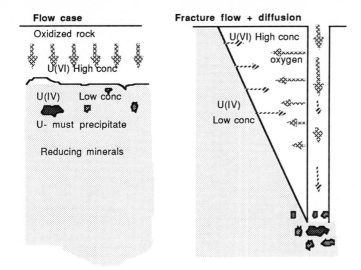

Figure 5.   Advective movement of the front (left) and diffusive mode (right).

The overall reaction formula for the pyrite oxidation is

$$FeS_2(s) + 15/4\ O_2(aq) + 7/2\ H_2O = Fe(OH)_3\ (s) + 4\ H^+ + 2\ SO_4^{2-}$$

This is the reaction which determines the overall movement of the redox front. There are secondary reactions induced by this reaction. One is the neutralization of the sulfuric acid formed by reactions with feldspars, the dissolution and re-precipitation of uranium at the redox front and several others which will be described later.

The movement of the redox front is determined by the mass of the intruding oxygen and by the mass of the pyrite. This reaction is fast. It is thus essentially a single reaction system and can be computed by simple analytical solutions. Zhu (1988) has done this for the one-dimensional advective and diffusive transport and Romero et al. (1991) have modelled the two-dimensional development of the redox wedges including competition with erosion.

Several groups have modelled the advective case using coupled codes. Zhu (1988) attempted using the CHEMTRN family of codes CHEMTRN, THCC and CHMTRNS. This was not successful because of numerical difficulties with the extreme changes in oxygen concentration at the redox front. Several attempts to change the primary variables by choosing other components were unsuccessful. Lichtner (1991) used MPATH which is based on the pseudo-stationary state approach but had to introduce an automatic procedure by which to switch variables between upstream and downstream of the various fronts. Lichtner´s results, obtained with the EQ3/6 database differ somewhat from those of Cross et al. (1991) who used CHEQMATE. In these simulations data were taken from the HATCHES database (Cross and Ewart, 1991).

In the latter calculations the following basic data were used. Initially the rock contains 30 ppm uraninite and 2% pyrite, with the remainder consisting of K-feldspar and kaolinite. The incoming rainwater has a pH 5.1 due to dissolved CO_2 and 10 ppm of dissolved oxygen (saturation with air). The water flux is assumed to be 0.1 m^3/m^2·annum. The pore diffusivity is $1.2 \cdot 10^{-10}$ m^2/s. The dispersion length is 0.1 m. In the computations 20 cells of equal size were used. The amount of minerals were scaled down by a factor of 100. This increases the rate at which the fronts move through the system by the same factor without changing the chemistry of the problem. Computation times are decreased accordingly.

Figure 6 shows the mineral composition profile after 38 000 years. The redox front has progressed to about 0.8 m. The neutralization of the protons by reaction with K-feldspar takes place forming a hydrolysis front where the feldspar reacts to kaolinite and possibly some silica mineral. Uranium oxide also accumulates at these closely spaced fronts. A second hydrolysis front arises from the reaction of feldspar with intruding H_2CO_3 at 0.25 m.

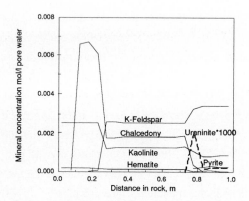

Figure 6. Mineral profiles obtained by CHEQMATE simulations after 38 000 years.

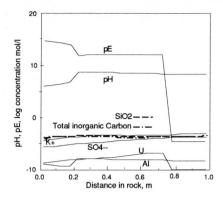

Figure 7. Concentration profiles of components after 38 000 years.

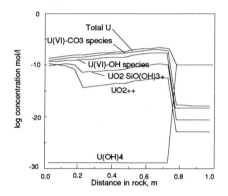

Figure 8. Concentration profiles of selected uranium species after 38 000 years.

Figures 7 and 8 show the concentrations of the components and some selected uranium species in solution.

At the redox front oxygen fugacity drops very sharply. Uranium upstream of the front is mainly U(VI) and U(V), whereas downstream it is in the tetravalent state U(IV) and thus has very low solubility. The mineral profiles obtained in the simulation were in good agreement with observations at the mine, except for the predicted silica, which is not present. With slightly different thermodynamic data for the feldspar and kaolinite, no silica formation and a different pH downstream of the front were predicted. The predicted water composition was otherwise in good agreement with that observed in boreholes at the mine (Cross et al., 1991).

In the simulations, it was noted that pseudo-stationary states develop, after the redox front has passed a few cells (reactors). This was expected and is one of the indicators which show that the mineral scaling is permissible.

It was not possible to simulate the development of all other expected fronts which develop nearer to the surface because these are much slower than the redox front. This problem can be circumvented by allowing the fast fronts to move out of the system and then rescaling the system by either again decreasing the mineral amounts or by simultaneously increasing the flowrate of water and the dispersivity.

The pseudo-stationary state method circumvents this problem. Lichtner (1991) successfully simulated the development of many different simultaneously moving fronts over long times. A possible drawback is that the present pseudo-stationary state codes cannot handle diffusion. This may be an important mechanism at very sharp and narrow fronts as pointed out by Lichtner (1991).

Simulations were also performed with CHEQMATE, in which diffusion is the only mechanism to introduce oxygen into the porous rock, as would be the case at a point in the fracture where water flows and oxygen diffuses into the rock and reacts with the pyrite forming the redox front, as shown in Figure 5. As the front moves further into the rock the uranium which was present in the rock dissolves and diffuses towards the fracture where it is swept downstream by the incoming uranium-lean water.

A theoretical analysis showed that no uranium will precipitate just inside the redox front. In this case, it will all diffuse out to the water in the fracture. The numerical simulations, however, showed that uranium should accumulate at the front. A further mathematical analysis of the condition at the front showed that the use of a discrete simulator such as the finite difference method used in CHEQMATE, will show an accumulation which depends directly on the size of the finite cell. The smaller the cell the less will accumulate. This effect must be kept in mind when using finite difference or finite element codes.

DISCUSSION AND CONCLUSIONS

The basic problems of simulating solids-water interaction have been approached with several different methods in recent years. Analytical tools have been developed for single reactions for both advective and diffusive transport with both slow and fast reactions. These methods have long been used in chemical engineering applications of catalysts, ion exchangers and sorbents. They still have a very important role to play in that they can be used to verify numerical codes. They have been successfully used in a class of two-dimensional problems found in fractured media, where such problems are often difficult to simulate with numerical techniques.

The one-step codes may have advantages because they use an implicit problem formulation, which should allow larger time steps and give higher accuracy than two-step codes. The number of equations that must be solved simultaneously is larger in one-step than in two-step codes. This has been one of the arguments for using two-step codes, especially for problems in two dimensions, in which the number of equations would otherwise be prohibitively large. The two-step codes have demonstrated their capability of handling large two- dimensional simulations.

Both the one-step and especially the two-step codes have limitations on the time steps which can be used. This becomes a serious problem in systems where the mass of the components is mainly in the solids. It has been found that some such problems can be scaled in time, to prevent computing times from becoming prohibitively large. Such problems have also been found to exhibit the properties of pseudo-stationary states. This has been used to speed up the computations for such systems.

There are several interesting problems which are not well handled by the available codes and techniques. These include transport in dual-porosity fractured media, problems with very different time scales and with bio-mediated reactions.

The thermodynamic databases are well developed for many substances but are lacking for other. It has also been found that some problems are very sensitive to seemingly small differences in thermodynamic data. Kinetic data are scarce. The problems can sometimes, but not always, be circumvented if the time spans of interest are long, compared with the time constants of the reactions.

NOTATION

a_r	Specific surface	m^2/m^3
C_i	Concentration of complex i	mol/m^3 water
C_j	Concentration of component j	mol/m^3 water
D_L	Dispersion coefficient	m^2/s
D_p	Pore diffusion coefficient	m^2/s
I	Number of complexes and components	-
J	Number of components	-
k_r	Mineral reaction rate for mineral r	$mol/m^2 \cdot s$
K_i	Equilibrium constant for complex i	-
K_r	Equilibrium constant for mineral r	-
$N_{complexes}$	Number of complexes	-
N_c	Number of space discretization steps (cells)	-
N_t	Number of time steps	-
R	Number of minerals	-
t	Time	s
T_j	Total concentration of component j	mol/m^3 water
u	Water velocity	m/s
v_f	Front velocity	m/s
V	Water flux	$m^3/m^2 \cdot s$
x	Coordinate	m
X	Concentration of mineral	mol/m^3 water
Y	Total concentration of dissolved species	mol/m^3 water
γ	Activity coefficient	-
θ	Porosity	-
ν	Stoichiometric coefficient	-

REFERENCES

Bird RB, Stewart WE, Lightfoot EN (1960) Transport Phenomena, Wiley

Bryant SL, Schechter RS, Lake LW (1987) Mineral sequences in precipitation/dissolution waves. AIChE J **33**:1271-1287

Carnahan CL (1986) Simulation of uranium transport with variable temperature and oxidation potential: The computer program THCC. Rep. 21639, Lawrence Berkeley Laboratory, Berkeley, California

Cederberg GA (1985) TRANQL: A groundwater mass–transport and equilibrium chemistry model for multicomponent systems. Ph.D. Dissertation Stanford University, Stanford

Cooper RS, Liberman DA (1970) Fixed-bed adsorption kinetics with pore diffusion control. Ind Eng Chem Fundam **9**: 620

Cross JE, Ewart FT (1991) HATCHES, A thermodynamic database and management system. Radiochimica Acta **52/53**:421-42227

Cross JE, Haworth A, Neretnieks I, Sharland SM, Tweed CJ (1991) Modelling of Redox Front and Uranium Movement in a Uranium Mine at Poços de Caldas. Radiochimica Acta **52/53**: 445-451

Dria MA, Bryant SL, Schechter RS, Lake LW (1987) Interacting precipitation waves: The movement of Inorganic Contaminants in Groundwater. Water Resources Res **23**: 2076-2090

Haworth A, Sharland SM, Tasker PW, Tweed CJ (1988) A guide to the coupled chemical equilibria and transport code CHEQMATE. Harwell Laboratory Report NSS-R113

Helfferich FG (1989) The theory of precipitation/dissolution waves. AIChE J **35**:75-87

Lapidus L, Pinder GF (1982) Numerical solution of partial differential equations in science and engineering. John Wiley, NY

Lichtner PC, Oelkers EK, Helgeson HC (1986) Interdiffusion with multiple precipitation/dissolution reactions: transient model and the steady-state limit. Geochim Cosmochim Acta **50**:1951-1966

Lichtner PC (1988) The quasi-stationary state approximation to coupled mass transport and fluid rock interactions in a porous medium. Geochim Cosmochim Acta **52**:143-165

Lichtner PC (1991) Redox front geochemistry and weathering: Theory with application to the Osamu Utsumi mine, Poços de Caldas, Brazil. Chemical Geology, Special issue on Poços de Caldas , in print

Liu CW, Narasimhan TN (1989) Redox controlled multiple species reactive chemical transport 1, Model development. Water Resources Research **25**:869-882

Liu CW, Narasimhan TN (1989) Redox controlled multiple species reactive chemical transport 2, Verification and application. Water Resources Research **25**:883-910

Miller CW (1983) CHEMTRN user's manual. Lawrence Berkeley Laboratory report, LBL-16152

Neretnieks I (1983) The movement of a redox front downstream from a repository for nuclear waste. Nuclear Technology **62**

Neretnieks I, Abelin H, Birgersson L, Moreno L, Rasmuson A, Skagius K (1985) Chemical Transport in Fractured Media. Proceedings of the "Fundamentals of Transport Phenomena in Porous Media". NATO/ASI Symposium:475-550, Delaware July 14-23

Neretnieks I, Arve S, Moreno L, Rasmuson A, Zhu M (Dec. 1987) Degradation of concrete and transport of radionuclides from SFR-Repository for low- and intermediate level waste. SKB Progress report SFR 87-11

Noorishad J, Carnahan CL, Benson LV (1987) Development of the non-equilibrium reactive chemical transport code CHMTRNS. Lawrence Berkeley Laboratory, University of California, LBL-22361

Novak CF, Schechter RS, Lake LW (1988) Rule-based mineral sequences in geochemical flow processes. AIChE J **34**:1607-1614

Nyman C, Ozolins V, Moreno L, Neretnieks I (1991) Development of a model for handling the movement of redox fronts and other sharp reaction fronts. Paper presented at the MRS meeting in Strasbourg, Nov 4-7, Proceedings in print

Parkhurst DL, Thorstenson DC, Plummer (1985) PHREEQE - A computer program for geochemical calculations. U.S. Geological Survey, Water Resources Investigation 80-96

Rhee HK, Aris R, Amundson NR (1970) On the theory of multicomponent chromatography. Phil Trans Roy Soc London **267**:419-455

Rhee HK, Amundson NR (1970) An analysis of an adsorption column: Part 1. Theoretical development. The Chem Eng J **1**:241-254

Romero L, Neretnieks I, Moreno L (1991) Release of radionuclides from the near field by various pathways - The influence by the sorption properties of materials in the near field. Paper presented at the MRS meeting in Strasbourg Nov 4-7, Proceedings in print

Steefel IS, Cappellen PV (1990) A new kinetic approach to modeling water-rock interaction: The role of nucleation, precursors and Ostwald ripening. Geochim Cosmochim Acta **54**: 2657-2677

Stumm W, Morgan JJ (1981) Aquatic chemistry 2nd Ed Wiley and sons

Schechter RS, Bryant SL, Lake LW (1987) Isotherm free chromatography: propagation of precipitation/dissolution waves. Chem Eng Comm **58**:353-376

Walsh MP, Bryant SL, Schechter RS, Lake LW (1984) Precipitation and dissolution of solids attending flow through porous media. AIChE J **30**(2):317

Westall J (June 1979a) MICROQL I, A chemical equilibrium program in BASIC. EAWAG, Swiss Federal Institute of Technology, Dübendorf

Wolery TJ (1983) EQ3NR, A computer program for geochemical aqueous speciation-solubility calculations: Users guide and documentation. UCRL-53414, Lawrence Livermore Laboratory, Livermore

Yeh G-T, Tripathi VS (1989) A critical evaluation of recent developments of hydrogeochemical transport models of reactive multicomponent components. Water Resources Research **25**: 93-108

Yeh G-T, Tripathi VS (1991) A model for simulating transport of reactive multispecies components: Model development and demonstration. Water Resources Research **27**: 3075-3094

Zhu Ming (1988) Some aspects of modelling of the migration of chemical species in groundwater system. Licentiate thesis Dept Chem Eng Royal Institute of Technology, Stockholm, Sweden

Impact of Small–Scale Spatial Variability upon the Transport of Sorbing Pollutants

Albert J. Valocchi and Hernán A. M. Quinodoz
Department of Civil Engineering
University of Illinois
205 N. Mathews Ave.
Urbana, IL 61801
USA

It is now widely recognized that groundwater aquifers exhibit significant three–dimensional, small–scale variability in their hydraulic properties and that this variability controls the migration and dispersion of contaminants at the field scale. Quantitative study of the impact of small–scale variability upon field–scale transport has been a central theme of groundwater research in recent years; this research has been motivated by a host of important questions. How can properties measured on small samples in the laboratory be extrapolated to larger scales? Are fundamental constitutive relations derived from studies at the laboratory scale valid at field scales? How can we quantify the inherent uncertainty in our information on spatially varying soil properties? What is the effect of this uncertainty upon the reliability of model predictions?

In order to address these and related questions, many investigators have advocated the use of stochastic methods. There is much available overview literature on this topic, including Dagan (1986, 1989) and Gelhar (1986). The stochastic approach provides a rational explanation for the observation that field–scale dispersivities not only tend to be much larger than their laboratory–scale counterparts, but that they also tend to increase with travel distance. Moreover, predictions from stochastic theory are in reasonable agreement with data from a limited number of natural gradient field experiments with nonreactive solutes (Freyberg, 1986; LeBlanc et al., 1991).

Most of the available literature on spatial variability considers only nonreactive solutes. The purpose of this chapter is to examine the impact of small–scale variability in soil hydraulic and chemical properties upon the transport of reactive solutes. For the sake of brevity, we will focus upon the transport of a single solute undergoing a kinetically–controlled reversible sorption reaction. Nonequilibrium sorption has been documented in several field studies (Semprini et al., 1990; Wood et al., 1990; Bahr, 1989) General results from the literature will be reviewed, but we will emphasize original results we have reported in Valocchi (1989) and Quinodoz and Valocchi (1992). As a result of kinetic sorption the plume–scale velocity will decrease and the dispersion will increase. The interaction between the two dispersion processes——one due to spatial variability of the soil hy-

NATO ASI Series, Vol. G 32
Migration and Fate of Pollutants in Soils and Subsoils
Edited by D. Petruzzelli and F. G. Helfferich
© Springer-Verlag Berlin Heidelberg 1993

draulic properties, and the other due to the kinetics of the reaction−−is the primary focus of this chapter. Spatial variability of soil hydraulic properties can also have a significant effect upon solutes undergoing biodegradation reactions; the reader is referred to Mac-Quarrie and Sudicky (1990) and Schaefer and Kinzelbach (1992) for further information.

MODELS OF SPATIAL VARIABILITY

There have been several recent field experiments which have documented the existence and implications of the spatial variability of the hydraulic conductivity (K). Results for the Borden (Sudicky, 1986) and Cape Cod (Hess, 1989) sites indicate that K measured for small soil samples varies in an erratic or random fashion. Hence, most theoretical work (see, e.g., Gelhar, 1986; Dagan, 1989; Neuman and Zhang, 1990) has assumed that K is a realization of a second–order stationary, lognormally distributed random space function (random field). A two dimensional example of such a random field is shown in Figure 1 which is a contour plot of $Y=\log(K)$ for an exponential, isotropic spatial covariance function. The original field was generated on a 1000 x 500 grid using the computer code TURN developed by Tompson et al. (1989) and has a correlation length $\lambda=5$ grid units. For presentation purposes, the original field was smoothed with a running average; the smoothed field is shown in Figure 1. This random field model will be adopted in this chapter, although when necessary we will resort to the conceptually simpler model of an ideal stratified aquifer in which K is only a function of the vertical location.

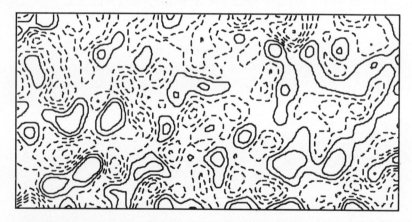

Figure 1. Contour plot of realization of two−dimensional random log hydraulic conductivity field (smoothed). Solid contours: higher than mean; dashed contours: lower than mean.

One key feature of the commonly adopted stationary random field model is that the $\log(K)$ variance, σ^2, and correlation length, λ, are fixed and finite. Thus, as a plume travels and increases in size due to dispersion, it will eventually attain a size large enough so that the entire statistical range of K variability has been sampled. In contrast some investigators have proposed models in which there are "nested" or "evolving" heterogeneities, so that a plume is constantly encountering larger scales of K variability as it increases in size. One such case, discussed by Neuman (1990), assumes a fractal model for K.

Although there have been relatively few field investigations of the spatial variability of reaction parameters, it is reasonable to expect that soil chemical properties would also exhibit spatial variability. At the Borden field site Robin et al. (1991) present evidence that equilibrium distribution coefficients (K_d) exhibit spatial variability which can be described mathematically as a stationary random field, but there does not appear to be any significant correlation between K and K_d.

GOVERNING FLOW AND TRANSPORT EQUATIONS

Porous media flow and solute transport has been examined extensively at the scale of homogeneous soil columns. These examinations are typically done with artificial porous media or natural material that has been sieved and repacked into laboratory columns or reactors. Hence, we will refer to this as the laboratory or "Darcy" scale. The appropriate governing flow and transport equations at this scale are contained in many standard textbooks (Bear, 1979; deMarsily, 1986). For the case of incompressible flow of a uniform–density fluid, Darcy's Law can be expressed as

$$v = -\frac{1}{n} K \nabla h \tag{1}$$

where v is the pore water velocity vector (also called the average linear velocity), n is the porosity, K is the hydraulic conductivity tensor, and h is the piezometric head. For the special case of steady state flow without any fluid sources or sinks, conservation of mass requires that $\nabla \cdot n v = 0$, which, after substituting (1) becomes

$$\nabla \cdot (K \nabla h) = 0 \tag{2}$$

Given appropriate boundary conditions, we can solve (1) and (2) to determine the spatially varying velocity field. For arbitrary heterogeneity in K numerical techniques are usually needed. However, as will be presented further below, an approximate analytical stochastic solution is possible in certain cases for which K is assumed to be a realization of a random space function. Of course, for the simplified model of a stratified aquifer, an exact solution of (1) and (2) is trivial and the result for $v_x(z)$ can be obtained directly.

In this chapter we will only consider the transport of a single solute undergoing linear reversible adsorption reactions. Furthermore, since most observations indicate that the porosity exhibits much less spatial variability than the hydraulic conductivity, we will also assume that n is constant. In this case, the solute transport equation can be written as

$$\frac{\partial c}{\partial t} + \frac{\partial s}{\partial t} = \nabla \cdot (D \nabla c) - \nabla \cdot (v c) \tag{3}$$

where c and s are the solute concentration (mass solute/volume fluid) in the dissolved and sorbed phase, respectively, and D is the hydrodynamic dispersion tensor. Since (3) is written at the Darcy scale, the hydrodynamic dispersion tensor accounts for mixing due to molecular diffusion and pore–scale spatial variability of the fluid velocity. This can be expressed as

$$D = D_m \, T + D^M \tag{4}$$

where T is the tortuosity tensor, D_m is the molecular diffusivity, and D^M is the mechanical dispersion tensor. D^M is related to the pore water velocity v according to the classical model described in the texts by Bear (1979), de Marsily (1986) and others. The principal directions of D^M are parallel and transverse to a streamline. The parallel (longitudinal) and transverse components are denoted $D_L{}^M$ and $D_T{}^M$, respectively. For homogeneous and isotropic porous media, these terms are given as $D_L{}^M = \alpha_L \, |v|$ and $D_T{}^M = \alpha_T \, |v|$, where α_L and α_T are the longitudinal and transverse dispersivities of the porous medium.

In addition to the transport equation (3), an equation describing the rate of adsorption, $\partial s / \partial t$, is required. Following previous work by Quinodoz and Valocchi (1992), we use the simple linear reversible rate equation

$$\frac{\partial s}{\partial t} = k_f \, c - k_r \, s = k_r \, (K_d \, c - s) \tag{5}$$

where k_f and k_r are the forward and reverse rate coefficients, respectively ($[\text{time}]^{-1}$), and $K_d = k_f / k_r$ is the equilibrium distribution coefficient. Note that K_d is dimensionless since c and s in (3) are expressed in the same units. In many other studies, c is expressed in units of mass per volume of solution and s in units of mass per mass of solids, so that K_d would have units of volume of solution per mass of solids. Several investigators (Nkedi–Kizza et al., 1984; van Genuchten, 1985; Valocchi, 1990a; Sardin et al., 1991] have demonstrated that equation (5) can be an accurate approximation to more complicated diffusion–based mass transfer models.

SPATIAL MOMENT ANALYSIS

The use of spatial moment analysis has a distinguished tradition in the fluid mechanics and chemical engineering literature. The early pioneering studies by Taylor (1953) and Aris (1956) focused upon shear flow dispersion in conduits, while later extensions by Brenner and co–workers (Brenner, 1980; Brenner and Adler, 1982; Shapiro and Brenner, 1988) encompassed arbitrary flow fields and chemically reactive flows. The motivation behind moment analysis is that it is usually very difficult to obtain the complete solution of the transport problem; in the context of this chapter, complete solution of equations (3) and (5) for $c(x,y,z,t)$ and $s(x,y,z,t)$ is only feasible for very simplified flow conditions. Spatial moment analysis seeks a solution for the lower–order spatial moments of the dependent variables. Such a solution is often relatively easy to obtain and, moreover, yields substantial physical insight into the transport process. In particular, spatial moment analysis has proven to be an invaluable tool for examining the effect of small–scale spatial variability upon large–scale transport behavior. Hence, we will employ these techniques extensively throughout this chapter.

As further illustration, consider the two dimensional dissolved–phase contaminant plume shown in Figure 2. The mean flow direction is from left to right, in the positive x direction. This figure was generated by tracking particles along streamlines generated by numerical solution of the groundwater flow equation in the heterogeneous K field shown in Figure 1. At any time, it is possible to define spatial moments of the plume by

$$M_{ijk}(t) = \int x^i y^j z^k \, c(x,y,z,t) \, dx \, dy \, dz \qquad (6)$$

It is clear that the lower–order spatial moments provide succinct, physically meaningful measures of the plume distribution. For example, M_{000} measures the total aqueous–phase mass in the plume, M_{100} measures the center of mass of the plume in the flow direction, and M_{200} provides a measure of the spread of the plume in the flow direction. As discussed extensively by several investigators (e.g., Freyberg, 1986; Guven et al, 1984; Dagan, 1990), the first and second spatial moments contain key information about the large–scale velocity and dispersion. In this chapter, we will use the following specific aqueous–phase moment definitions:

$$M_0(t) = M_{000}(t) = \int c(x) \, dx \qquad (7)$$

$$\overline{M}(t) = \frac{1}{M_0} \int c(x,t) \, x \, dx \qquad (8)$$

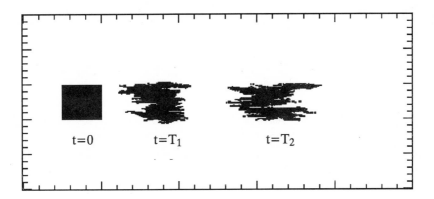

Figure 2. General illustration of transport of plume in randomly heterogeneous K field.

$$S(t) = \frac{1}{M_0} \int c(x,t) \, (x - \overline{M}(t)) \, (x - \overline{M}(t))^T \, dx \tag{9}$$

In (7) through (9), the Cartesian coordinate location is represented by the vector $x = (x,y,z)^T$, where the superscript T signifies the vector transpose. Equation (7) defines the total aqueous–phase mass in the plume, the vector $\overline{M}$ defined in (8) is the center of mass of the solute plume, and the tensor S defined by (9) is the variance of the solute plume. It is also possible to define equivalent moments of the sorbed–phase concentration $s(x, t)$, although they will not be utilized in this chapter. Related to (8) and (9) are definitions of the effective velocity of the plume, $V^e(t)$, and the effective dispersion tensor, $D^e(t)$:

$$V^e(t) = \frac{d\,\overline{M}}{dt} \qquad\qquad D^e(t) = \frac{1}{2} \frac{d\,S}{dt} \tag{10}$$

Another reason why spatial moment analysis is so popular is due to its natural connection with the Lagrangian, or particle–based, description of solute transport. As written above in (6), the spatial moment is simply an average of the concentration field; the concentration field is the solution of equations (3) and (5), which are written in the standard Eulerian framework. However, as described by Taylor (1921), Dagan (1989), Kinzelbach (1987), Tompson and Dougherty (1988), and Dagan and Neuman (1991), solute transport can also be examined from a Lagrangian, or particle–based, perspective. This entails representation of a solute plume by a large number of indivisible "particles" of finite mass; these particles are then transported according to some equations, which may include both deterministic and random components. Spatial moments of the particle displacements can then be interpreted identically to the spatial moments of the solute plume defined by (6). The particle tracking approach is described in greater detail in the next section.

Although spatial moment analysis will be used exclusively in this chapter, it should be noted that temporal moment analysis of solute breakthrough curves is a complementary and powerful tool for studying the influence of various processes and sources of uncertainty. Valocchi (1990a) and Goltz and Roberts (1987) discuss applications to nonequilibrium sorption reactions for homogeneous soils. Dagan and Nguyen (1989) and Cvetkovic and Shapiro (1990) present applications to transport in randomly heterogeneous aquifers.

PARTICLE TRACKING (LAGRANGIAN) APPROACH

In this section we summarize the main features of the particle tracking approach to solute transport in aquifers. The first subsection presents results of Kinzelbach (1987) and Tompson and Dougherty (1988) for the simple case of nonreactive solutes and deterministic velocity fields. The second subsection presents results of Dagan (1989) for nonreactive transport with the hydraulic conductivity field taken as a sample realization of a second–order stationary random field. The third subsection, based upon recent results of Quinodoz and Valocchi (1992), presents a particle–based model of the adsorption–desorption reaction (5).

Deterministic Velocity Field

A general treatment of particle tracking methods as an approximate numerical technique for solving the advection–dispersion equation has been provided by Tompson and Dougherty (1988). Here we consider nonreactive transport and assume that the aqueous–phase concentration $c(x,y,z,t)$ can be represented by a sum of discrete particles each having a finite mass; the position of any arbitrary particle at time t is denoted by the vector $X(t)$. Note that in the Cartesian coordinate system adopted here $X(t) = [X(t), Y(t), Z(t)]^{\mathrm{T}}$. The use of upper case for $X(t)$ signifies that the coordinates are associated with the position of a particle; this can be interpreted as a Lagrangian coordinate. In contrast, the lower case vector $x = (x,y,z)^{\mathrm{T}}$ signifies a fixed position in the Cartesian coordinate system; that is, x is the Eulerian coordinate. A general equation to move the particle over one time step Δt is given by Tompson and Dougherty (1988) as

$$X(t + \Delta t) = X(t) + A(X(t), t)\, \Delta t + B(X(t), t) \cdot Z \sqrt{(\Delta t)} \tag{11}$$

where A is a deterministic forcing vector, B is a deterministic square scaling tensor, and the vector Z contains three independent random numbers with zero mean and unit variance.

If a finite number of particles are moved according to (11), the density of particles at time t lying in some small volume centered at x is a measure of the concentration, $c(x, t)$. Tompson and Dougherty (1988) and Kinzelbach (1987) show that selection of

$$A(X,t) \equiv v(X(t)) + \nabla \cdot D(v(X(t))) \tag{12}$$

$$B(X,t) \cdot B^{T}(X,t) \equiv 2\, D(v(X(t))) \tag{13}$$

guarantees that the concentration field is a solution of the nonreactive advection–dispersion equation (i.e., equation (3) without the source/sink term due to sorption reactions). The random step in (11) is clearly associated with dispersion. The correction term $\nabla \bullet D$ becomes important for cases of highly nonuniform flow, as demonstrated by Kinzelbach (1987). If the dispersion term is neglected, the deterministic step in (11) is associated with advection.

We remark again that there is a direct relationship between spatial moments of the solute plume defined by (7) through (9) and the averages of the positions $X(t)$ of all the particles. If, at time t there are N_p aqueous–phase particles each having mass m, then Tompson and Gelhar (1990) show that

$$M_0 \approx mN_p \tag{14}$$

$$\overline{M}(t) \approx \frac{1}{N_p} \sum_p X_p(t) \tag{15}$$

$$S(t) \approx \frac{1}{N_p} \sum_p X_p(t)\, X_p(t)^T - \overline{M}(t)\, \overline{M}(t)^T \tag{16}$$

Spatially Random Velocity Field

As described earlier, one of the most prominent developments in groundwater science over the past decade has been the recognition of the important influence of small–scale spatial variability of soil permeability upon field–scale transport and mixing processes. An important consequence of this variability is that predictions of field–scale solute transport will be uncertain because it is not possible to make enough measurements to map deterministically the spatial distribution of hydraulic conductivity (K). One of the most commonly used quantitative approaches to this problem is based upon assuming that the K field is a realization of a second–order stationary spatial stochastic process. This approach has spawned an entire new sub–area of stochastic groundwater hydrology. We will only present a brief summary of the stochastic approach here, and we refer the reader to the excellent reviews by Gelhar (1986), Dagan (1986), Dagan (1989), ASCE (1990), and Sudicky and Huyakorn (1991) for details.

We will use the schematic depiction of transport in a spatially random K field shown in Figure 3. Using a random field generator such as the Turning Bands Method (Tompson et al., 1989; Mantoglou and Wilson, 1982), we can generate a large number of equally likely realizations of the spatially variable K field; it is also possible to "condition" these realizations upon available point measurements of K (ASCE, 1990). For each realization, the groundwater flow and solute transport equations can be solved to determine the head, velocity, and concentration fields. This is the procedure demonstrated schematically in Figure 3, which shows the contaminant plume at time t resulting from an instantaneous input over a finite–sized source area at time 0. The flow is assumed to be steady, and the ensemble–averaged flow field is one–dimensional. It is possible to perform averages over the ensemble; ensemble–averaged quantities will be denoted by angle brackets $<>$. Quantities of interest include the average head field, $<h(x)>$, the average velocity field, $<v(x)>$, and the average concentration field, $<c(x, t)>$. We are also interested in the variability among the individual realizations, or, equivalently, the difference between the ensemble–average results and the individual realization results. This variability is measured by the ensemble variance or covariance. For example, the ensemble covariance of the head field is defined by

$$C_h\ (x_1, x_2)\ =\ <\ [h(x_1) - < h(x_1) >]\ [h(x_2) - < h(x_2) >]\ > \tag{17}$$

Other important covariances include $C_{v_i v_j}(x_1, x_2)$, where v_i refers to a particular component of the pore water velocity vector $v(x)$, and $C_c(x_1, x_2)$, where $c(x, t)$ is the concentration field at time t.

As discussed in the previous section, we recognize the difficulty in solving directly for the concentration plume, so we are content to determine only its lower–order spatial moments. Consequently, we can consider applying the definition (6) to compute the spatial moments of the solute plume in each realization of Figure 3, as well as the spatial moments

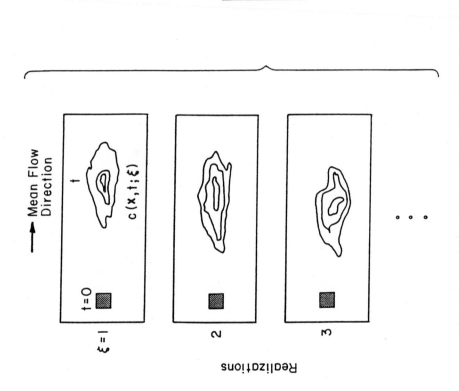

Figure 3. Schematic illustration of stochastic approach to transport in heterogeneous aquifers.

of the ensemble–averaged plume. Then it becomes possible to compute an ensemble variance of the spatial moments. As discussed by Dagan (1990), these variances are typically smaller than those of the point values of the concentration field. In other words, prediction of global features of the solute plume (as measured by spatial moments) is much less prone to error than prediction of point concentration values.

Most of the theoretical studies of solute transport in randomly heterogeneous aquifers has focused upon determination of the ensemble–averaged concentration field, $<c(x,t)>$, or, more commonly, upon its first two spatial moments. Recent work by Dagan (1990) and Vomvoris and Gelhar (1990) has addressed the important and difficult problem of concentration uncertainty. Without quantifying the uncertainty, it is not possible to ascertain the applicability of more easily obtained ensemble–average results to a single realization. (Keep in mind that in the stochastic approach the actual field site represents one particular realization of the K field.) Recent numerical studies by Graham and McLaughlin (1989) and Rubin (1991) have examined concentration uncertainty.

Under some conditions there is very little difference between individual realizations and the ensemble–averaged results; in this case we say that ergodic conditions hold, or that ergodicity can be assumed. Dagan (1990, 1991) and Quinodoz and Valocchi (1990) show that this normally requires the dimensions of the solute plume to be much greater than the length scale characterizing the velocity field variability. Referring to Figure 3, ergodicity requires that the transverse width (in the z–direction) of the input zone be much greater than the spatial correlation scale of the K field random process. In this case the solute plume in each individual realization encounters "identical" (in a statistical sense) velocity variability. Two–dimensional analyses reported by Dagan (1991) and Quinodoz and Valocchi (1990) suggest that the input zone width may need to be as much as 100 times greater than the log (K) correlation scale. Black and Freyberg (1987) have also examined rigorously the requirements for ergodicity in two–dimensional problems involving transport in perfectly stratified aquifers. The case of non–ergodic transport is beyond the scope of our work, and in this chapter we focus upon ensemble–averaged results for the problem of sorbing solute transport. In particular, we will determine the zeroth, first, and second spatial moments of $<c(x, t)>$.

We next summarize the key results from Dagan's work on nonreactive transport in randomly heterogeneous porous media. The book by Dagan (1989) succinctly presents his methodology and results. The hydraulic conductivity is assumed to be a lognormally distributed, second–order stationary random field having an exponential spatial correlation structure. The first step in Dagan's approach is to solve the groundwater flow problem given by (1) and (2) in order to determine the statistical properties of the head and velocity fields. The major results of this step which are used subsequently in the solute transport step are determination of $<v(x)>$ and $C_{v_i v_j}(x_1, x_2)$. Dagan obtains these results through a linearization and small perturbation procedure; therefore, the results are only valid for small values of σ^2, the variance of the log (K) field.

The second step utilizes the theory of diffusion by continuous movements (Taylor, 1921) to find the mean and covariance of particle displacement, using the velocity statistics obtained in the first step. As in equation (11), the total displacement is assumed to be the result of two additive terms: an advective and a dispersive component. The dispersive term is a consequence of pore–scale dispersion and can be represented by the zero–mean Brownian motion process given in (11) and (13). Here we disregard pore–scale dispersion, since in field–scale situations its contribution is usually orders of magnitude smaller than that of macrodispersion. Therefore, particle displacement becomes a purely advective process. Without loss of generality, we consider that the initial position of the particle (at time $t=0$) is located at the origin $x=0$. Thus, the total cumulative particle displacement and the particle position at time t are equivalent.

Almost all of the analytical studies of stochastic groundwater hydrology in the literature has been restricted to the case of mean uniform flow; that is, $<v(x)> = Ui$, where U is the modulus of the (spatially uniform) mean velocity, and i is the unit vector along the x axis, which is the assumed direction of the mean flow. This is the case depicted schematically in Figures 2 and 3, and is the case that will be considered in this chapter. Therefore, we decompose the random velocity field as $v(x) = <v(x)> + v'(x) = Ui + v'(x)$, where v' is a zero–mean spatially random fluctuation. Then the particle position $X(t)$ at time t is given by

$$X(t) = U t i + \int_0^t v'[X(t')] \, dt' \tag{18}$$

Equation (18) is an integrated version of the single Δt step particle tracking equation (11) (without the random dispersion step). Equation (18) is a nonlinear integral equation since the unknown displacement $X(t)$ appears on both sides. The first–order solution used by Dagan approximates the displacement X in the argument of the integral by its ensemble average: $<X(t)> = U t i$. Note that because we assume ergodicity $<X(t)> = \overline{M}(t)$, where $\overline{M}(t)$ is defined by (15). The total particle displacement can also be decomposed as $X(t) = <X(t)> + X'(t)$, where $X'(t)$ is a zero–mean displacement fluctuation. Since $<X(t)> = U t i$, the fluctuation $X'(t)$ equals the integral term in (18). Therefore the final result for the displacement covariance tensor is

$$C_{X_j X_k}(t) = < X'_j \, X'_k > = \int_0^t \int_0^t C_{v_j v_k}[< X(t') >, < X(t'') >] \, dt' \, dt'' \tag{19}$$

Equation (19) is a fundamental result which relates the particle displacement $(C_{X_j X_k})$ and velocity $(C_{v_j v_k})$ covariance functions. Note that because of ergodicity $C_{X_j X_k} = S(t)$, where

$S(t)$ is defined by (16). Dagan solves (19) for two– and three–dimensional flows for the case in which the velocity covariance is given by the small–perturbation solution of the groundwater flow problem for an exponential isotropic log (K) covariance function. Some of these solutions will be discussed further in section VII below.

Linear Reversible Sorption Kinetics

One of the objectives of this chapter is to use the Lagrangian approach outlined in the previous subsections to obtain results for the transport of sorbing solutes in randomly heterogeneous aquifers. These results will be presented in Section VII. Hence we need a particle–based model for the adsorption–desorption reaction (5). Fortunately, a stochastic particle–based analog to the rate equation (5) has been developed in the chromatography literature (Keller and Giddings, 1960; McQuarrie, 1963). Valocchi and Quinodoz (1989) discuss the numerical implementation of the stochastic model developed by Keller and Giddings (1960), which is a particular case of a two–state Markov Chain.

The two–state Markov Chain model considers a single indivisible particle having a fixed mass (e.g., a molecule). At any instant of time the particle can be in either the dissolved or sorbed state; the time spent in each state (prior to making a transition to the other state) is an exponentially distributed random variable having a mean equal to the inverse of the corresponding reaction rate (i.e., the time spent in the dissolved state has mean k_f^{-1}, and the time spent in the sorbed phase has mean k_r^{-1}). Based upon the Markov chain model, it is possible to compute several important types of probability estimates, including the probability that a particle is in a given state at time t conditioned upon its state at time 0. In the Lagrangian approach a given mass of solute is represented by a large number of particles. Since the stochastic process for each particle is independent of every other particle, ergodicity is valid and the ensemble–based probabilities given by the Markov chain model are equivalent to estimates based upon particle statistics. For example, Valocchi and Quinodoz (1989) show that the probability that a particle is in a particular state can be estimated by the number of particles in each state (i.e., phase), which is equivalent to the phase concentration (c or s in equation (5)).

Keller and Giddings (1960) used the two–state Markov Chain model and found an exact expression for the probability density function of the fraction of time $\beta(t)$ spent in the dissolved phase by a solute particle, conditioned on the particle state at the beginning and end of the time interval t. In this chapter we will focus on the behavior of the dissolved–phase plume for the case in which the solute mass is input initially in the dissolved phase. Therefore, we use the probability density function for the case in which the particle is in the dissolved phase at the beginning and end of the time interval. The appropriate probability density function is given by Quinodoz and Valocchi (1992) as

$$p(\beta) = [\, g(\beta) + \exp(-b)\, \delta(\beta - 1)\,]\, \frac{a + b}{a + b\exp(-a - b)} \qquad (20)$$

where

$$g(\beta) = \sqrt{\frac{ab\beta}{1 - \beta}} \; \exp[-a(1 - \beta) - b\beta] \; I_1[\sqrt{4ab\beta(1 - \beta)}] \qquad (21)$$

$\delta(\ )$ is Kronecker's delta function, $a = k_r t$, $b = k_f t$, $I_1[\ ]$ is the modified Bessel function of order one, and $0 \leq \beta \leq 1$.

HOMOGENEOUS SOIL COLUMN

In this section we use the particle–based approach to obtain results for the simple case of one–dimensional flow in a spatially uniform soil column with constant rate parameters. We do this in order to focus upon the impact of sorption kinetics without the added complications of spatial heterogeneity. Goltz and Roberts (1987) studied this deterministic case using classical Taylor–Aris moment analysis. For simplicity, we will neglect pore–scale dispersion processes. We assume that a given mass of solute is input instantly in the dissolved phase at time zero. The solute plume advects downgradient at the constant pore water velocity v and undergoes linear reversible adsorption–desorption.

Since in the Lagrangian approach the total solute mass is represented by a large number of particles, the dissolved–phase mass in the plume (M_0) is given by (14). As stated previously, N_p, the number of particles in the dissolved phase at time t, is proportional to the probability that a particle is in the dissolved phase at time t, conditioned on the particle being in the dissolved phase at time 0. The result is (Valocchi, 1988; Quinodoz and Valocchi, 1992):

$$M_0 = \frac{M_T}{R} \; [1 + K_d \exp(-k_r R t)] \qquad (22)$$

where M_T is the total mass input at time 0, and $R = 1 + K_d$ is the retardation factor. Figure 4, which is a plot of equation (22), shows that all of the mass is in the dissolved phase at early time and is partitioned between the dissolved and sorbed phases according to local equilibrium at late time; that is, M_0 decays exponentially from M_T at early time to M_T/R at late time.

The position of any particle comprising the dissolved–phase solute mass is given by

$$X(t) = \int_0^{t_m(t)} v \; dt' = v \; t_m(t) = v \; t \; \beta(t) \qquad (23)$$

where $t_m(t)$ is the time that the particle spends in the mobile (i.e. dissolved) phase during the time interval from 0 to t, and $\beta(t)$ is the fraction of the time interval that the particle is in the dissolved phase; $\beta(t)$ is a random variable having the probability density function given by (20). Equation (23) is a consequence of the fact that the particle is stationary when it is in the sorbed phase and advects with uniform velocity v when it is in the dissolved phase. The

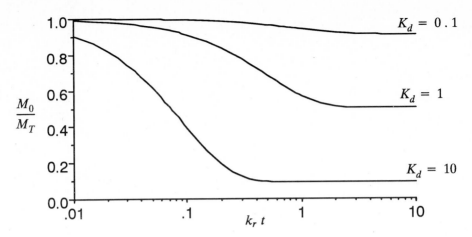

Figure 4. Dissolved–phase mass in plume versus time.

mean location of the aqueous–phase plume, defined by (8), can be calculated according to (15) as the average of the spatial location of all the particles in the aqueous phase. By the ergodicity argument above, the mean plume location (i.e., the location of the plume center–of–mass) equals the ensemble average of (23):

$$\overline{M}_x(t) = \; < X(t) > \; = v \, t \; < \beta(t) > \tag{24}$$

The mean of the β process is reported by Quinodoz and Valocchi (1992) as

$$< \beta > = \frac{1 + K_d^2 \exp(-\gamma) + 2K_d[1 - \exp(-\gamma)]\gamma^{-1}}{R \, [1 + K_d \exp(-\gamma)]} \tag{25}$$

where $\gamma = k_r R t$. The effective velocity of the plume center–of–mass defined by (10) is plotted in Figure 5. At early times V^e is equal to the nonreactive pore water velocity v because there has not yet been sufficient time for particles to adsorb. At later times the effective velocity asymptotically tends to the retarded velocity v/R.

The variance of the particle cloud can be computed by substituting the above results into the definition (16). The result is

$$S(t) = v^2 \, t^2 \; Var[\beta(t)] \tag{26}$$

where $Var[\beta(t)] = \; <\beta^2> - (<\beta>)^2$ and $<\beta^2>$ is a lengthy expression determined by Quinodoz and Valocchi (1992) which will not be reported here for sake of brevity. The effective dispersion coefficient of the plume defined by (10) is plotted in Figure 6. This figure shows that $D^e(t)$ asymptotically approaches a constant value. This asymptotic limit is:

$$D^e(t \to \infty) = (K_d \, v^2) \, / \, (k_r R^3) \tag{27}$$

Also, for large values of K_d, $D^e(t)$ has a peculiar behavior with an early peak that can be much larger than the asymptotic value. As discussed by Bellin et al. (1991) this behavior is associated with bimodal solute profiles resulting from fast adsorption and slow desorption.

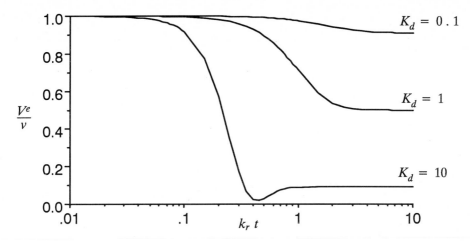

Figure 5. Effective velocity of dissolved–phase plume versus time

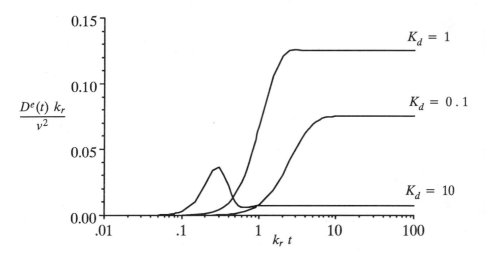

Figure 6. Effective dispersion coefficient of dissolved phase plume versus time.

RANDOMLY HETEROGENEOUS HYDRAULIC CONDUCTIVITY WITH UNIFORM REACTION PARAMETERS

Here we use the results from the previous two sections and combine the Lagrangian approach of Dagan for particle transport in randomly heterogeneous aquifers with the particle–based model of adsorption–desorption kinetics. Since we assume that the reaction parameters are constant, the particle displacement stochastic process (due to the random velocity field) is independent of the particle phase change stochastic process. Hence particle displacement by the random velocity field occurs only when the particle is in the dissolved phase, and so the basic results (18) and (19) hold if the time interval t is replaced by the random variable $t_m(t) = \beta(t)\, t$.

Quinodoz and Valocchi (1992) show in detail how the mean and variance of the particle displacement can be calculated in two steps using the notion of conditional expected values. In the first step $t_m(t)$ is considered constant and the expected values are over the ensemble of the velocity process. In the second step the expected values are with respect to the random process $t_m(t)$. The final result for the mean particle displacement is simply

$$< X(t) > = U\,t\ < \beta(t) > i \qquad or \qquad < X(t) > = U\,t\ < \beta(t) > \tag{28}$$

where we have used the scalar $X(t)$ to denote the particle position along the x–direction, which is the direction of the mean flow. It is interesting to note that (28) is identical with result (24) for the case of a homogeneous soil column; therefore, a plot of the effective velocity of the plume center–of–mass (V^e) would look similar to Figure 5, except the pore water velocity v would be replaced by the ensemble average velocity U.

For the second moment results, we will only report the particle displacement variance in the x–direction, which is denoted $S_x(t)$; additional results for moments in the directions transverse to flow are reported by Quinodoz and Valocchi (1992). The total variance is a sum of two terms

$$S_x(t) = S1(t) + S2(t) \tag{29}$$

The first term in (29) is the expected value of the particle displacement variance over the random process $t_m(t) = \beta t$; that is,

$$S1(t) = < C_{XX}(t_m) > = \int_0^1 C_{XX}(\beta t)\, p(\beta)\, d\beta \tag{30}$$

where the function C_{XX} is given by (19) and the probability density function $p(\beta)$ is given by (20). The second term, $S2(t)$, is identical to (26), which is the variance for the case of uniform velocity. This term will be denoted the "kinetics–only" variance, whereas the first term will be denoted the "mixed macrodispersion–kinetics" variance. Similarly, the effective dispersion coefficient $D_x^e(t)$ (see equation (10)) can be expressed as the sum of two terms

$$D_x^e(t) = D1(t) + D2(t) \tag{31}$$

Since the $D2(t)$ term in (31) represents the contribution of kinetics for the case of uniform velocity, it is identical to the result presented in Section VI and plotted in Figure 6. The $D1(t)$ term depends upon the particular model of the K field variability adopted, as can be seen from equations (30) and (19). Because the particle displacement covariance function in the integrand of (30) is generally a nonlinear function of t_m, it is necessary to use numerical quadrature to evaluate $S1(t)$ and $D1(t)$. Typical results are shown in Figure 7 for the case of a three–dimensional log (K) random process having an isotropic exponential covariance function. Due to the added complexity of the chemical reaction, a plot of the dimensionless dispersion coefficient $(D1 / \sigma^2 U \lambda)$ versus dimensionless time (Ut / λ) will depend upon the reaction parameters K_d and k_r. The plots shown in Figure 7 are for fixed $K_d = 1$ and for varying values of the dimensionless rate or Damkohler number, Da, which is defined as $Da = k_r \lambda / U$. As shown in Quinodoz and Valocchi (1992) and Figure 7, the large time asymptote of the D1(t) term is the same for all values of the chemical parameters; the large–time result is:

$$D1(t \to \infty) = \sigma^2 U \lambda R^{-1} \qquad (32)$$

where σ^2 and λ are the variance and correlation length, respectively, of the log (K) field and $R = 1 + K_d$ is the retardation factor. Figure 7 also shows that $D1(t)$ is always bounded by the nonreactive and local equilibrium limits; these bounds are attained for the case of very slow and very fast reaction rates. These bounds have already been presented by Dagan (1989), and they demonstrate the well–known "scale effect" whereby the effective dispersion coefficient increases with travel distance. Note that in the local equilibrium limit (i.e., $k_r \to \infty$ and $Da \to \infty$), $D2(t) = 0$ and $D_x^e(t) = D1(t)$.

In order to plot conveniently the total effective dispersion coefficient defined by (31) it is necessary to pick particular values for the parameters σ^2 and K_d. Figure 8 is a plot of $D_x^e(t) \ (U\lambda)^{-1}$ versus Ut / λ for $K_d = 1$ and $\sigma^2 = 0.5$. This figure shows that the $D_x^e(t)$ curve for fixed Da displays three different kinds of behavior. At early time $D_x^e(t)$ follows the nonreactive curve because most of the initially input solute mass is still in the dissolved phase. At very large time $D_x^e(t)$ reaches its asymptotic limit, given by the sum of equations (32) and

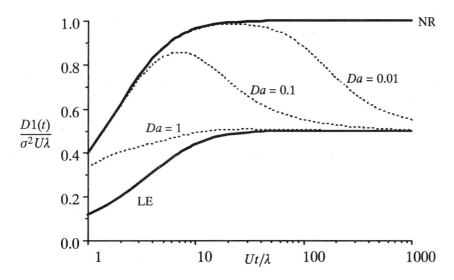

Figure 7. Temporal evolution of "mixed macrodispersion–kinetics" portion of effective dispersion coefficient of dissolved–phase plume; $K_d = 1$. LE: local–equilibrium case; NR: nonreactive case.

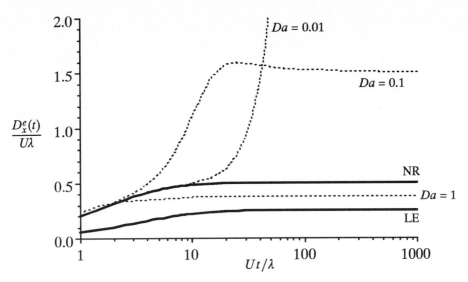

Figure 8. Temporal evolution of total effective dispersion coefficient of dissolved–phase plume; $K_d=1$, $\sigma^2=0.5$.

(27) (with U replacing v in (27)). At intermediate times $D_x^e(t)$ changes from its nonreactive to asymptotic value, and may exhibit a maximum depending upon the value of Da.

Deviations from Local Equilibrium

The analytical solution presented above provides a simple and useful tool for understanding the combined effect of sorption kinetics and macrodispersion upon large–scale mixing processes. One particularly important application is to determine conditions for which the local equilibrium assumption is valid. Of the many possible ways to measure the difference between kinetics–based and equilibrium–based models of sorbing solute transport, we adopt the approach advocated by Valocchi (1989) and quantify the importance of kinetic sorption by its relative effect upon the effective dispersion coefficient. The particular measure we use is the ratio of the "kinetics–only" term $D2(t)$ to the "mixed macrodispersion–kinetics" term $D1(t)$. Here we simplify the analysis and just consider the large–time limit of this ratio, which, after using (32) and (27), becomes

$$\Delta_\infty \equiv \Delta(t \to \infty) = \frac{D2(t \to \infty)}{D1(t \to \infty)} = \frac{K_d}{R^2 k_r} \frac{U}{\sigma^2 \lambda} \tag{33}$$

This asymptotic relationship shows that deviations from local equilibrium not only diminish with increasing reaction rate (k_r) for fixed hydraulic parameter values (U, σ^2, λ), but also diminish with increasing values of the product $\sigma^2\lambda$, for fixed mean velocity (U) and reaction parameters (k_r, K_d). If we set the expression (33) to be less than an arbitrary small number, we obtain a practical condition for validity of the local equilibrium assumption. Quinodoz and Valocchi (1992) examine the behavior of $\Delta(t)$ at early times.

Numerical Simulations

We have conducted numerical simulations in order to verify the approximate analytical solution presented above. The numerical simulations involve the following steps: (1) generation of a single large realization of a randomly heterogeneous K field using the code TURN (Tompson and Gelhar, 1990); (2) solution of the groundwater flow equation using a block–centered finite difference model; (3) simulation of reactive particle transport using the method described by Valocchi and Quinodoz (1989). Further information about the numerical experiments is given by Valocchi (1990b). Figure 9 shows that the numerical results for the longitudinal variance of the plume agrees very closely with the theory as given by equation (29). The parameters used for Figure 9 are: $U=10^{-2}$ m/day, $\sigma^2=0.5$, $\lambda=5$ m, $K_d=2$, and $k_r=0.00016$ day^{-1}.

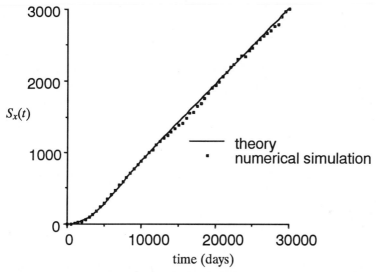

Figure 9. Theoretical and numerical results for longitudinal variance of dissolved–phase plume.

SPATIALLY VARIABLE REACTION PARAMETERS

As indicated in Section II, there is limited field evidence that field soils and aquifers exhibit spatial variability in their reactive, as well as hydraulic, properties. Therefore there have been many recent efforts to analyze the impact of small–scale variability of sorption parameters (e.g., K_d, k_r) upon large–scale transport behavior. In this section we review briefly a portion of the published literature and we discuss the main features of the available results.

Several investigators have considered the simplified case of reaction parameter heterogeneity for a hydraulically homogeneous porous medium (e.g., Chrysikopoulos et al.,

1992). These studies are typically performed as an intermediate step to obtain results for the more general case with space–varying velocity.

Large–Time Asymptotic Results

Transport of reactive solutes in a porous medium having spatially variable velocity and reaction parameters is a formidable mathematical problem. Accordingly, nearly all of the published analyses of reaction heterogeneity have focused upon large–time asymptotic behavior, when the effective velocity and dispersion tensor of the solute plume become constants. A notable exception is the work of Kabala and Sposito (1991); these authors considered a solute undergoing linear equilibrium adsorption with the pore water velocity and retardation factor modeled as random space functions. They used the cumulant expansion method to derive an ensemble average transport equation valid for all times. In another interesting study oriented to vadose zone transport, van der Zee and van Riemsdijk (1987) modeled the soil as a large number of parallel, non–interacting homogeneous columns, each with randomly selected velocity and sorption parameters. Although van der Zee and van Riemsdijk (1987) neglected sorption kinetics, they did consider a nonlinear (Freundlich) equilibrium isotherm.

A three–dimensional, spatially varying velocity field resulting from the second–order stationary random field model for K was adopted by Garabedian (1987) and Dagan (1989). Garabedian (1987) assumed that K_d is linearly related to K thereby also giving K_d as a random field. He assumed the sorption reaction was governed by local chemical equilibrium and used the spectral small–perturbation techniques of Gelhar (1986) to derive large–time ensemble–averaged results for the first and second spatial moments of the solute plume. One of Garabedian's most important results is that negative correlation between K and K_d tends to increase dispersion. Similar results were reported by Dagan (1989, p. 344), who also invoked ergodicity and assumed local chemical equilibrium with a linear correlation between K and K_d. Dagan used the Lagrangian approach outlined above in Section V; although his formulation is general for all time, he introduces several large–time approximations in order to obtain the final results.

Valocchi (1989) and Chrysikopoulos et al. (1991, 1992) adopted a deterministic model of the spatially varying velocity and reaction parameters. Valocchi considered steady, horizontal flow through a perfectly stratified aquifer of constant thickness with arbitrary vertical variation in the pore water velocity, dispersion coefficients, and reaction parameters. Chrysikopoulos et al. considered the spatially periodic model of a porous medium used by Brenner (1980) and Brenner and Adler (1982). In the spatially periodic model the hydrodynamic and reaction parameter values repeat themselves with a certain period in each spatial direction. Although the spatially periodic model assumes a deterministic pattern of variability, Kitanidis (1990) explores some extensions to nonperiodic and random media.

In order to obtain results for the large–time spatial moments, both Valocchi and Chrysikopoulos et al. employ the same mathematical technique––a generalization of the classical Taylor spatial moment analysis sometimes referred to as generalized Taylor–Aris–Brenner moment analysis (Frankel and Brenner, 1989). The generalized moment analysis provides a rigorous methodology for spatial averaging of the small–scale variability and it is related to other so–called "homogenization" techniques (Ene, 1990). We will not review the methodology here, instead we will summarize briefly the results of Valocchi (1989); those of Chrysikopoulos et al. (1991) are similar though somewhat more general.

The mathematical model of Valocchi (1989) consists of equations (3) and (5) for the case of steady, horizontal flow through a perfectly stratified aquifer; for this model $v=(V(z),0,0)^T$, and D_x, D_y, D_z, K_d, and k_r are all functions of z only. Valocchi solved for the large–time mean and variance in the x–direction (the direction of flow); that is, only for $\overline{M}_x$ in (8) and S_x in (9). Note that the spatial integration over the z direction in definitions (8) and (9) is bounded and is taken over the vertical thickness of the aquifer, which is assumed constant and is denoted H. Based upon the first moment results, the effective velocity of the plume center–of–mass defined by (10) is

$$V^e_{x\,\infty} = <V> / <R> \qquad (34)$$

where the subscript ∞ refers to the large–time asymptotic result, $R=1+K_d$ is the retardation factor, and the angle brackets denote depth–averaged quantities. That is,

$$<V> = \frac{1}{H} \int_0^H V(z)\ dz \qquad <R> = \frac{1}{H} \int_0^H R(z)\ dz \qquad (35)$$

where $z=0$ and $z=H$ denote the aquifer's bottom and top, respectively. Note that the angle brackets were used previously to represent ensemble–averaged quantities in Sections V and VII.

The final result for the large–time effective dispersion coefficient is

$$D^e_{x\,\infty} = D^L_{x\,\infty} + D^T_{x\,\infty} + D^K_{x\,\infty} \qquad (36)$$

In (36) $D^L_{x\,\infty} = <D_x>/<R>$ is the contribution due to "local" hydrodynamic dispersion, $D^T_{x\,\infty}$ is the contribution due to the so–called "Taylor" mechanism, which is related to differential advection caused by the vertically varying velocity and retardation factor, and $D^K_{x\,\infty}$ is the contribution due to the kinetic adsorption reaction. The Taylor dispersion coefficient can be written as

$$D^T_{x\,\infty} = \frac{H^2 <V>^2}{<R>^3} \int_0^1 \left[\int_0^\eta \chi(\xi)\ d\xi \right]^2 \frac{1}{D_z(\eta)}\ d\eta \qquad (37)$$

where $\eta = z/H$ and the function $\chi(\eta)$ is defined as

$$\chi(\eta) = \frac{V(\eta) - R(\eta)[<V>/<R>]}{<V>/<R>} \tag{38}$$

The expression for $D_{x\infty}^K$ is

$$D_{x\infty}^K = \frac{<V>^2}{<R>^3} <K_d/k_r> \tag{39}$$

The results above show that sorption kinetics enhances the asymptotic longitudinal spreading of the plume in a simple additive way; this agrees with the large–time results in Section VII. However, (37) indicates that $D_{x\infty}^T$ depends upon the vertical variation of both $V(z)$ and $R(z)$. In particular, Valocchi (1989) shows that the magnitude of $D_{x\infty}^T$ will be greater if $V(z)$ and $R(z)$ are negatively correlated than if they are positively correlated. This is in agreement with the results of Chrysikopoulos et al. (1991) for a spatially periodic medium and Garabedian (1987) for a random medium.

REFERENCES

ASCE Task Committee on Geostatistical Techniques (1990) Review of geostatistics in geo-hydrology. I: basic concepts. J Hydraul Eng 116:612–632

Bahr JM (1989) Analysis of nonequilibrium desorption of volatile organics during a field test of aquifer decontamination. J Contam Hydrol 4:205–222

Bear J (1979) Hydraulics of groundwater, McGraw–Hill, New York

Bellin A, Valocchi AJ, Rinaldo A (1991) Double peak formation in reactive solute transport in one–dimensional heterogeneous porous media. Quaderni del dipartimento IDR 1/1991, Dipartimento di Ingegneria Civile ed Ambientale, Università degli Studi di Trento

Black TC, Freyberg DL (1987) Stochastic modeling of vertically averaged concentration uncertainty in a perfectly stratified aquifer. Water Resour Res 23:997–1004

Brenner H (1980) A general theory of Taylor dispersion phenomena, Physicochem Hydrodyn 1:91–123

Brenner H, Adler PM (1982) Dispersion resulting from flow through spatially periodic porous media, II, Surface and intraparticle transport, Philos Trans R Soc London Ser A 307:149–200

Chrysikopoulis CV, Kitanidis PK, Roberts PV (1991 submitted) Macrodispersion of sorbing solutes in heterogeneous porous formations with spatially periodic retardation factor and velocity field. Water Resour Res

Chrysikopoulis CV, Kitanidis PK, Roberts PV (1992) Generalized Taylor–Aris moment analysis of the transport of sorbing solutes through porous media with spatially–periodic retardation factor. Transport in Porous Media 7:163–185

Cvetkovic VD, Shapiro A (1990) Mass arrival of sorptive solute in heterogeneous porous media. Water Resour Res 26:2057–2068.

Dagan G (1986) Statistical theory of groundwater flow and transport: Pore to laboratory, laboratory to formation, formation to regional scale. Water Resour Res 22:120s–134s

Dagan G (1989) Flow and transport in porous formations. Springer–Verlag, Berlin

Dagan G (1990) Transport in heterogeneous porous formations: spatial moments, ergodicity, and effective dispersion, Water Resour Res 26:1281–1290

Dagan G (1991) Dispersion of a passive solute in non–ergodic transport by steady velocity fields in heterogeneous formations. J Fluid Mech 233:197–210

Dagan G, Nguyen V (1989) A comparison of travel time and concentration approaches to modeling transport by groundwater, J Contam Hydrol 4:79–92

Dagan G, Neuman SP (1991) Nonasymptotic behavior of a common Eulerian approximation for transport in random velocity fields, Water Resour Res 27:3249–3256

de Marsily G (1986) Quantitative hydrogeology, Academic Press, Orlando Florida

Ene H (1990) Applicatiopn of the homogenization method to transport in porous media. In: Cushman JH (ed) Dynamics of fluids in hierarchical porous media. Academic Press, London, p 223–241

Frankel I, Brenner H (1989) On the foundations of generalized Taylor dispersion theory. J Fluid Mech 204:97–119

Freyberg D (1986) A natural gradient experiment on solute transport in a sand aquifer, 2. Spatial moments and the advection and dispersion of nonreactive tracers. Water Resour Res 22: 2031–2046

Garabedian SP (1987) Large–scale dispersive transport in aquifers: field experiments and reactive transport theory. Ph.D. dissertation, Massachusetts Institute of Technology, Cambridge MA

Gelhar LW (1986) Stochastic subsurface hydrology from theory to applications. Water Resour Res 22:135s–145s

Goltz M, Roberts PV (1987) Using the method of moments to analyze three–dimensional diffusion–limited solute transport from temporal and spatial perspectives. Water Resour Res 23:1575–1585

Graham W, McLaughlin D (1989) Stochastic analysis of nonstationary subsurface transport, 2, conditional moments. Water Resour Res 25:23312355

Guven O, Molz F, Melvill JG (1984) An analysis of dispersion in a stratified aquifer, Water Resour Res 10:1337–1354

Hess K (1989) Use of a borehole flowmeter to determine spatial heterogeneity of hydraulic conductivity and macrodispersion in a sand and gravel aquifer, Cape Cod, Massachussets. In: Molz FJ, Melville JG, Guven O (eds) Proceedings of the conference on new field techniques for quantifying the physical and chemical properties of heterogeneous aquifers. National Water Well Association, Dublin OH, p 497–508

Kabala ZJ, Sposito G (1991) A stochastic model of reactive solute transport with time–varying velocity in a heterogeneous aquifer. Water Resour Res 27:341–350

Keller RA, Giddings JC (1960) Multiple zones and spots in chromatography. J Chromatog 3:205–220

Kinzelbach W (1987) The random walk method in pollutant transport simulation. In:Custodio E, et al. (eds) Groundwater flow and quality modeling. D Reidel, Dordrecht, p 227–245

Kitanidis, P.K. (1990) Effective hydraulic conductivity for gradually varying flow. Water Resour Res 26:1197–1208

LeBlanc DR, Garabedian SP, Hess KM, Gelhar LW, Quadri RD, Stollenwerk KG, Wood WW (1991) Large–scale natural gradient test in sand and gravel, Cape Cod, Massachusetts, 1, experimental design and observed tracer movement. Water Resour Res 27:895–910

MacQuarrie KTB, Sudicky EA (1990) Simulation of biodegradable organic contaminants in groundwater, 2, Plume behavior in uniform and random flow fields. Water Resour Res 26:223–240

Mantoglou A, Wilson J (1982) The turning bands method for simulation of random fields using line generation by a spectral method. Water Resour Res 18:1379–1394

McQuarrie DA (1963) On the stochastic theory of chromatography. J Chem Phys 38:437–435

Neuman SP (1990) Universal scaling of hydraulic conductivities and dispersivities in geologic media. Water Resour Res 26:1749–1758

Neuman SP, Zhang YK (1990) A quasi–linear theory of non–Fickian subsurface dispersion, 1, Theoretical analysis with application to isotropic media. Water Resour Res 26:887–902

Nkedi–Kizza P, Biggar JW, Selim HM, van Genuchten M Th,Wierenga PJ, Davidson JM, Nielsen DR (1984) On the equivalence of two conceptual models for describing ion exchange during transport through an aggregated oxysol, Water Resour Res 20:1123–1130

Quinodoz HAM, Valocchi AJ (1990) Macrodispersion in heterogeneous aquifers: numerical experiments. In: Moltyaner G (ed) Transport and mass exchange processes in sand and gravel aquifers: field and modelling studies. Atomic Energy of Canada Limited, Chalk River Ontario, p 455–468

Quinodoz HAM, Valocchi AJ (1992 submitted) Stochastic analysis of the transport of kinetically adsorbing solutes in randomly heterogeneous aquifers. Water Resour Res

Robin MJL, Sudicky EA, Gillham RW, Kachanoski RG (1991) Spatial variability of strontium distribution coefficients and their correlation with hydraulic conductivity in the CFB Borden aquifer. Water Resour Res 27:2619–2632

Rubin Y (1991) Transport in heterogeneous porous media: prediction and uncertainty. Water Resour Res 27:1723–1738

Sardin M, Schweich F, Leij FJ, van Genuchten M Th (1991) Modeling the nonequilibrium transport of linearly interacting solutes in porous media: a review. Water Resour Res 27:2287–2307

Schaefer W, Kinzelbach W (1992 in press) Stochastic modeling of in situ bioremediation in heterogeneous aquifers. J Contam Hydrol

Semprini L, Roberts PV, Hopkins GD, McCarty PL (1990) A field evaluation of in–situ biodegradation of chlorinated ethenes: part 2, results of biostimulation and biotransformation experiments. Ground Water 28:715–727

Shapiro M, Brenner H (1988) Dispersion of a chemically reactive solute in a spatially periodic model of a porous medium, Chem Eng Sci 43:551–571

Sudicky EA (1986) A natural gradient tracer experiment on solute transport in a sand aquifer: Spatial variability of hydraulic conductivity and its role in the dispersion process. Water Resour Res 22:2069–2082

Sudicky EA, Huyakorn P (1991) Contaminant migration in imperfectly known heterogeneous groundwater systems. In: Reviews of Geophysics, Supplement, April, p. 240–253

Taylor GI (1921) Diffusion by continuous movements, Proc London Math Soc 2:196–212

Taylor GI (1953) The dispersion of matter in a solvent flowing slowly through a tube, Proc R Soc London Ser A 219:189–203

Tompson AFB, Dougherty DE (1988) On the use of particle tracking methods for solute transport in porous media. In: Celia MA, et al. (eds) Computational methods in water resources, Vol. 2, Numerical methods for transport and hydrologic processes. Elsevier, Amsterdam, p 227–232

Tompson AFB, Gelhar LW (1990) Numerical simulation of solute transport in three–dimensional, randomly heterogeneous porous media, Water Resour Res 26:2541–2562

Tompson AFB, Ababou R, Gelhar LW (1989) Implementation of the three–dimensional Turning Band random field generator. Water Resour Res 25:2227–2244

Valocchi AJ (1988) Theoretical analysis of deviations from local equilibrium during sorbing solute transport through idealized stratified aquifers. J Contam Hydrol 2:191–207

Valocchi AJ (1989) Spatial moment analysis of the transport of kinetically adsorbing solutes through stratified aquifers. Water Resour Res 25:273–279

Valocchi AJ (1990a) Use of temporal moment analysis to study reactive solute transport in aggregated porous media. Geoderma 46:233–247

Valocchi AJ (1990b) Numerical simulation of the transport of adsorbing solutes in hetero-geneous aquifers. In: Gambolati G, et al. (eds) Computational methods in subsurface hydrology. Springer–Verlag, Berlin, p. 373–382

Valocchi AJ, Quinodoz HAM (1989) Application of the random walk method to simulate the transport of kinetically adsorbing solutes. In: Abriola LM (ed) Groundwater con-tamination, IAHS Publ. No. 185, p 35–42

van der Zee SEATM, van Riemsdijk WH (1987) Transport of reactive solutes in spatially variable soil systems. Water Resour Res 23:2059–2069

van Genuchten M Th (1985) A general approach for modeling solute transport in struc-tured soils. In: Hydrology of rocks of low permeability. Proceedings of the 17th interna-tional congress. Intl Assoc Hydrogeol 17:513–526

Vomvoris E, Gelhar LW (1990) Stochastic analysis of the concentration variability in a three–dimensional, randomly heterogeneous porous media. Water Resour Res 26:2591–2602

Wood WW, Kraemer TF, Hearn P (1990) Intragranular diffusion: an important mechanism influencing solute transport in clastic aquifers? Science 247:1569–1572

LIST OF SYMBOLS

$A(X, t)$	deterministic forcing vector in particle tracking equation (L/T)
$B(X, t)$	deterministic square scaling tensor in particle tracking equation (L/T)
$C_c(x_1, x_2)$	ensemble–covariance of concentration field c at locations x_1 and x_2
$C_h(x_1, x_2)$	ensemble–covariance of head field h at locations x_1 and x_2
$C_{Xi\,Xj}(t)$	ensemble–covariance of components i and j of the particle displacement X at time t
$C_{vi\,vj}(x_1, x_2)$	ensemble–covariance of components i and j of the spatially random velocity field v' at locations x_1 and x_2
c	dissolved–phase solute concentration (M/L^3)
Da	Damkohler number $= k_r \lambda / U$ (dimensionless)
D_m	molecular diffusion coefficient (L^2/T)
D	hydrodynamic dispersion tensor (L^2/T)
$D^e(t)$	effective dispersion tensor of the plume (L^2/T)
D^M	mechanical dispersion tensor (L^2/T)
$D_L{}^M$	longitudinal component of mechanical dispersion tensor (L^2/T)
$D_T{}^M$	transverse component of mechanical dispersion tensor (L^2/T)
$D^L_{x\infty}$	contribution to the large–time effective dispersion coefficient due to local hydrodynamic dispersion for the case of horizontal flow in a stratified aqui-fer
$D^T_{x\infty}$	contribution to the large–time effective dispersion coefficient due to the vertically varying velocity (the "Taylor" mechanism) for the case of horizon-tal flow in a stratified aquifer
$D^K_{x\infty}$	contribution to the large–time effective dispersion coefficient due to sorp-tion kinetics for the case of horizontal flow in a stratified aquifer
$D1(t)$	mixed macrodispersion–kinetics term of longitudinal effective dispersion coefficient (L^2/T)
$D2(t)$	kinetics–only term of longitudinal effective dispersion coefficient (L^2/T)
h	piezometric head (L)
i	unit vector along x–axis
k_f	forward adsorption rate coefficient (T^{-1})
k_r	reverse adsorption rate coefficient (T^{-1})

$I_1[\ ]$	modified Bessel function of order 1
H	aquifer thickness
$\boldsymbol{K}$	hydraulic conductivity tensor (L/T)
K	hydraulic conductivity (point value) (L/T)
K_d	equilibrium distribution coefficient (dimensionless)
$\overline{\boldsymbol{M}}(t)$	center−of−mass of dissolved solute plume (vector) (L)
$M_{ijk}(t)$	spatial moment of order ijk; where i,j,k indicate order along three coordinate axes
$M_0(t)$	zeroth−order spatial moment of the aqueous−phase concentration (M)
M_T	total mass input in the aqueous phase at time zero (M)
n	porosity
N_p	number of particles in mobile phase
R	retardation coefficient $= 1+K_d$ (dimensionless0
s	sorbed−phase solute concentration (M/L^3)
$S(t)$	spatial variance of aqueous−phase solute plume (tensor)
t	time
$\boldsymbol{T}$	tortuosity tensor
$t_m(t)$	time from 0 to t spent by a particle in the aqueous phase $= \beta(t)\ t$
U	modulus of mean velocity (L/T)
$\boldsymbol{v}$	pore water velocity vector (L/T)
$\boldsymbol{v}'(\mathbf{x})$	zero−mean spatially−variable velocity fluctuation
$\mathbf{V}^e(t)$	effective velocity of the aqueous−phase plume−center−of−mass (L/T)
x	fixed position in Eulerian reference frame (L)
$X(t)$	vector denoting the position of a particle at time t (L)
$X'(t)$	zero−mean particle displacement fluctuation (L)
Y	ln (K)
$\mathbf{Z}$	three−component random number vector
α_L	longitudinal dispersivity
α_L	transverse dispersivity
$\beta(t)$	fraction of time t spent in dissolved phase
$\delta(\ )$	Kronecker delta function
Δt	time step used in particle−tracking
Δ_∞	large−time asymptotic value of ratio D2(t) / D1(t)
λ	correlation length of the ln (K) random field
σ^2	variance of the ln (K) random field

Numerical Modeling of Contaminant Transport in Groundwater

Giuseppe Gambolati, Claudio Paniconi, and Mario Putti
Dipartimento di Metodi e Modelli Matematici per le Scienze Applicate
University of Padua - Italy

Contamination of groundwater is a major concern in the management of water resources. Sources of groundwater pollution include chemical compounds in industrial,
agricultural, and urban settings. Contaminants entering groundwater systems through
the unsaturated zone may degrade soils, aquifers (through leaching and recharge), and
streams (through surface and subsurface runoff and leakage). Transported by groundwater, contaminants may also pollute withdrawal sites at pumping wells, and they may
reappear at the surface, emerging at springs and seepage faces.

Contamination can occur at point sources (e.g., isolated spills; leaking storage tanks;
sanitary landfills; septic tanks; radioactive waste disposal; waste tailings from mining
operations) or at nonpoint sources (e.g., herbicides, pesticides, and fertilizers used in
agriculture; atmospheric deposition such as acid rain). The contaminants can be organics, trace metals, or radionuclides, and can interact with the fluid and solid phases
of the porous medium through a wide variety of processes, including chemical diffusion,
mechanical dispersion, advection, chemical reaction, decay, and biodegradation. These
processes are complicated by numerous factors. For instance, due to heterogeneity and
anisotropy, dispersion occurs at very different spatial scales, ranging from a few meters
(local scale) to several kilometers (basin or regional scale). Also, chemical and biochemical reactions are usually very complex and may act at time scales different from
those characterizing groundwater transport. There are three types of chemical reaction
which may influence the fate of a contaminant [Mangold and Tsang, 1991]:

1. ion complexation in the aqueous solution (liquid phase);

NATO ASI Series, Vol. G 32
Migration and Fate of Pollutants in Soils and Subsoils
Edited by D. Petruzzelli and F. G. Helfferich
© Springer-Verlag Berlin Heidelberg 1993

2. precipitation/dissolution of solutes (solid phase);

3. sorption onto solid grains (interphase boundaries).

Ion complexation between two or more reacting species is controlled by the law of mass action and involves thermodynamic equilibrium constants. Dissolution or precipitation of solutes is also governed by the mass action equations for the participating species. An important mechanism accounting for sorption is the exchange of ions between groundwater and mineral or colloidal surfaces. In addition to these three basic types of reaction, a contaminant may also undergo decay. Decay occurs independently of all other processes, and may be modeled using the half-life period of the species.

Mathematical models can be effectively used to study the migration and fate of contaminants and their effects on water resources, and to design and analyze remediation strategies. The models are based on the partial differential equations of fluid and solute continuity, and may be solved by finite element methods. If the physical characteristics of the subsurface system are given, including the initial state of the system and the boundary conditions of the region, the numerical simulation will yield information on the fluid pressure and contaminant concentration distributions and on the direction and rate of transport of the pollutants. The numerical procedure consists of first solving the flow equation to obtain the pressure and velocity distributions. Using the velocity values thus obtained, the transport equation is then solved to obtain the concentration distribution. If the flow and transport processes are coupled (e.g., saltwater intrusion in coastal aquifers), the equations are to be solved simultaneously, using ad hoc techniques to deal with the nonlinearities due to coupling.

In realistic subsurface settings the numerical system of equations can become very large and complex given the variety of contaminant sources and interaction mechanisms possible, and the large temporal and spatial dimensions which may be involved. In addition, difficulties may arise due to nonlinear processes and the high degree of variability inherent in porous media. For these reasons it is important to use cost-effective numerical techniques to solve the subsurface flow and transport equations.

In this chapter the basic equations governing groundwater flow and contaminant transport are first presented, followed by consideration of chemical reaction and decay processes. We then discuss the coupling of flow and transport phenomena, finite element solution methods, and efficient techniques for solving the coupled flow and transport equations. Some illustrative examples are then presented, and we conclude with a summary and some recommendations.

Basic Equations

In a porous medium the processes of groundwater flow and chemical transport and the stresses in the solid skeleton are coupled in a complex manner. To simulate the migration and fate of contaminants in groundwater it is necessary to capture the essential features of these processes in a mathematical formulation. In order to obtain tractable equations in the formulation, a number of simplifying assumptions are made:

- the water and contaminant are perfectly miscible;

- Darcy's law applies to the mixture;

- the water and porous medium are elastic (compressible);

- deformations in the porous medium are small;

- the flow field is not affected by the contaminant—that is, the flow and transport processes are assumed to be uncoupled (this assumption will be relaxed later);

- the dispersion coefficient accounts for both molecular diffusion and mechanical dispersion (mixing);

- the concentration of contaminant is given as weight per unit volume;

- the hydraulic potential head h of the mixture is expressed as $h = z + p/\gamma$ where z [L] is the elevation above a datum, p [F/L^2] is the pressure of water, and γ [F/L^3] is the specific weight of water.

Using indicial notation where the indices i and j denote summation over the three coordinate dimensions $(i, j = 1, 2, 3)$, the equation for groundwater flow can be written as [Bear, 1972; Gambolati, 1973; Bear, 1979]:

$$\frac{\partial}{\partial x_i}\left(k_{ij}\frac{\partial h}{\partial x_j}\right) = S_s\frac{\partial h}{\partial t} - q \tag{1}$$

where:

$k_{ij} = k_{ij}^*\gamma/\mu$ is the hydraulic conductivity tensor [L/T];

k_{ij}^* is the intrinsic permeability tensor [L^2];

μ is the dynamic viscosity of water [FT/L^2];

x_i is the ith Cartesian coordinate $(x_3 = z)$;

$S_s = \gamma(\alpha + n\beta)$ is the specific elastic storage of the medium $[L^{-1}]$;

$\alpha = (1 + \nu)(1 + 2\nu)/E(1 - \nu)$ is the vertical compressibility of the medium $[L^2/F]$;

E is Young's modulus for the medium $[F/L^2]$;

ν is Poisson's ratio for the medium (dimensionless);

β is the volumetric compressibility of water $[L^2/F]$;

n is the porosity of the medium (dimensionless);

t is time;

q represents distributed source or sink terms (volumetric flow rate per unit volume) $[T^{-1}]$.

Initial and boundary conditions for equation (1) can be expressed as:

$$h(x_i, 0) = h_o(x_i) \tag{2a}$$
$$h(x_i, t) = \overline{h}(x_i, t) \qquad \text{on } \Gamma_1 \tag{2b}$$
$$v_i n_i = -q_n(x_i, t) \qquad \text{on } \Gamma_2 \tag{2c}$$

where h_o is the potential head at time $t = 0$, $\overline{h}$ is the prescribed head on boundary Γ_1, and v_i is the Darcy velocity

$$v_i = -k_{ij}\frac{\partial h}{\partial x_j} \tag{3}$$

on the complementary boundary Γ_2, with n_i being the direction cosine of the outward normal to the boundary and q_n being the prescribed flux on Γ_2. We use the sign convention of q_n positive for an inward flux and negative for an outward flux. Boundary condition (2b) is said to be of the first, principal, or Dirichlet type while (2c) is said to be of the second, natural, or Neumann type.

Using the Darcy velocities v_i given by (3) we can write the weight conservation equation for the contaminant. With the dispersion part of the contaminant velocity expressed as $-(D_{ij}/\gamma)(\partial c/\partial x_j)$, we obtain the equation describing transport of a non-reactive solute [Bredehoeft and Pinder, 1973; Pinder and Gray, 1977; Bear, 1979; Neuman, 1984; Burnett and Frind, 1986; Kinzelbach, 1986]:

$$\frac{\partial}{\partial x_i}(D_{ij}\frac{\partial c}{\partial x_j}) - \frac{\partial}{\partial x_i}(v_i c) = n\frac{\partial c}{\partial t} - qc_1 - f \tag{4}$$

where $D_{ij} = n\tilde{D}_{ij}$, $\tilde{D}_{ij}$ is the dispersion tensor $[L^2/T]$ as defined by Bear [1979, p.234], c is the concentration of the solute $[F/L^3]$, c_1 is the concentration of solute injected (withdrawn) with the fluid source (sink) $[F/L^3]$, and f is the distributed flow rate of the solute per unit volume $[F/L^3T]$.

If the flow field is steady, that is the velocity satisfies the equation $\partial v_i/\partial x_i = q(x_i)$, then (4) can be equivalently written as (e.g., Huyakorn and Pinder [1983], Huyakorn *et al.* [1985; 1986]):

$$\frac{\partial}{\partial x_i}\left(D_{ij}\frac{\partial c}{\partial x_j}\right) - v_i\frac{\partial c}{\partial x_i} = n\frac{\partial c}{\partial t} + q(c - c_1) - f \tag{5}$$

If we are considering a well which pumps out contaminated water and q is the pumping rate distributed over the wellbore volume, then in equations (4) or (5) we set $c_1 = c$. Consequently the sink term disappears from equation (5) despite the fact that the contaminant is removed from the system at rate qc. If the well is injecting clean water into a polluted aquifer then $c_1 = 0$ and in equation (5) we have the apparent source term qc, even though no contaminant is added to the system.

The dispersion tensor for a two-dimensional porous medium is given by [Bear, 1979]:

$$D_{ij} = \alpha_L\,|v|\,\delta_{ij} + (\alpha_L - \alpha_T)\frac{v_iv_j}{|v|} + D_o\,n\,\tau\,\delta_{ij} \tag{6}$$

where α_L is the longitudinal dispersivity $[L]$, α_T is the transverse dispersivity $[L]$, δ_{ij} is the Kronecker delta, τ is the tortuosity (dimensionless, ≤ 1), D_o is the molecular diffusion coefficient $[L^2/T]$, and $|v| = \sqrt{v_1^2 + v_2^2}$. Usually we assume $\tau = 1$.

For a three-dimensional system dispersivity is a third-rank tensor and is isotropic in the plane orthogonal to the Darcy velocity. If the Darcy velocity is parallel to the coordinate axis x_1 equation (6) simplifies to:

$$
\begin{aligned}
D_{11} &= D_L = \alpha_L\,|v_1| + D_o\,n\,\tau \\
D_{22} &= D_T = \alpha_T\,|v_1| + D_o\,n\,\tau \\
D_{33} &= D_{22} \\
D_{12} &= D_{13} = D_{23} = 0
\end{aligned}
$$

The initial and boundary conditions for equations (4) or (5) can be expressed as:

$$
\begin{aligned}
c(x_i, 0) &= c_o(x_i) & &\tag{7a}\\
c(x_i, t) &= \bar{c}(x_i, t) & \text{on } \Gamma_3 &\tag{7b}\\
D_{ij}\frac{\partial c}{\partial x_j}n_i &= q^d(x_i, t) & \text{on } \Gamma_4 &\tag{7c}\\
\left(v_i c - D_{ij}\frac{\partial c}{\partial x_j}\right)n_i &= q^c(x_i, t) & \text{on } \Gamma_5 &\tag{7d}
\end{aligned}
$$

where c_o is the initial concentration, $\bar{c}$ is the prescribed concentration on the Dirichlet boundary Γ_3, q^d is the prescribed dispersive flux normal to the Neumann boundary Γ_4 (positive inward), and q^c is the prescribed flux of solute across the Cauchy or Rubin boundary Γ_5. Boundary condition (7c) is usually imposed along the outflow boundary with $q^d = 0$, that is $D_{ij}(\partial c/\partial x_j)n_i = 0$. Along an impermeable boundary we have $v_i n_i = 0$, and a zero dispersive flux implies $q^c = 0$ as well. For a detailed discussion of the numerical treatment of boundary conditions for the transport equation the reader is referred to Galeati and Gambolati [1989].

If the saturated porous system is confined above by a free surface, the gravity drainage equation holds at the water table:

$$k_{33}\frac{\partial h}{\partial x_3} = I - S_y\frac{\partial h}{\partial t} \tag{8}$$

where S_y (dimensionless) is the specific yield of the upper aquifer and I [L/T] is the net vertical infiltration rate per unit horizontal area.

Reactive Solutes

Assume that the contaminant is subject to radioactive or biodegradation decay. If $T_{1/2}$ is the half-life constant, the decay constant λ [T^{-1}] is defined as $\lambda = \ln 2/T_{1/2}$. The convection-dispersion equation (4) with decay is written as:

$$\frac{\partial}{\partial x_i}\left(D_{ij}\frac{\partial c}{\partial x_j}\right) - \frac{\partial}{\partial x_i}(v_i c) = n\frac{\partial c}{\partial t} + n\lambda c - qc_1 - f \tag{9}$$

and is subject to the same initial and boundary conditions as (4).

Local Sorption Equilibrium

If a fraction of the solute is adsorbed onto the solid phase an additional term is to be added to equation (9). Let S (dimensionless) denote the adsorbed contaminant (mass adsorbed per mass of dry sediment). Then the transport equation accounting for decay and adsorption becomes [Bear, 1979]:

$$\frac{\partial}{\partial x_i}\left(D_{ij}\frac{\partial c}{\partial x_j}\right) - \frac{\partial}{\partial x_i}(v_i c) = n\frac{\partial c}{\partial t} + \gamma_s(1-n)\frac{\partial S}{\partial t} + \lambda\left[nc + (1-n)\gamma_s S\right] - qc_1 - f \tag{10}$$

where γ_s [F/L^3] is the specific weight of dry sediment.

The importance of sorption/desorption on the transport of an organic contaminant depends on many factors, including the physical and chemical properties of the soil and contaminant, the presence of competing solutes, and the nature of any background organic matter in solution. These factors may, for instance, retard the movement of the contaminant by orders of magnitude relative to the rate of groundwater movement [Miller and Weber Jr., 1984]. If the contaminant moves slowly enough that chemical equilibrium is achieved and the conditions of the local equilibrium assumption (LEA) hold, then the relationship between S and c can be approximated by the Freundlich isotherm:

$$S = k_d c^m \tag{11}$$

where k_d is the distribution coefficient and is equal to the (relative) adsorbed weight for unit concentration. In the case $m = 1$ the sorption process is described by the linear isotherm $S = k_d c$.

Defining the retardation factor

$$R = 1 + \frac{1-n}{n} \gamma_s k_d > 1$$

equation (10) can be written as:

$$\frac{\partial}{\partial x_i} \left(D_{ij} \frac{\partial c}{\partial x_j} \right) - \frac{\partial}{\partial x_i}(v_i c) = R n \frac{\partial c}{\partial t} + \lambda n R c - q c_1 - f \tag{12}$$

Equation (12) holds for the case of reversible adsorption with a linear isotherm [Anderson, 1979]. Note that R increases with k_d, that is the contaminant front moves more slowly for a larger distribution coefficient.

It is well known that the adsorption isotherm may depend on several factors, including surface charge of the sorbing phase, ionic strength, solution pH, competing counter-ions and their concentrations, and the concentration of the adsorbed phase. In some cases the "adsorption" isotherm may be different from the "desorption" isotherm (chemical hysteresis).

For the case of a linear reversible isotherm, k_d is given by:

$$k_d = \frac{f_s}{f_w} \frac{u}{w}$$

where u is the solution volume, w is the weight of the solid phase, f_s is the percentage of solute adsorbed in the solid phase, and f_w is the percentage of solute in solution.

Some authors (e.g., Weber Jr. [1972], Brusseau and Rao [1989]) have used nonlinear isotherms, especially for cases where concentrations vary over a wide range. In addition to the nonlinear Freundlich isotherm given earlier (equation (11) with $m \neq 1$), another commonly used nonlinear isotherm is the Langmuir isotherm:

$$S = \frac{S^\circ b c}{1 + b c}$$

where S^o is the limiting concentration of adsorbed solute onto the solid phase and b is the adsorption constant.

Lack of Local Sorption Equilibrium

The local equilibrium assumption is not always satisfied for transient flow. In this case van Genuchten and Wierenga [1976] have postulated the existence of two mechanisms:

1. instantaneous processes in the largest pores (related to the mobile fraction of water);

2. diffusion-controlled (retarded) processes in complex aggregates and in dead-end pores (where water is stagnant).

These two mechanisms can be accounted for by the so-called "dual porosity model" which is able to describe the typical "tailing effect" observed in experimental concentration profiles. van Genuchten and Wierenga [1976] proposed reaction kinetics that tend towards equilibrium to account for the intra-pore diffusive process. The resulting model is the dual porosity model (for earlier references see also Lapidus and Amundson [1952] and Coats and Smith [1964]):

$$\frac{\partial}{\partial x_i}\left(D_{ij}\frac{\partial c_m}{\partial x_j}\right) - \frac{\partial}{\partial x_i}(v_i c_m) = n_m\frac{\partial c_m}{\partial t} + n_{im}\frac{\partial c_{im}}{\partial t} + \dots \tag{13a}$$

$$n_{im}\frac{\partial c_{im}}{\partial t} = \alpha(c_m - c_{im}) \tag{13b}$$

where c_m is the concentration in the mobile phase, c_{im} is the concentration in the immobile phase, n_m is the pore volume fraction occupied by the mobile phase, n_{im} is the pore volume fraction occupied by the immobile phase, and α is the mass transfer coefficient $[T^{-1}]$. For the sake of simplicity we have not included decay and distributed contaminant sources in equation (13). More realistically c_m in equation (13b) should be replaced by the equilibrium concentration $f(c_m)$ based on the Freundlich or Langmuir isotherm. Finite volume numerical methods [Putti et al., 1990] are particularly suitable for solving non-LEA equations involving complex chemical processes.

Equation (13b) is a first order linear differential equation of the kind $y' + f_1(x)y + f_2(x) = 0$ which possesses the general solution:

$$y = \exp[-\int f_1(x)\,dx]\left\{k - \int [f_2(x)\exp\int f_1(x)\,dx]\,dx\right\}, \qquad k \text{ constant} \tag{14}$$

Writing equation (13b) as

$$\frac{dc_{im}}{dt} + \frac{\alpha}{n_{im}}c_{im} - \frac{\alpha}{n_{im}}c_m = 0$$

and applying solution (14) we get:

$$c_{im} = \exp\left(-\frac{\alpha t}{n_{im}}\right)\left\{k + \frac{\alpha}{n_{im}}\int_0^t \exp\left(\frac{\alpha}{n_{im}}t\right)c_m(t)\,dt\right\} \tag{15}$$

Setting $\beta = \alpha/n_{im}$ and substituting equation (15) into equation (13a) we obtain an integro-differential equation for the mobile phase concentration c_m:

$$\frac{\partial}{\partial x_i}\left(D_{ij}\frac{\partial c_m}{\partial x_j}\right) - \frac{\partial}{\partial x_i}(v_i c_m) = n_m\frac{\partial c_m}{\partial t} + \alpha\left[c_m - \beta\exp\left(-\beta t\right)\int_0^t \exp\left(\beta\tau\right)c_m(\tau)\,d\tau\right] \tag{16}$$

The constant of integration has been set equal to zero, that is we assume $c_{im} = 0$ at time $t = 0$.

The solution to equation (16) lies between the solutions to the limiting equations:

$$\frac{\partial}{\partial x_i}\left(D_{ij}\frac{\partial c_m}{\partial x_j}\right) - \frac{\partial}{\partial x_i}(v_i c_m) = n_m\frac{\partial c_m}{\partial t} \qquad (\alpha = 0)$$

$$\frac{\partial}{\partial x_i}\left(D_{ij}\frac{\partial c_m}{\partial x_j}\right) - \frac{\partial}{\partial x_i}(v_i c_m) = (n_m + n_{im})\frac{\partial c_m}{\partial t} \qquad (\alpha = \infty)$$

The first equation with $\alpha = 0$ describes the case where no mass exchange takes place between the mobile and immobile phases. The second equation with $\alpha = \infty$ describes the case where LEA holds, that is $c_{im} = c_m$ with instantaneous chemical equilibrium between the two phases. Reality lies somewhere between these two limiting cases which differ only by a time scaling factor. Hence the tailing effect may be fully accounted for by the integro-differential equation (16). For a more complete review of other possible sorption mechanisms, see Miller and Weber Jr. [1984] and Brusseau and Rao [1989].

Column and Field Experiment Examples

Now let us examine a typical laboratory experiment involving reactive transport in a one-dimensional vertical porous column of length L. Define the effluent pore volume T as $T = v_z t/(nL)$ where v_z is the vertical Darcy velocity. Assume that sorption is controlled by a linear isotherm and that at time $t = 0$ the contaminant concentration is instantaneously set equal to $c_o =$ constant at the top of the column. The "breakthrough curve" is a plot of c/c_o versus T at the bottom of the column ($z = L$). Typical breakthrough curves are shown in Figure 1a. Profile 1 is for a non-reactive contaminant, and the slope of the concentration front is related to the magnitude of dispersion. If the dispersion coefficient is zero the front maintains a vertical slope while moving through the column. Profile 2 describes a reactive contaminant subject to LEA and for which a linear isotherm applies. The difference between profiles 1 and 2 is a simple translation in

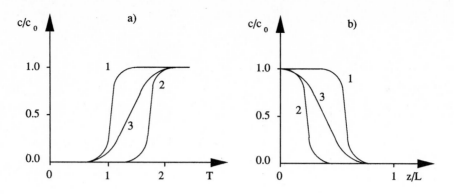

Figure 1: Typical breakthrough curves (a) and concentration front profiles (b) for LEA and non-LEA contaminants in laboratory column experiment

time (i.e. a time scaling) determined by the retardation factor. Profile 3 is typical of a reactive contaminant for which LEA does not hold. In this case the process is described by linear kinetics tending towards equilibrium ($\alpha \neq 0$ and $\alpha \neq \infty$). The tailing effect in profile 3 is easily recognized in Figure 1a. Qualitatively similar results are obtained if we plot the position of the contaminant front at a fixed time t (Figure 1b).

Pickens *et al.* [1981] suggest incorporating the tailing effect into the dispersivity term. They propose the following field experiment (Figure 2a). A non-reactive contaminant and the reactive contaminant under study are injected into an aquifer at a rate Q. Breakthrough curves are obtained from an observation borehole at a distance r from the injection well. Under steady state flow conditions the average water velocity in the pores is $v_m = Q/(2\pi b r n)$ where b is the aquifer thickness. We also have $Q\bar{t} = \pi r^2 b n$ where $\bar{t}$ is the time at which a relative concentration $c/c_o = 0.5$ is observed in the borehole. The average distance $\bar{r}$ of the contaminant front at time $\bar{t}$ is therefore given by $\bar{r} = \sqrt{Q\bar{t}/(\pi b n)}$. The dispersivity α_L can be computed as [Pickens *et al.*, 1981]:

$$\alpha_L = \frac{3r}{16\pi} \left(\frac{\Delta t}{\bar{t}}\right)^2$$

where Δt is the projection onto the t-axis of the tangent to the breakthrough curve at the point where $c/c_o = 0.5$ (Figure 2b). The retardation factor R is then computed as [Pickens *et al.*, 1981]:

$$R = \frac{V_{NR}}{V_R} = \frac{\bar{t}_R}{\bar{t}_{NR}}$$

where V_{NR} is the mean pore velocity of the non-reactive contaminant, V_R is the mean

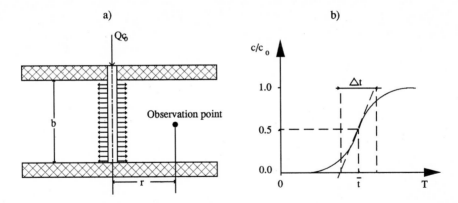

Figure 2: Field experiment of Pickens *et al.* [1981] (a) and related breakthrough curve (b)

velocity of the reactive contaminant under study, $\bar{t}_R$ is the time at which $c/c_o = 0.5$ for the reactive contaminant, and $\bar{t}_{NR}$ is the time at which $c/c_o = 0.5$ for the non-reactive contaminant. We then obtain the distribution coefficient

$$k_d = (R - 1)\frac{n}{(1 - n)\gamma_s} = \left(\frac{\bar{t}_R}{\bar{t}_{NR}} - 1\right)\frac{n}{(1 - n)\gamma_s}$$

Coupled Processes (Flow and Transport)

When solute density (or specific weight) is variable the flow and transport processes become "coupled". This is frequently the case with saltwater intrusion in coastal aquifers where the density of the solute is relatively higher than that of water and the contaminant plume tends to migrate downwards. The coupled problem can be formulated in terms of a continuity equation for the fluid mixture, a continuity equation for the contaminant, and a constitutive equation relating specific weight of the mixture to solute concentration.

The continuity equation for the mixture may be written either in terms of pressure [Voss, 1984; Sanford and Konikow, 1985; Souza and Voss, 1987] or in terms of equivalent freshwater head h [Frind, 1982; Huyakorn *et al.*, 1987; Galeati *et al.*, 1992] defined as:

$$h = \frac{p}{\gamma_o} + z$$

where p is pressure, γ_o is the reference freshwater specific weight, and z is elevation.

Defining the specific weight difference ratio as:

$$\varepsilon = \frac{\gamma_{\max} - \gamma_o}{\gamma_o}$$

where $\gamma_{\max}$ is the specific weight corresponding to the maximum concentration of solute, the constitutive relation is written as:

$$\gamma = \gamma_o(1 + \varepsilon c^*) \tag{17}$$

where c^* is relative (dimensionless) concentration (actual divided by maximum). Implicit in (17) are assumptions that conditions are isothermal and that the mixture is incompressible. Assuming further that the contaminant does not undergo chemical reaction and that its concentration is not high enough to affect viscosity or to invalidate Darcy's and Fick's laws [Hassanizadeh and Leijnse, 1988], the continuity equation (1) for the fluid mixture becomes:

$$\frac{\partial}{\partial x_i}\left[k_{ij}\left(\frac{\partial h}{\partial x_j} + \varepsilon c^* \eta_j\right)\right] = S_s \frac{\partial h}{\partial t} + n \frac{\gamma_o}{\gamma} \varepsilon \frac{\partial c^*}{\partial t} - q \tag{18}$$

where $\eta_j = 0$ along horizontal directions and $\eta_j = 1$ along the vertical.

Solute continuity is again described by equation (4), writing c^* instead of c, and the Darcy flux in this case is given by [Bear, 1979; Kinzelbach, 1986]:

$$v_i = -k_{ij}\left(\frac{\partial h}{\partial x_j} + \varepsilon c^* \eta_j\right) \tag{19}$$

From equations (18) and (19) we may write

$$\frac{\partial v_i}{\partial x_i} = -S_s \frac{\partial h}{\partial t} - n \frac{\gamma_o}{\gamma} \varepsilon \frac{\partial c^*}{\partial t} + q \tag{20}$$

Using equation (20) and differentiating the convective product term $v_i c$ in equation (4) we obtain:

$$\frac{\partial}{\partial x_i}\left(D_{ij}\frac{\partial c^*}{\partial x_j}\right) - v_i \frac{\partial c^*}{\partial x_i} = n\left(1 - \frac{\gamma_o}{\gamma}\varepsilon c^*\right)\frac{\partial c^*}{\partial t} - c^*\left(S_s \frac{\partial h}{\partial t} - q\right) - qc_1 - f \tag{21}$$

where now c_1 is the relative (dimensionless) concentration of solute injected or withdrawn and f is the relative distributed flow rate $[1/T]$. Initial and boundary conditions (2) and (7) with c^* substituted for c apply to equations (18) and (21) respectively.

Equations (18) and (21) are nonlinearly coupled through the ε term. The strength of the coupling, and hence of the nonlinearity, is dependent upon the magnitude of ε. For saltwater intrusion typical ε values range between 0.025 and 0.040. For ε values in this range the nonlinearity is moderately strong and we require ad hoc techniques to solve the coupled problem. The coupling from the $\partial c^*/\partial t$ term in equation (18) is weak and can be neglected [Galeati *et al.*, 1992]. Actually, since $c^* \leq 1$ we have $1 - (\gamma_o/\gamma)\varepsilon c^* \approx 1$ and equation (21) can be written in the form:

$$\frac{\partial}{\partial x_i}\left(D_{ij}\frac{\partial c^*}{\partial x_j}\right) - v_i\frac{\partial c^*}{\partial x_i} = n\frac{\partial c^*}{\partial t} - c^*\left(S_s\frac{\partial h}{\partial t} - q\right) - qc_1 - f \tag{22}$$

If we neglect $\partial c^*/\partial t$ in equation (18) and the elastic storage of the aquifer ($S_s = 0$), we have the simplified equations which are often used in practice:

$$\frac{\partial}{\partial x_i}\left[k_{ij}\left(\frac{\partial h}{\partial x_j} + \varepsilon c^*\eta_j\right)\right] = -q \tag{23a}$$

$$\frac{\partial}{\partial x_i}\left(D_{ij}\frac{\partial c^*}{\partial x_j}\right) - v_i\frac{\partial c^*}{\partial x_i} = n\frac{\partial c^*}{\partial t} + q(c^* - c_1) - f \tag{23b}$$

Solution by Finite Elements

The finite element solution of the flow and transport equations is described for the more general coupled problem, equations (18) and (21). The numerical equations for the uncoupled problem are readily obtained from the coupled ones by setting $\varepsilon = 0$. For a detailed analysis of the finite element method as applies to the flow and transport equations the reader is referred to Pinder and Gray [1977], Huyakorn and Pinder [1983] and Kinzelbach [1986].

Define an approximate solution $\hat{h}$ for h in the following form:

$$h \approx \hat{h} = \sum_{i=1}^{l} \hat{h}_i(t)N_i(x_1, x_2, x_3) \tag{24}$$

where $N_i(x_1, x_2, x_3)$ are linear shape or basis functions for three-dimensional tetrahedral finite elements, $\hat{h}_i$ are the unknown nodal potential heads, and l is the number of nodes. Substituting $\hat{h}$ in equation (18) and prescribing an orthogonality condition between the residual

$$L(\hat{h}) = \frac{\partial}{\partial x_i}\left[k_{ij}\left(\frac{\partial \hat{h}}{\partial x_j} + \varepsilon c^*\eta_j\right)\right] - S_s\frac{\partial \hat{h}}{\partial t} - n\frac{\gamma_o}{\gamma}\varepsilon\frac{\partial c^*}{\partial t} + q$$

and each basis function yields the Galerkin integral:

$$\int_V L(\hat{h})N_i\, dV = 0, \qquad i = 1, \ldots, l$$

where V is the flow region or integration domain.

Assuming the coordinate directions x_j to be parallel to the principal directions of hydraulic anisotropy and applying Green's lemma we get:

$$-\int_V \left[k_{11} \frac{\partial \hat{h}}{\partial x_1} \frac{\partial N_i}{\partial x_1} + k_{22} \frac{\partial \hat{h}}{\partial x_2} \frac{\partial N_i}{\partial x_2} + k_{33} \left(\frac{\partial \hat{h}}{\partial x_3} + \varepsilon c^* \right) \frac{\partial N_i}{\partial x_3} \right] dV$$

$$+ \int_\Gamma \left[k_{11} \frac{\partial \hat{h}}{\partial x_1} n_1 + k_{22} \frac{\partial \hat{h}}{\partial x_2} n_2 + k_{33} \left(\frac{\partial \hat{h}}{\partial x_3} + \varepsilon c^* \right) n_3 \right] N_i \, d\Gamma -$$

$$\int_V S_s \frac{\partial \hat{h}}{\partial t} N_i \, dV - \int_V n \frac{\gamma_o}{\gamma} \varepsilon \frac{\partial c^*}{\partial t} N_i \, dV + \int_V q N_i \, dV = 0, \qquad i = 1, \ldots, l \qquad (25)$$

where n_i, $i = 1, 2, 3$ are the direction cosines of the outer normal to the boundary Γ.

Using equation (24), changing sign, and recalling the physical meaning of the surface integral in equation (25), we obtain for tetrahedral finite elements e of volume V^e:

$$\sum_{j=1}^l \hat{h}_j \left[\sum_e \int_{V^e} \left(k_{11} \frac{\partial N_i}{\partial x_1} \frac{\partial N_j}{\partial x_1} + k_{22} \frac{\partial N_i}{\partial x_2} \frac{\partial N_j}{\partial x_2} + k_{33} \frac{\partial N_i}{\partial x_3} \frac{\partial N_j}{\partial x_3} \right) dV^e \right] +$$

$$\sum_{j=1}^l \frac{\partial \hat{h}_j}{\partial t} \left(\sum_e \int_{V^e} S_s N_i N_j \, dV^e \right) + \sum_e \int_{V^e} k_{33} \varepsilon \bar{c} \frac{\partial N_i}{\partial x_3} dV^e +$$

$$\sum_e \int_{V^e} n \frac{\gamma_o}{\gamma} \varepsilon \frac{\partial \bar{c}}{\partial t} N_i \, dV^e - \sum_e \int_{V^e} q N_i \, dV^e - \sum_e \int_{\Gamma_2^e} q_n N_i \, d\Gamma = 0, \qquad i = 1, \ldots, l \quad (26)$$

where $\sum_e$ denotes summation over the elements which share node j and $\bar{c}$ represents the average relative concentration over V^e.

In matrix form equation (26) can be written as:

$$H\hat{h} + P\frac{\partial \hat{h}}{\partial t} + q^* = 0 \qquad (27)$$

where H is the stiffness matrix, P is the capacity matrix, and vector q^* accounts for the prescribed boundary flux, the withdrawal or injection rate, and the coupling with the transport equation. Integration in time of (27) by a weighted finite difference scheme yields:

$$\left(\nu H_{t+\Delta t} + \frac{\nu P_{t+\Delta t} + (1 - \nu)P_t}{\Delta t} \right) \hat{h}_{t+\Delta t} =$$

$$\left(\frac{\nu P_{t+\Delta t} + (1 - \nu)P_t}{\Delta t} - (1 - \nu)H_t \right) \hat{h}_t - \nu q_{t+\Delta t}^* - (1 - \nu)q_t^* \qquad (28)$$

where for stability reasons the weighting parameter ν must satisfy the condition $0.5 \leq \nu \leq 1$. Selecting $\nu = 0.5$ leads to the Crank-Nicolson scheme while $\nu = 1$ gives the fully implicit backward difference scheme. Note that if hydraulic conductivity and elastic storage are not time-dependent and if the flow region V does not deform in time then H and P are constant matrices.

If the aquifer system is bounded above by a free surface where equation (8) holds, then taking into account the modification of Darcy's law expressed by equation (19) we obtain on the free surface

$$k_{33}\left(\frac{\partial h}{\partial x_3} + \varepsilon c^*\right) = I - S_y\frac{\partial h}{\partial t} \tag{29}$$

Introducing equation (29) as a second-type (or Neumann) boundary condition into the Galerkin integral (26) we get:

$$\sum_e \int_{\Gamma_2^e}\left(k_{33}\frac{\partial \hat{h}}{\partial x_3} + \varepsilon c^*\right)N_i\,d\Gamma = \sum_e \int_{\Gamma_2^e}IN_i\,d\Gamma - \sum_e \int_{\Gamma_2^e}S_y\frac{\partial \hat{h}}{\partial t}N_i\,d\Gamma$$

where Γ_2^e are surface elements on the phreatic surface. If the free surface changes are large relative to the flow depth, deforming elements and an equation more general than (25) are to be used (see for instance Neuman and Witherspoon [1971] and Isaacs [1980]).

For the finite element integration of the transport equation (21) we assume, as we did for the flow equation, an approximate solution given in the form of a linear combination of linear basis functions $N_i(x_1, x_2, x_3)$, and we write c^* (or c) as:

$$c^* \approx \hat{c} = \sum \hat{c}_i(t)N_i(x_1, x_2, x_3) \tag{30}$$

Substitution of $\hat{c}$ in equation (21) yields the residual:

$$M(\hat{c}) = \frac{\partial}{\partial x_i}\left(D_{ij}\frac{\partial \hat{c}}{\partial x_j}\right) - v_i\frac{\partial \hat{c}}{\partial x_i} - n\left(1 - \frac{\gamma_o}{\gamma}\varepsilon\hat{c}\right)\frac{\partial \hat{c}}{\partial t} + \hat{c}\left(S_s\frac{\partial h}{\partial t} - q\right) + qc_1 + f$$

Imposing the condition that the residual $M(\hat{c})$ be orthogonal over V to the test functions $W_k(x_1, x_2, x_3)$, $k = 1, \ldots, l$, we get the weighted residual equation:

$$\int_V M(\hat{c})W_k\,dV = 0, \qquad k = 1, \ldots, l \tag{31}$$

If $W_k \equiv N_k$ the classical Galerkin method is obtained. If however we use test functions W_k which are different from the basis functions N_k we obtain a different method, for example the upwind finite element scheme (e.g., Huyakorn and Taylor [1976], Huyakorn et al. [1987], Pini et al. [1989]) where nonsymmetric test functions are used for the convective component of the transport equation. Upwinding is used to increase the stability of the numerical solution.

Application of Green's lemma to both the dispersive and convective components of integral (31) is attractive in that it would allow us to incorporate Cauchy boundary conditions directly into the integral. Unfortunately this approach leads to unstable numerical solutions, as was noticed by Gureghian [1983], Huyakorn and Pinder [1983], and Huyakorn *et al.* [1985] and recently demonstrated by Galeati and Gambolati [1989]. Therefore a weak formulation which applies Green's lemma to the dispersive component only is followed in the present analysis. This yields (assuming $S_s = 0$ and neglecting the $(\gamma_o/\gamma)\varepsilon\hat{c}$ term):

$$-\int_V \left(D_{ij}\frac{\partial \hat{c}}{\partial x_j}\frac{\partial W_k}{\partial x_i} + v_i\frac{\partial \hat{c}}{\partial x_i}W_k \right) dV + \int_\Gamma \left(D_{ij}\frac{\partial \hat{c}}{\partial x_j} \right) n_i W_k \, d\Gamma$$

$$-\int_V n\frac{\partial \hat{c}}{\partial t}W_k \, dV + \int_V [(c_1 - \hat{c})q + f] W_k \, dV = 0, \qquad k = 1,\ldots,l \qquad (32)$$

Substituting the approximate solution (30) into equation (32), changing sign, and incorporating boundary conditions (7c) and (7d), we obtain:

$$\sum_{m=1}^{l} \hat{c}_m \left[\sum_e \int_{V^e} \left(D_{ij}\frac{\partial N_m}{\partial x_j}\frac{\partial W_k}{\partial x_i} + v_i\frac{\partial N_m}{\partial x_i}W_k \right) dV^e \right] +$$

$$\sum_{m=1}^{l} \frac{\partial \hat{c}_m}{\partial t} \left(\sum_e \int_{V^e} n N_m W_k \, dV^e \right) + \sum_{m=1}^{l} \hat{c}_m \left(\sum_e \int_{V^e} q N_m W_k \, dV^e \right) -$$

$$\sum_e \int_{V^e} (qc_1 + f) W_k \, dV^e - \sum_e \int_{\Gamma_4^e} q^d W_k \, d\Gamma$$

$$-\sum_e \int_{\Gamma_5^e} q^c W_k \, d\Gamma - \sum_{m=1}^{l} \hat{c}_m \left(\sum_e \int_{\Gamma_5^e} (v_i n_i) N_m W_k \, d\Gamma \right) = 0, \qquad k = 1,\ldots,l \qquad (33)$$

Equation (33) represents a system of l equations for the unknown concentrations $\hat{c}_m$ $(m = 1,\ldots,l)$, and is coupled with the flow equation (26) through the ε term. Setting

$$a_{km} = \sum_e \int_{V^e} D_{ij}\frac{\partial N_m}{\partial x_j}\frac{\partial W_k}{\partial x_i} \, dV^e$$

$$b_{km} = \sum_e \int_{V^e} v_i\frac{\partial N_m}{\partial x_i} W_k \, dV^e$$

$$c_{km} = \sum_e \int_{V^e} n N_m W_k \, dV^e$$

$$d_{km} = -\sum_e \int_{\Gamma_5^e} (v_i n_i) N_m W_k \, d\Gamma$$

equation (33) becomes, in matrix form,

$$(A + B + D)\hat{c} + C\frac{\partial \hat{c}}{\partial t} + r^* = 0 \qquad (34)$$

where the vector r^* accounts for point and distributed contaminant sources and sinks. Vector r^* can also incorporate the unknown term [Huyakorn *et al.*, 1987]:

$$\sum_{m=1}^{l} \hat{c}_m \sum_e \int_{V^e} q N_m W_k \, dV^e \qquad (35)$$

However this contribution to the transport equation only appears if clean water is injected into the aquifer (in which case we have $c_1 = 0$ in equation (33)). In the more general case we have $c_1 = c^* \approx \hat{c}$, (for instance if contaminated groundwater is withdrawn), (35) cancels out with the term $-\sum_e \int_{V^e} q c_1 W_k \, dV^e$ from equation (33), and there is no contribution to r^* of equation (34).

Integration in time of equation (34) is again performed using a weighted finite difference scheme:

$$\left[\nu(A + B + D)_{t+\Delta t} + \frac{\nu C_{t+\Delta t} + (1 - \nu)C_t}{\Delta t} \right] \hat{c}_{t+\Delta t} =$$
$$\left[\frac{\nu C_{t+\Delta t} + (1 - \nu)C_t}{\Delta t} - (1 - \nu)(A + B + D)_t \right] \hat{c}_t - \nu r^*_{t+\Delta t} - (1 - \nu)r^*_t \qquad (36)$$

where as before $1/2 \leq \nu \leq 1$. Unlike the flow equation, the transport equation is quite sensitive to the value of the weighting parameter ν. A value close to $1/2$ leads to accurate but unstable solutions while values close to 1 yield good stability but with larger numerical dispersion [Peyret and Taylor, 1983].

Note that in equation (36) matrices A, B, and D are nonlinearly dependent on the Darcy velocity, and therefore on the hydraulic potential head via equation (19). The solution to the flow equation (28) in turn depends on concentration through the term q^*. Thus equations (28) and (36) exhibit a nonlinear coupling and require a special iterative strategy for their solution. In the uncoupled case ($\varepsilon = 0$), iteration is not required and equations (28) and (36) are solved independently at each time step.

A different approach for solving the coupled flow and transport equations was presented by Galeati et al. [1992] who developed an Eulerian-Lagrangian model where the flow equation is solved by finite elements and the transport equation by a combination of a method of characteristics for the convective part and an Eulerian grid method (such as finite elements) for the dispersive part. A modified method of characteristics (MMOC) is used which is implicit, unconditionally stable, and tracks particles backwards according to a relatively efficient single-step reverse algorithm of the kind described by Neuman [1981], Douglas Jr. and Russell [1982], Baptista et al. [1984], Casulli [1987], and others. One advantage of this approach is that it requires only the solution of symmetric positive definite systems of equations, for which accurate and cost-effective preconditioned conjugate gradient schemes may be used [Gambolati and Perdon, 1984]. Another advantage is its suitability to parallel computation, since each particle can be tracked independently on separate processors. The nonlinearity is handled through an iterative procedure conceptually similar to the one described in the next section.

Efficient Solution Schemes

Solution with Full Coupling

In solving the coupled equations (28) and (36) we must remember that vector q^* of equation (28) contains averaged concentration values and that matrices A, B, and D of equation (36) depend on $\hat{h}$ (through the Darcy velocity). In some cases C and r^* may also be $\hat{h}$-dependent, for instance if $c_1 \neq c$ in the pumping wells, or if q represents a distributed source with $c_1 = 0$, or if the numerical model includes the terms which have been dropped from equation (32), i.e. the contributions:

$$\int_V \frac{\gamma_0}{\gamma} \varepsilon \hat{c} \frac{\partial \hat{c}}{\partial t} W_k \, dV \tag{37}$$

$$\int_V \hat{c} S_s \frac{\partial \hat{h}}{\partial t} W_k \, dV \tag{38}$$

For ε values in the range $0.025 \leq \varepsilon \leq 0.040$ mentioned earlier, the dependence of the flow solution on transport is moderately strong. In this case an iterative Picard scheme may be appropriate to solve the nonlinear equations. Convergence can be accelerated with the relaxation procedure described later. Since H and P in equation (28) are symmetric positive definite matrices, the flow equation is conveniently solved using the preconditioned conjugate gradient method with incomplete Cholesky factorization of the coefficient matrix [Gambolati and Perdon, 1984]. By contrast the coefficient matrix in equation (36) is nonsymmetric and its coefficients change at each Picard iteration. The matrix topology, however, is fixed for any given problem, and hence a preliminary reordering can be performed in order to minimize the computational cost of using a direct method of solution. The reordered matrix is first symbolically factorized into its lower triangular part L and its upper triangular part U, and then numerically factorized at each Picard iteration. The use of this solution strategy for the transport equation allows for a marked saving of computer memory and time since the preliminary reordering minimizes the number of nonzero elements in L and U, and the only operations to be repeated at each iteration are the numerical computation of L and U and the subsequent backward and forward substitutions.

The iterative solution of the fully coupled flow and transport equations can be summarized in the following steps:

1. at the start of each time step use the potential head and concentration values from the previous time step as initial estimates $\hat{h}_i^{r=0}$ and $\hat{c}_i^{r=0}$, where r denotes iteration level and i node number.

2. Solve equation (28) for $\hat{h}_i^{r+1}$ using $\hat{c}_i^r$ in the evaluation of $q_{t+\Delta t}^*$.

3. Compute the Darcy velocities from equation (19) and the dispersion coefficients from equation (6) using $\hat{h}_i^{r+1}$ and $\hat{c}_i^r$.

4. Update matrices A, B, and D of equation (36) using the velocity values determined in step 3. If needed, matrix C and the term $r_{t+\Delta t}^*$ can also be updated by computing equations (37) and (38) using $\hat{c}_i^r$ and evaluating $S_s(\partial \hat{h}/\partial t)$ numerically over each element.

5. Solve equation (36) for $\hat{c}_i^{r+1}$.

6. Check for convergence. If the differences $\mid \hat{c}_i^{r+1} - \hat{c}_i^r \mid$ and $\mid \hat{h}_i^{r+1} - \hat{h}_i^r \mid$, $i = 1, \ldots, l$ are below a prescribed tolerance the iterative procedure has converged and we proceed to the next time step. Otherwise assign $\hat{h}_i^{r+1}$ to $\hat{h}_i^r$ and $\hat{c}_i^{r+1}$ to $\hat{c}_i^r$ and repeat steps 2–5. If convergence does not occur within a pre-set maximum number of iterations the time step size is reduced and the procedure is repeated for the current time step.

Convergence should be rapid if the time step size Δt is sufficiently small. The optimal value of Δt for a given simulation will depend on the problem parameters and boundary conditions. Frind [1982] suggested however that the time step size for a simulation is dictated more by the need to reduce numerical dispersion than by the requirements for convergence of the Picard iteration, and he reported convergence of the Picard procedure in 2–4 iterations for a two-dimensional simulation. For the three-dimensional simulations reported in Huyakorn et al. [1987], on the other hand, it was necessary to accelerate Picard convergence by means of the relaxation formula $\psi_i^{r+1} = (1 - \omega)\psi_i^r + \omega \psi_i^{r+1}$, where ψ_i represents either $\hat{h}_i$ or $\hat{c}_i$ and ω $(0 \leq \omega \leq 1)$ is an under-relaxation parameter defined as $\omega = (3 + \xi)/(3 + |\xi|)$ for $\xi \geq -1$ and $\omega = 0.5/|\xi|$ for $\xi < -1$. ξ is a parameter defined as $\xi = 1$ for $r = 0$ and $\xi = e_{r+1}/(\xi_{old} e_r)$ for $r > 0$, where ξ_{old} is the previous iteration value of ξ and e_{r+1} is the largest absolute difference in $\hat{c}_i$ or $\hat{h}_i$ values between the current and previous iterations.

Solution with Partial Coupling

In the solution procedure for the partially coupled case Picard iterations are not performed [Galeati et al., 1992]. Equation (28) is solved only once at time level $k + 1$ using the $\hat{c}_i^k$ values at time level k and setting $\hat{c}_i^{k+1} = \hat{c}_i^k$ in the computation of $q_{t+\Delta t}^*$. This

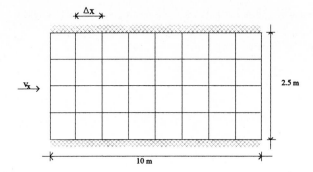

Figure 3: Geometry for dual porosity transport example

allows us to evaluate $\hat{h}_i^{k+1}$ which is then used to update the transport matrices (and possibly also vector r^*) as described in step 4 earlier. Equation (36) is then solved only once to provide $\hat{c}_i^{k+1}$. The rationale for this simplified procedure is that the dependence of $\hat{h}$ on $\hat{c}$ is weak for ε values smaller than 0.04 [Galeati et $al.$, 1992]. Computational costs can be greatly reduced in this way.

Uncoupled Solution

This is the classical problem where flow and transport are uncoupled ($\varepsilon = 0$ in equations (28) and (36)). There is no need to iterate in the solution procedure: the output produced from solving the flow problem is used in the calculation of Darcy velocities which become part of the input data for the transport code. If the flow field is steady the coefficient matrices A, B, and D do not need to be updated at each time step.

Illustrative Examples

Transport with the Dual Porosity Model

We present a solution to equation (16) (the dual porosity model with non-LEA conditions). The numerical example is of two-dimensional flow and transport (Figure 3) with $v_x = 1$ m/s, $v_y = 0$, and $n_m = n_{im} = 0.2$. The condition $c = 1$ is set on the left boundary at time $t = 0$ and the behavior of the concentration is examined on a horizontal cross section of the region for different values of the dispersivities α_L and α_T and the transfer coefficient α. Recall that for $\alpha = 0$ or $\alpha = \infty$ the dual porosity equations reduce to the classical transport equation (4) with $n = n_m$ or $n = n_{im} + n_m$ respectively, that is the controlling processes are advection and dispersion.

The concentration profiles after 20 time steps are shown in Figure 4 for $\alpha_L = \alpha_T = 10^{-2}$ m. Note the pronounced tailing effect for intermediate α values. For this example the dispersivity values are small and thus the concentration profiles display overshooting which is characteristic of instabilities arising from advection dominated transport. In fact the Peclet number, defined as $Pe = \Delta x/\alpha_L$, is much higher than the theoretical value $Pe = 2$ [Frind, 1987] that defines the stability region for one-dimensional parabolic equations.

Overshooting is decreased when Pe is reduced, as we see in Figure 5 where the dispersivities were increased by a factor of 10, thus yielding a Peclet number $Pe = 2.5$.

Coupled Flow and Transport for a Passive Contaminant

We present a solution to the coupled flow and transport problem described earlier. Equations (18) and (22) are discretized using the finite element procedure, yielding equations (28) and (36).

Two iterative methods, a Picard scheme and a partial Newton scheme, are used to solve the nonlinearly coupled system. The Picard approach is the 6-step iterative procedure described earlier for solving the fully coupled problem. For the partial Newton scheme we follow the same 6-step procedure, except that we express the Darcy velocities in steps 3 and 4 at $\hat{c}_i^{r+1}$ instead of at $\hat{c}_i^r$, and we linearize the transport equation using a Newton procedure.

The problem domain is shown in Figure 6. The following parameters are used: $k_1 = k_2 = 0.01$ m/s, $n = 0.35$, $S_s = 10^{-2}$ m^{-1}, $\epsilon = 0.025$, and $\alpha_L = \alpha_T = 0.035$ m. Zero flux boundary conditions are imposed on the top and bottom of the aquifer for both the flow and transport equations. On the right boundary there is an incoming flux of fresh water $q_n = 6.6 \times 10^{-5}$ m/s, and zero relative concentration. The equivalent freshwater head is set to hydrostatic conditions along the left boundary. The upper part (20 m) of the left boundary has a zero concentration flux condition while in the lower 80 m we impose a relative concentration equal to 1 (sea water). The two-dimensional problem domain is discretized using a three-dimensional grid of unit width with 693 nodes and 2400 tetrahedra.

In Figure 7 we show the steady state potential head contours together with the velocity field, and the steady state concentration contour lines. The effect of the density difference between freshwater and saltwater can be easily seen at the left boundary, where an inversion of the flux occurs at the bottom portion of the boundary. In Figure 8

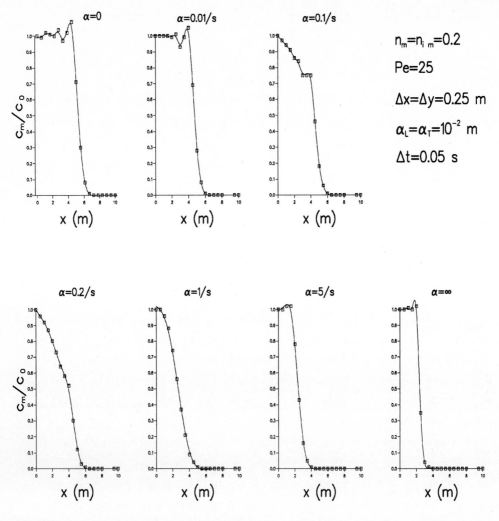

Figure 4: Concentration vs. x at $t = 20\Delta t$ ($\Delta t = 0.05$ s) for $Pe = 25$ and full range of α-values

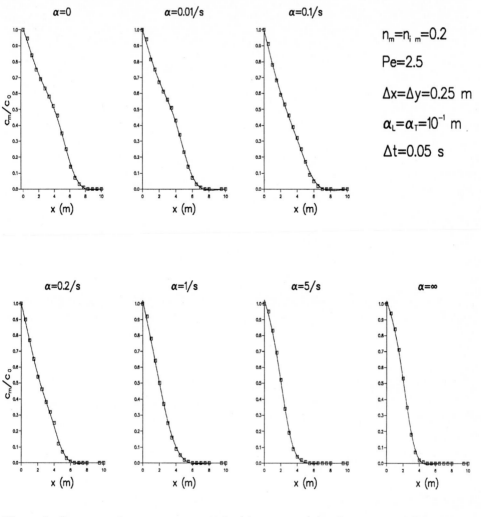

Figure 5: Concentration vs. x at $t = 20\Delta t$ ($\Delta t = 0.05$ s) for $Pe = 2.5$ and full range of α-values

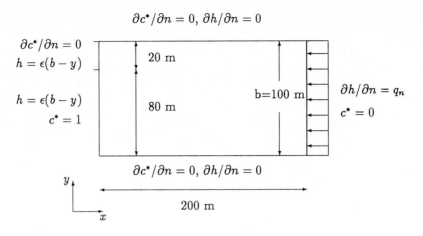

$$\partial c^*/\partial n = 0, \ \partial h/\partial n = 0$$

$\partial c^*/\partial n = 0$
$h = \epsilon(b - y)$

$h = \epsilon(b - y)$
$c^* = 1$

20 m

80 m

$b{=}100$ m

$\partial h/\partial n = q_n$
$c^* = 0$

$$\partial c^*/\partial n = 0, \ \partial h/\partial n = 0$$

200 m

Figure 6: Geometry for coupled flow and transport example

we compare the behavior of the Picard and partial Newton schemes for this steady state problem. Without the relaxation procedure described in a previous section the Picard scheme fails to converge, whereas with relaxation it converges in 26 iterations. Better behavior is observed for the partial Newton scheme, which converges in 9 iterations without relaxation and in 10 iterations with relaxation. The convergence criterion used for this problem is that the maximum norm of the solution difference between the current and previous iterations should be less than 10^{-2}. In Figure 8 we plot only the concentration norm, since the potential head solution converges much more rapidly than the concentration solution.

Conclusions

We have provided a brief overview of some of the mathematical and numerical principles pertaining to the modeling of groundwater flow and chemical transport in porous media. We have limited our discussion to the case of a perfectly miscible contaminant partially adsorbed onto the solid phase, and we have focused on finite element techniques for numerically solving the coupled flow and transport systems. For the case of reactive contaminants we considered sorption as modeled under LEA (local equilibrium assumption) and non-LEA conditions. Procedures for solving the nonlinearly coupled systems were presented in some detail for the case of moderately strong (full) coupling and for the case of weak (partial) coupling. Their application to the simpler uncoupled problem is straightforward.

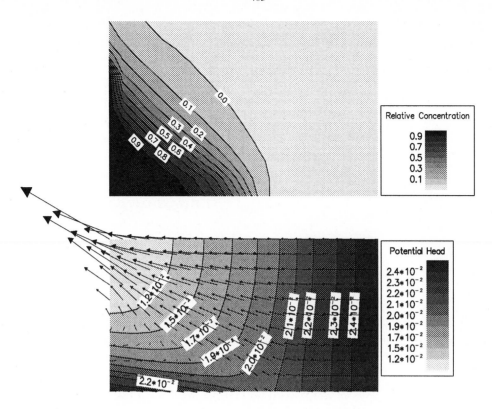

Figure 7: Steady state potential head, velocity, and concentration fields for coupled flow and transport example

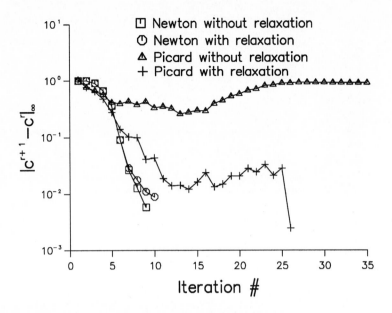

Figure 8: Convergence profiles for Picard and partial Newton schemes for coupled flow and transport example

Illustrative numerical examples related to non-LEA chemical transport of a reactive contaminant and to coupled flow and transport of a passive contaminant were presented and briefly discussed. For the coupled problem the convergence behavior of the Picard and partial Newton procedures were compared.

There are many issues related to the physics and modeling of groundwater flow and chemical transport processes which remain unresolved, and many fertile areas of continuing research. More realistic models of transport processes such as dispersion and complex chemical reaction are required. On the numerical side, there is a need for robust and efficient algorithms to solve the nonsymmetric systems arising from the discretization of the transport equation, and alternative methods for dealing with the nonlinearities and the coupling of the flow, transport, and chemical equations need to be investigated. For more details on recent progress in the modeling of flow and transport processes, the reader is referred to recent reviews by Nielsen *et al.* [1986], Pinder and Abriola [1986], Gee *et al.* [1991], Mangold and Tsang [1991], Sardin *et al.* [1991], and Sudicky and Huyakorn [1991].

Acknowledgments This work has been partially supported by the Italian CNR, Gruppo Nazionale per la Difesa delle Catastrofi Idrogeologiche, linea di Ricerca n. 4.

References

Anderson MP (1979) Using models to simulate the movement of contaminant through groundwater flow systems. *Crit Rev Env Contr* 9:97–156

Baptista AM, Adams EE, Stolzenback KD (1984) The 2-D unsteady transport equation solved by the combined use of the finite element method and the method of characteristics. In: *Proc V Int Conf on Finite Elements in Water Resources* Springer-Verlag, New York, pp 353–362

Bear J (1972) *Dynamics of Fluids in Porous Media.* Elsevier, New York

Bear J (1979) *Hydraulics of Groundwater.* McGraw Hill, New York

Bredehoeft JD, Pinder GF (1973) Mass transport in flowing groundwater. *Water Resour Res* 9:194–210

Brusseau ML, Rao PSC (1989) Sorption nonideality during organic contaminant transport in porous media. *Crit Rev Env Contr* 19(1):33–99

Burnett RD, Frind EO (1986) A comparison of transport simulation in two and three dimensions using alternating direction Galerkin techniques. In: *Proc VI Int Conf on Finite Elements in Water Resources* Springer-Verlag, New York, pp 459–475

Casulli V (1987) Eulerian-Lagrangian methods for hyperbolic and convection dominated parabolic problems. In: Taylor C, Owen DRJ, Hinton EE (eds) *Computational Methods for Nonlinear Problems*, Pineridge, Swansea, pp 239–269

Coats KH, Smith BD (1964) Dead-end pore volumes and dispersion in porous media. *Soc Pet Eng J* 4:73–84

Douglas, Jr. J, Russell TF (1982) Numerical methods for convection dominated diffusion problems based on combining the method of characteristics with finite element or finite difference procedures. *SIAM J Num Anal* 19:871–885

Frind EO (1982) Simulation of long-term transient density-dependent transport in groundwater. *Adv Water Res* 5:73–88

Frind EO (1987) Modelling of contaminant transport in groundwater–an overview. In: *The Canadian Soc Civil Eng Centennial Symp on Management of Waste Contamination of Groundwater*, Montreal, pp 1–30

Galeati G, Gambolati G (1989) On boundary conditions and point sources in the finite element integration of the transport equation. *Water Resour Res* 25:847–856

Galeati G, Gambolati G, Neuman SP (1992) Coupled and partially coupled Eulerian-Lagrangian model of freshwater-seawater mixing. *Water Resour Res* 28(1):149–165

Gambolati G (1973) Equation for one-dimensional vertical flow of groundwater. 1. The rigorous theory. *Water Resour Res* 9:1022–1028

Gambolati G, Perdon AM (1984) The conjugate gradients in flow and land subsidence modeling. In: Bear J, Corapcioglu Y (eds) *Fundamental of Transport Phenomena in Porous Media* NATO-ASI Series, Applied Sciences 82, Martinus Nijoff B.V., The Hague, pp 953–984

Gee GW, Kincaid CT, Lenhard RJ, Simmons CS (1991) Recent studies of flow and transport in the vadose zone. In: *U.S. National Report to International Union of Geodesy and Geophysics 1987–1990: Contributions in Hydrology* American Geophysical Union, Washington, DC, pp 227–239

Gureghian AB (1983) TRIPM, a two-dimensional finite element model for the simultaneous transport of water and reacting solutes through saturated and unsaturated porous media. Technical Report ONWI-465 Off. of Nuclear Water Isolation Columbus, Ohio

Hassanizadeh SN, Leijnse T (1988) On the modeling of brine transport in porous media. *Water Resour Res* 24:326–330

Huyakorn PS, Taylor C (1976) Finite element models for coupled groundwater and convective dispersion. In: *Proc I Int Conf on Finite Elements in Water Resources* Pentech, London, pp 1131–1151

Huyakorn PS, Pinder GF (1983) *Computational Methods in Subsurface Flow.* Academic Press, London

Huyakorn PS, Mercer JW, Ward DS (1985) Finite element matrix and mass balance computational schemes for transport in variably saturated porous media. *Water Resour Res* 21:346–358

Huyakorn PS, Geoffrey B, Andersen PF (1986) Finite element algorithms for simulating three dimensional groundwater flow and solute transport in multilayer systems. *Water Resour Res* 22:361–374

Huyakorn PS, Andersen PF, Mercer JW, White HO (1987) Saltwater intrusion in aquifers: Development and testing of a three dimensional finite element model. *Water Resour Res* 23:293–312

Isaacs LT (1980) Location of seepage free surface in finite element analysis. *Civil Eng Trans Inst Eng Aust* CE22(1):9–17

Kinzelbach W (1986) *Groundwater Modeling.* Elsevier, Amsterdam

Lapidus L, Amundson NR (1952) Mathematics of adsorption in beds, VI. The effect of longitudinal diffusion in ion exchange and chromatographic columns. *J Phys Chem* 56:984–988

Mangold DC, Tsang CF (1991) A summary of subsurface hydrological and hydrochemical models. *Rev Geophys* 29:51–79

Miller CT, Weber Jr. WJ (1984) Modeling organic contaminant partitioning in groundwater systems. *Ground Water* 22:584–592

Neuman SP (1981) A Eulerian-Lagrangian numerical scheme for the dispersion convection equation using conjugate space time grids. *J Comp Phys* 41:270–294

Neuman SP (1984) Adaptive Eulerian-Lagrangian finite element method for advection-dispersion. *Int J Numer Methods Eng* 20:321–337

Neuman SP, Witherspoon PA (1971) Analysis of nonsteady flow with a free surface using the finite element method. *Water Resour Res* 7:611–623

Nielsen DR, van Genuchten MT, Biggar AJW (1986) Water flow and solute transport processes in the unsaturated zone. *Water Resour Res* 22(9):89S–108S

Peyret R, Taylor TD (1983) *Computational Methods for Fluid Flow*. Springer-Verlag, New York

Pickens JF, Jackson RE, Inch KJ, Merrit WF (1981) Measurement of distribution coefficients using a radial injection dual-tracer test. *Water Resour Res* 17:529–544

Pinder GF, Gray WG (1977) *Finite Element Simulation in Surface and Subsurface Hydrology*. Academic Press, San Diego

Pinder GF, Abriola LM (1986) On the simulation of nonaqueous phase organic compounds in the subsurface. *Water Resour Res* 22(9):109S–119S

Pini G, Gambolati G, Galeati G (1989) 3-D finite element transport models by upwind preconditioned conjugate gradients. *Adv Water Res* 12:54–58

Putti M, Yeh WWG, Mulder WA (1990) A triangular finite volume approach with high resolution upwind terms for the solution of groundwater transport equations. *Water Resour Res* 26(12):2865–2880

Sanford WE, Konikow LF (1985) A two constituent solute transport model for groundwater having variable density. Technical Report 4279 U.S. Geol. Surv. Water Resour. Invest.

Sardin M, Schweich D, Leij FJ, van Genuchten MT (1991) Modeling the nonequilibrium transport of linearly interacting solutes in porous media: A review. *Water Resour Res* 27(9):2287–2307

Souza WR, Voss CI (1987) Analysis of an anisotropic coastal aquifer system using variable-density flow and solute transport simulation. *J Hydrol* 92:17–41

Sudicky EA, Huyakorn PS (1991) Contaminant migration in imperfectly known heterogeneous groundwater systems. In: *U.S. National Report to International Union of Geodesy and Geophysics 1987–1990: Contributions in Hydrology* American Geophysical Union, Washington, DC, pp 240–253 van Genuchten MT, Wierenga PJ (1976) Mass transfer studies in sorbing porous media: 1. analytical solutions. *Soil Science Soc Amer J* 40:463–480

Voss CI (1984) SUTRA: A finite-element simulation model for saturated-unsaturated, fluid-density-dependent groundwater flow with energy transport of chemically-reactive simple-species solute transport. Technical Report 4369 U.S. Geol. Surv. Water Resour. Invest.

Weber Jr. WJ (1972) *Physicochemical Processes for Water Quality Control.* Wiley-Interscience, New York

PART III. SPECIFIC PROBLEMS AND APPLICATIONS

GROUNDWATER RECHARGE WITH RECLAIMED MUNICIPAL WASTEWATER

Takashi Asano
Department of Civil and Environmental Engineering
University of California at Davis
Davis, CA 95616
U.S.A.

Natural replenishment of the vast supply of underground water occurs very slowly; therefore, excessive continued exploitation of groundwater at a rate greater than this replenishment causes declining groundwater levels in the long term and if not corrected, leads to eventual mining of groundwater. To increase the natural supply of groundwater, artificial recharge of groundwater basins is becoming increasingly important in groundwater management and particularly in situations where the conjunctive use of surface water and groundwater resources is being considered.

Groundwater recharge with reclaimed municipal wastewater is an approach to water reuse that results in the planned augmentation of groundwater for various beneficial uses. The beneficial uses are the many ways water can be used, either directly by people, or for their overall benefit. Groundwater is used as a source of water supply, and its major beneficial uses include municipal water supply, agricultural irrigation, and industrial water supply. The purposes of artificial recharge of groundwater have been (Bouwer, 1978; Todd, 1980; Asano, 1985):

o   To reduce, stop, or even reverse declines of groundwater levels
o   To protect underground freshwater in coastal aquifers against saltwater intrusion from the ocean
o   To store surface water, including flood or other surplus water, and reclaimed municipal wastewater for future use.

Groundwater recharge is also incidentally achieved in land treatment and disposal of municipal and industrial wastewater via percolation and infiltration.

There are several advantages to storing water underground:

o   The cost of artificial recharge may be less than the cost of equivalent surface water reservoirs

NATO ASI Series, Vol. G 32
Migration and Fate of Pollutants in Soils and Subsoils
Edited by D. Petruzzelli and F. G. Helfferich
© Springer-Verlag Berlin Heidelberg 1993

o   The aquifer serves as an eventual distribution system in underground and may eliminate the need for transmission pipelines or canals for surface water

o   Water stored in surface reservoirs is subject to evaporation, potential taste and odor problems due to algae and other aquatic productivity, and to pollution; these may be avoided by underground storage

o   Suitable sites for surface water reservoirs may not be available or environmentally acceptable

o   The inclusion of groundwater recharge in a water reuse project may provide psychological and esthetic benefits as a result of the transition between reclaimed municipal wastewater and groundwater. This aspect is particularly significant when a possibility exists in the wastewater reclamation and reuse plan to augment substantial portions of domestic or potable water supplies.

Technique of Groundwater Recharge

Two types of groundwater recharge are commonly used with reclaimed municipal wastewater: surface spreading or percolation, and direct injection.

Groundwater Recharge by Surface Spreading. Direct surface spreading is the simplest, oldest, and most widely applied method of artificial recharge (Todd, 1980). In surface spreading, recharge waters such as treated municipal wastewater percolates from spreading basins through the unsaturated groundwater (vadose) zone. Infiltration basins are the most favored methods of recharge because they allow efficient use of space and require only simple maintenance. In general, infiltration rates are highest where soil and vegetation are undisturbed.

Where hydrogeological conditions are favorable for groundwater recharge with spreading basins, wastewater reclamation can be implemented relatively simply by the soil-aquifer treatment (SAT) process. The necessary treatment can be obtained by the filtration process as the wastewater percolates through the soil and the vadose zone, down to the groundwater and then some distance through the aquifer. Recommended pretreatment for municipal wastewater for the SAT process includes primary treatment or stabilization pond. Pretreatment processes that leave high algal concentrations in the recharge water should be avoided. Algae can severely clog the soil of infiltration basins. While renovated wastewater from the SAT process is much better water quality than the influent wastewater, it could be lower quality than the native groundwater. Thus, the SAT process should be designed and managed to avoid encroachment into the native groundwater and to use only a portion of the aquifer. The distance between infiltration basins and wells or drains should be as large as possible, usually at least 50-100 m to give adequate soil-aquifer treatment (Bouwer, 1978 and 1988).

The advantage of groundwater recharge by surface spreading is:

o   Groundwater supplies may be replenished in the vicinity of metropolitan and agricultural areas where groundwater overdraft is severe

o   Surface spreading provides the added benefits of the treatment effect of soils and transporting facilities of aquifers.

Direct Injection to Groundwater.   Direct subsurface recharge is achieved when water is conveyed and emplaced directly into an aquifer. In direct injection, generally, highly treated wastewater is pumped directly into the groundwater zone, usually into a well-confined aquifer. Groundwater recharge by direct injection is practiced:

o   Where groundwater is deep or where the topography or existing land use makes surface spreading impractical or too expensive

o   When direct injection is particularly effective in creating freshwater barriers in coastal aquifers against intrusion of saltwater.

Both in surface spreading and direct injection, locating the extraction wells as great a distance as possible from the spreading basins or the injection wells increases the flow path length and residence time of the recharged water. These separations in space and in time contribute to the mixing of the recharged water and the other aquifer contents, and the loss of identity of the recharged water originated from municipal wastewater. The latter is an important consideration in successful wastewater reuse from the public acceptance point of view.

In arid climates where the practice of groundwater recharge is most imperative, recharge will occur through such means as dry river beds and spreading basins, and in most situations there will be an unsaturated zone between the surface and the aquifer. The fundamental principles that apply to the saturated flow can also be extended to unsaturated flow. Unsaturated flow processes, however, are complicated by complex relationships between water content, pressure head, and hydraulic conductivity. The driving forces for flow under saturated conditions are total head gradients that include positive pressure heads; whereas for flow under unsaturated conditions, driving forces are total head gradients that include negative pressure heads.

Pretreatment for Groundwater Recharge

Four water quality factors are particularly significant in groundwater recharge with reclaimed wastewater: (1) microbiological quality, (2) total mineral content (total dissolved solids), (3) presence of toxicant of the heavy metal type, and (4) the concentration of stable organic substances. Thus, groundwater recharge with reclaimed wastewater presents a wide spectrum of technical and health challenges that must be carefully evaluated. Some basic questions that need to be addressed include (Asano and Wassermann, 1980; Roberts, 1980):

o   What treatment processes are available for producing water suitable
       for groundwater recharge?
o   How do these processes perform in practice?
o   How does water quality change during infiltration-percolation
       and in the groundwater zone?
o   What do infiltration-percolation and groundwater passage contribute
       to the overall treatment system performance and reliability?
o   What are the important health issues?
o   How do these issues influence groundwater recharge regulations
       at the points of recharge and extraction?
o   What benefits and problems have been experienced in practice?

Pretreatment requirements for groundwater recharge vary considerably, depending on purpose of groundwater recharge, sources of reclaimed wastewater, recharge methods, and location. Although the surface spreading method of groundwater recharge is in itself an effective form of wastewater treatment, a certain degree of pretreatment must be provided to untreated municipal wastewater before it can be used for groundwater recharge.

A variation of the SAT process in Ben Sergao (a suburb of Agadir, Morocco) is described below. The on-going pilot study is of interest not only for the Greater Agadir where water resources are limited but also for a number of cities in Morocco where reuse of treated wastewater constitutes an essential option in wastewater treatment and disposal.

***

Wastewater Treatment of Greater Agadir, Morocco (Bennani, *et al.*, 1992)

Today with its population of over 350,000, the rapidly growing Greater Agadir faces need for wastewater treatment and increased demand for water supply. The two main discharges of raw sewage, one into the port area, the other into the bed of the Souss wadi, at a few kilometers of its mouth, become less and less compatible with valuable tourist attraction.

In the Moroccan-French cooperation project, pilot wastewater treatment through dune sand infiltration-percolation is underway at Ben Sergao (a suburb of Agadir). After treatment by anaerobic stabilization pond (chemical oxygen demand, COD, of raw sewage = 1,190 mg/L), the pilot wastewater treatment by infiltration-percolation plant treats 1,000 m^3/d of highly concentrated effluents in five infiltration basins of 1,500 m^2 each, consisting of two- meter thick eolian sand. The anaerobic stabilization pond (1,500 m^3 for a theoretical residence time of 2 days; depth of the pond: 3-4 m) is used to reduce suspended solids (40-50 %) and organic matter (50-60 %); thus, reducing the surface areas necessary for the infiltration basin. The basin is submerged for 8 hours and stays dry for 16 hours.

The wastewater was infiltrated at the rate of one meter per day. With this process nearly 100 percent of suspended solids and 95 percent of COD are removed, and 85 % of nitrogen is in oxidized forms and 56 % removed. Microbiological quality of raw sewage, pond effluent, and percolated water are shown below:

	Raw Sewage	Pond Effluent	Percolated Water	Overall Removal Efficiency
Bacteria:				
Fecal coliforms (Geometric mean counts/100 mL)	6×10^6	5×10^5	327	4.26 logs
Fecal streptococci (Geometric mean counts/100 mL)	2×10^7	1.6×10^6	346	4.78 logs
Parasites:				
Nematode egg (Numbers/L)	139	32	0	100 %
Cestode egg (Numbers/L)	75	18	0	100 %
Total helminths egg (Numbers/L)	214	47	0	100 %

The percolated water will be used in growing tomatoes (a vegetable extensively cultivated in the Agadir region), public gardens, and future golf courses.

***

Inasmuch as recharged groundwater may be an eventual source of potable water supply, groundwater recharge with reclaimed municipal wastewater may often involve treatment beyond the conventional secondary wastewater treatment. In the past, prior to the recent concerns about protozoan cysts, enteric viruses, and trace organics in drinking water, several apparently successful groundwater recharge projects were developed and operated using secondary effluents in spreading basins. However, because of the increasing concerns for these contaminants, groundwater recharge with reclaimed wastewater normally entails further treatment following conventional secondary treatment. For example, for surface spreading operations practiced in the United States, common wastewater reclamation processes include primary and secondary wastewater treatment, and tertiary granular-medium filtration followed by chlorine disinfection.

The groundwater recharge operation at the Whittier Narrows spreading basins in Los Angeles County, California is an example where the reclaimed wastewaters from both San Jose Creek and Whittier Narrows Wastewater Treatment Plants, operated by the County Sanitation Districts of Los Angeles County, are used as the recharge water along with the stormwater runoff, river water, and imported water from northern California and Arizona as shown in the following section.

***

Groundwater Recharge in the Water Resources Planning in Southern California

Like many semiarid regions of the United States, Southern California does not receive sufficient water from local sources to support 18 million people residing in the area. Almost two thirds of the water supply is imported 300 to 800 Km from the point of use. The remainder is derived from local groundwater basins. In some areas, the occurrence of overdraft conditions and saltwater intrusion has led to the adjudication of groundwater extractions and the implementation of artificial groundwater recharge. Water sources used for groundwater replenishment include stormwater runoff, imported water from northern California and Arizona, and reclaimed municipal wastewater.

The Rio Hondo spreading facility is the one of the largest recharge facilities in the U.S.A., having a total of 1.8 Km^2 of wetted areas available for spreading. The San Gabriel River spreading basins have some 0.5 Km^2 wetted areas available for recharge, with an additional 0.5 Km^2 of river bottom that can be used when needed. The Rio Hondo and San Gabriel River spreading grounds are subdivided into individual basins that range in size from 1.6 - 8.1 ha (Nellor, et al., 1985).

1. Recharge Operation. Under normal operating conditions, batteries of the basins are rotated through a 21-day cycle consisting of: (1) a 7-day flooding period during which the basins are filled to maintain a constant 1.2 m depth, (2) a 7-day draining period during which the flow to the basins is terminated and the basins are allowed to drain, and (3) a 7-

day drying period during which the basins are allowed to dry out thoroughly. This wetting and drying operation serves several purposes, including maintenance of aerobic conditions in the upper soil strata and vector control.

Infiltration rates during the flooding period average about 0.6 m/d. The capacity of the spreading grounds during normal operation is about 8.5 m^3/s. During the winter storm period, when all the basins are in use and filled to a depth of about 2.5 m, the capacity increases to 17 m^3/s.

2. Source of Recharge Water. The available percolation capacity of the recharge facilities is utilized only during a small portion of the year for spreading stormwater runoff, thus allowing for replenishment by water from other sources. This recharge practice first began in the 1950's with the purchase of Colorado River water imported by the Metropolitan Water District of Southern California.

The other water source used for replenishment in the Montebello Forebay has been reclaimed municipal wastewater purchased from the County Sanitation Districts of Los Angeles County. Reclaimed wastewater was first made available in 1962 following the completion of the Whittier Narrows Water Reclamation Plant and later in 1973 when the San Jose Creek Water Reclamation Plant was placed in service. Reclaimed water was also been incidentally supplied by the Pomona Water Reclamation Plant. Effluent from the Pomona plant that is not put to beneficial use is discharged into San Jose Creek, a tributary of the San Gabriel River, and ultimately becomes a source of recharge for the Montebello Forebay. As the Pomona effluent becomes fully utilized for irrigation and industrial applications within the Pomona area, this source of recharge is expected to diminish.

Through 1977-78, the reclaimed wastewater used for replenishment was disinfected activated sludge secondary effluent. Since that time, all three treatment plants have been upgraded to include either dual-media filtration (Whittier Narrows and San Jose Creek Water Reclamation Plants) or activated carbon filtration (Pomona Water Reclamation Plant) prior to chlorine disinfection. The final effluents produced by each treatment facility comply with primary drinking water standards and meet average coliform and turbidity effluent discharge requirements of less than 2.2 per 100 ml and less than 2 NTU, respectively.

During the 20-year period following the completion of the Whittier Narrows Water Reclamation Plant, over 542 x 10^6 m^3 of reclaimed wastewater have infiltrated into the Montebello Forebay. On an annual basis, the amount of reclaimed wastewater entering the Forebay averages about 33 x 10^6 m^3/y, or 16 percent of the total inflow to the basin. An arbitrary upper limit for reclaimed wastewater of 40 x 10^6 m^3/y has been established based on historical spreading operations. The Los Angeles Regional Water Quality Control Board has approved that the current level of use of reclaimed wastewater for groundwater recharge in this location be expanded to a maximum of 62×10^6 m^3/y, or approximately 30 percent of the total inflow to the Montebello Forebay.

***

In the above example, the reclaimed water produced for groundwater recharge by surface spreading by each treatment facility complies with the U.S. primary drinking water standards and meets total coliform and turbidity requirements of less than 2.2/100 mL and 2 NTU, respectively. Analysis of samples taken at three wastewater reclamation plants from October 1988 through September 1989 provides examples of reclaimed water quality shown in Tables 1 and 2 (Crook, *et al.*, 1990).

The pretreatment processes for direct injection where potable water supply is involved may include chemical oxidation, coagulation and flocculation, clarification, filtration, air stripping, ion exchange, activated carbon adsorption, and reverse osmosis or other membrane separation processes. The groundwater recharge with direct injection of reclaimed wastewater has been operated since 1976 at the Orange County Water District, Fountain Valley, California. The so-called, Water Factory 21 is the first such installation in the United States where highly treated municipal wastewater is directly injected into groundwater aquifers. The 57×10^3 m^3/d advanced wastewater treatment facilities include lime clarification, ammonia stripping, recarbonation, mixed media filtration, activated carbon adsorption, demineralization by reverse osmosis, and chlorination (Argo and Rigby, 1981; Asano, 1985). The original purpose of groundwater recharge in this case was to create a seawater intrusion barrier system, but many potable water wells exist in the vicinity of the affected aquifers and for all practical purposes it is a groundwater replenishment project.

Fate of Contaminants in Groundwater

The behavior of organic substances in soil-aquifer system is crucial in evaluating the feasibility of groundwater recharge using reclaimed wastewater. Treated effluents contain trace quantities of organic contaminants even when the most advanced treatment technology is used. The transport and fate of these substances in the subsurface environment are governed by various mechanisms which include biodegradation by microorganisms, chemical oxidation and reduction, sorption and ion exchange, filtration, chemical precipitation, dilution, volatilization, and photochemical reactions in spreading basins (McCarty, *et al.*, 1981; Ward, *et al.*, 1985).

Particulate Contaminants. Particulate contaminants in reclaimed water introduced into the ground are effectively retained by filtration. This applies to microorganisms as well as other particulate organic substances.

Table 1. Analysis of Recharge Waters (1988-89) - The Whittier Narrows Groundwater Recharge Project (After Crook, *et al.*, 1990).

Constituents	Units	San Jose Creek	Whittier Narrows	Pomona	Discharge Limits
Arsenic	mg/L	0.005	0.004	<0.004	0.05
Aluminum	mg/L	<0.06	<0.10	<0.08	1.0
Barium	mg/L	0.06	0.04	0.04	1.0
Cadmium	mg/L	ND	ND	ND	0.01
Chromium	mg/L	<0.02	<0.03	<0.03	0.05
Lead	mg/L	ND	ND	<0.05	0.05
Manganese	mg/L	<0.02	<0.01	<0.01	0.05
Mercury	mg/L	<0.0003	ND	<0.0001	0.002
Selenium	mg/L	<0.001	0.007	<0.004	<0.01
Silver	mg/L	<0.005	ND	<0.005	0.005
Lindane	μg/L	0.05	0.07	<0.03	4
Endrin	μg/L	ND	ND	ND	0.2
Toxaphene	μg/L	ND	ND	ND	5
Methoxyxhlor	μg/L	ND	ND	ND	100
2,4-D	μg/L	ND	ND	ND	100
2,4,5-TP	μg/L	<0.11	ND	ND	10
Suspended Solids	mg/L	<3	<2	<1	15
BOD	mg/L	7	4	4	20
Turbidity	NTU	1.6	1.6	1.0	2
Total coliform	No/100 mL	<1	<1	<1	2.2
Total dissolved solids	mg/L	598	523	552	700
Nitrate and nitrite	mg/L	1.55	2.19	0.69	10
Chloride	mg/L	123	83	121	250
Sulfate	mg/L	108	105	82	250
Fluoride	mg/L	0.57	0.74	0.50	1.6

[a]ND means not detected

Dissolved Inorganic and Organic Contaminants. In addition to the common dissolved mineral constituents, reclaimed wastewater also contains many dissolved trace elements. The physical action of filtration would not accomplish removal of these dissolved inorganic contaminants. For trace metals to be retained in the soil, physical, chemical, or microbiological reactions are required to immobilize the dissolved contaminants.

Table 2. Trace Organic Analyses of Recharge Waters (1988-89) - The Whittier Narrows Groundwater Recharge Project (After Crook, *et al.*, 1990).

Constituents	Average concentrations, $\mu g/L$			
	San Jose Creek	Whittier Narrows	Pomona	Discharge Limits
Atrazine	ND[a]	ND	ND	3
Simazine	ND	ND	ND	10
Methylene chloride	<2.1	8.6	<4.7	40
Chloroform[c]	5.0	4.6	5.5	100
1,1,1-Trichloroethane	<1.0	<1.6	<0.5	200
Carbon tetrachloride	<0.2	<0.3	ND	0.5
1,1-Dichloroethane	<0.2	ND	ND	6
Trichloroethylene	<0.2	ND	<0.3	5
Tetrachloroethylene	<0.8	<0.5	4.1	5
Bromodichloromethane[b]	0.7	<0.6	<0.9	10
Dibromochloromethane[b]	<0.4	<0.3	<0.5	10
Bromoform[b]	<0.3	ND	ND	10
Chlorobenzene	ND	ND	<0.3	30
Vinyl chloride	ND	ND	ND	0.5
o-Dichlorobenzene	<0.7	<0.5	ND	130
m-Dichlorobenzene	ND	ND	ND	130
p-Dichlorobenzene	<1.8	<1.8	ND	5
1,1-Dichloroethane	ND	<0.2	ND	5
1,1,2-Trichloroethane	ND	ND	ND	32
1,2-Dichloroethane	<0.2	<0.3	ND	0.5
Benzene	<0.2	<0.2	ND	1
Toluene	ND	<0.5	ND	100
Ethyl benzene	<0.2	<0.4	<0.3	680
o-Xylene	<0.4	<0.4	<0.4	1750
p-Xylene	<0.4	<0.7	<0.3	1750
Trans-1,2-dichloroethylene	ND	ND	ND	10
1,2-dichloropropane	ND	ND	ND	5
2 Cis-1,3-dichloropropene[c]	ND	ND	ND	0.5
Trans-1,3-dichloropropene[c]	ND	ND	ND	0.5
1,1,2,2-Tetrachloroethane	ND	ND	ND	1
Freon 11	ND	ND	ND	150
Pentachlorophenol	ND	ND	ND	30

[a]ND means not detected.
[b]Limit for total trihalomethanes is 100 $\mu g/L$.
[c]Limit for total of both isomers is 0.5 $\mu g/L$.

In a groundwater recharge system, the impact of microbial activities on the attenuation of inorganic microcontaminants is small. Physical and chemical reactions that are important to a soil's capability to react with trace metal elements include cation exchange reactions, precipitation, surface adsorption, and chelation and complexation. Although soils do not possess unlimited capability in attenuating inorganic microcontaminants, experimental studies have demonstrated that soils do have capacities for retaining large amounts of trace metal elements. Therefore, it is conceivable that a site used for groundwater recharge may be effective in retaining trace metals for extended periods of time (Chang and Page, 1985).

Removal of dissolved organic contaminants is affected primarily by biodegradation and adsorption during groundwater recharge operations. Biodegradation offers the potential of permanent conversion of hazardous organic substances into harmless products. The rate and extent of biodegradation is strongly influenced by the nature of the organic substances, as well as by the presence of electron acceptors such as dissolved oxygen and nitrate. There are strong indications that biodegradation of easily degradable substances takes place almost exclusively in the immediate vicinity of recharge, within a few meters of travel. However, the effects of slow degradation of materials less easily broken down over the long time periods are still poorly understood.

Among the end products of complete biodegradation of dissolved organic contaminants are carbon dioxide and water under aerobic conditions, or carbon dioxide, nitrogen, sulfide, and methane under anaerobic conditions. However, the degradation process does not necessarily proceed to completion. Degradation may terminate at an intermediate stage, leaving a residual organic product that under the particular conditions cannot be degraded further at an appreciable rate.

Pathogen removal. Groundwater contamination by pathogenic microorganisms has not received as much attention as surface water pollution. It has been generally assumed that groundwater has a good microbiological quality and is free of pathogenic microorganisms. However, a number of well-documented disease outbreaks have been traced to contaminated groundwater. For example, a total of 673 waterborne outbreaks affecting 150,268 persons occurred in the United States from 1946 to 1980. Of these, 295 (44 %) outbreaks involving 65,173 cases were attributed to contamination of groundwater (Craun, 1979).

The fate of pathogenic bacteria and viruses in the subsurface is determined by their survival and their retention by soil particles. Both survival and retention are largely determined by (1) climate, (2) nature of the soil, and (3) nature of microorganism. Climate will control two important factors in determining viral and bacterial survival: temperature and rainfall. At higher temperature, inactivation or dieoff is fairly rapid. In case of bacteria, and probably viruses, the dieoff rate is approximately doubled with each 10 C° rise in temperature between 5 C° and 30 C°.

The nature of the soil will also play a major role in determining survival and retention. Soil properties influence moisture-holding capacity, pH, and organic matter. All of these factors control the survival of bacteria and viruses in the soil. Resistance of microorganisms to environmental factors will vary among different species as well as strains. Bacteria are believed to be removed largely by filtration processes in the soil while adsorption is the major factor controlling virus retention (Gerba and Goyal, 1985).

With proper design, soil-aquifer treatment can be used as an effective method for reducing the number of pathogens in reclaimed wastewater. Table 3 summarizes several potential treatment practices useful in enhancing virus removal in groundwater recharge with municipal wastewater.

Table 3. Operating Practices Which Control Pathogen Transmission in Groundwater Recharge[a]

Operating practice	Remarks
Drying	Enhances pathogen inactivation and removal at soil surface
Spreading with wastewater after rain fall	Reduces virus movement
Addition of cation	Enhances virus adsorption
Spreading cycle	Shorter flooding-drying cycles limit virus penetration
Infiltration rate	Slower rate promotes virus retention and removal

[a] Adopted from reference (Gerba and Goyal, 1985)

Guidelines for Groundwater Recharge with Reclaimed Wastewater

As discussed in the previous sections, groundwater recharge with reclaimed municipal wastewater presents a wide spectrum of health concerns. It is essential that water extracted from a groundwater basin for domestic use be of acceptable physical, chemical, microbiological, and radiological quality. Main concerns governing the acceptability of groundwater recharge projects are that adverse health effects could result from the introduction of pathogens or trace amounts of toxic chemicals into groundwater that is eventually consumed by the public. Because of the increasing concern for long-term health effects every effort should be made to reduce the number of chemical species and concentration of specific organic constituents in the applied water (National Research Council, 1982; State of California, 1987).

A source control program to limit potentially harmful constituents entering the sewer system must be an integral part of any groundwater recharge project. Extreme caution is warranted because of the difficulty in restoring a groundwater basin once it is contaminated. Additional cost would be incurred if groundwater quality changes resulting from recharge necessitated the treatment of extracted groundwater and/or the development of additional water sources.

In the United States, federal requirements for groundwater recharge with reclaimed municipal wastewater have not been established. As a consequence, water reclamation and reuse requirements for groundwater recharge are established by the state agencies such as the California Regional Water Quality Control Boards with a case-by-case determination of the project. Considerably higher wastewater treatment prior to groundwater recharge is advocated in general because of the health concerns related to potential chronic effects of trace organics, and waterborne pathogens, particularly, enteric viruses upon human health.

Proposed California Groundwater Recharge Criteria

The proposed criteria for groundwater recharge with reclaimed municipal wastewater rightly reflect cautious attitude toward such short-term as well as long-term health concerns. Proposed Criteria (State of California, 1992) are shown in Table 4. The criteria rely on a combination of controls intended to maintain a microbiologically and chemically safe groundwater recharge operation. No single method of control would be effective in controlling the transmission and transport of contaminants of concern into and through the environment. Therefore, source control, wastewater treatment processes, treatment standards, recharge methods, recharge area, extraction well proximity, and monitoring wells are all specified.

The requirements in Table 4 are specified by "project category" which identify a set of conditions that constitute an acceptable project. An equivalent level of perceived risk is inherent in each project category when all conditions are met and enforced. Main concerns governing the acceptability of groundwater recharge projects with reclaimed municipal wastewater are that adverse health effects could result from the introduction of pathogens or trace amounts of toxic chemicals into groundwater that is eventually consumed by the public.

Microbiological Considerations. Of the known waterborne pathogens, enteric viruses have been considered most critical in wastewater reuse in California because of the possibility of contracting disease with relatively low doses and difficulty of routine examination of re-claimed wastewater for their presence. Thus, essentially virus-free effluent via the full treatment process (primary/secondary, coagulation/flocculation, clarification, filtration, and disinfection, cf. Table 4) is deemed necessary by the California Department of Health Services for reclaimed wastewater applications with higher potential exposures, e.g., spray irrigation of food crops eaten raw, or most of groundwater recharge applications (Project Categories I, II, and IV in Table 4).

Table 4. Proposed Requirement for Groundwater Recharge with Reclaimed Municipal Wastewater (State of California, 1992)

Treatment and Recharge Site Requirements	Project Category			
	Surface Spreading			Direct Injection
	I	II	III	IV
Level of Wastewater Treatment:				
Primary/Secondary	X	X	X	X
Filtration	X	X		X
Organics Removal	X			X
Disinfection	X	X	X	X
Max. Allowable Reclaimed Wastewater in Extracted Well Water (%)	50	20	20	50
Depth to Groundwater (m) at Initial Percolation Rate of:				
50 mm/min	3	3	6	na[1]
80 mm/min	6	6	15	na[1]
Retention Time Underground (Months)	6	6	12	12
Horizontal Separation[2] (m)	150	150	300	300

[1]Not applicable.
[2]From the edge of the groundwater recharge operation to nearest potable water supply well.

Enteric virus concentrations in unchlorinated secondary effluents are reported in Table 5. Enteric viruses were detected in 45 to 87 percent of samples and estimated geometric mean ranged from 2 to 200 vu/100 L. In the Pomona Virus Study (County Sanitation Districts, 1977), 35 samples were collected and analyzed for native enteric viruses. Virus concentrations detected in two positive samples were 30 vu/100 L and 111 vu/100 L with echo 1 and polio 1 viruses. In the seeded poliovirus study, there was no difference in the overall removal between the full treatment process and the contact filtration process when high chlorine residuals of approximately 10 mg/L were used. However, when low chlorine residuals of 5 mg/L were applied, a slight difference of 5.2 log removal vs. 4.7 log removal was observed (The log removal refers to the fraction of poliovirus remaining after treatment, thus one log removal is equivalent to 90 % removal; five log removal is 99.999 %.). The public health significance of this slight difference in overall removal of the seeded poliovirus is not known.

The wastewater treatment requirements in Table 4 are designed to provide assurance that reclaimed water is essentially pathogen-free prior to extraction from the groundwater. The pathogen, e.g., enteric viruses, removal capabilities of an individual or a combination of treatment processes have been estimated (Hultquist, *et al.*, 1991) and the virus removals achieved by various combinations of wastewater treatment is reported in Table 5.

In addition to the treatment processes shown in Table 5, passage through an unsaturated zone of significant depth (> 3 m) reduces organic constituents and pathogens in treated effluents. At low infiltration rates of less than 5 m/day in sands and sandy loams, the rates of virus removal are approximated by a semi-log plot (k = -0.007 log/cm) against infiltration rates, resulting approximately 99.2 % or 2.1 logs removal for 3 m depth soils. The overall estimates for the removal of enteric viruses by the treatment processes, unsaturated zone, and horizontal separation (retention time in groundwater) as specified in the proposed Criteria are shown in Table 6.

Quantifying the enteric virus concentration in the treated effluent is necessary for estimating the risk of infection upon exposure to reclaimed municipal wastewater. A risk assessment based on the following exposure scenarios was made using virus concentrations reported in Table 7. A groundwater recharge exposure scenario, based on California's proposed groundwater recharge regulations (cf. Table 4), assumes that the nearest domestic water well from a groundwater recharge site draws water that contains 50 percent reclaimed wastewater which has been groundwater for a period of six months after percolating through 3 m of unsaturated soil beneath the recharge basin. Any person drinking water from this well consumes 2 L per day for 70 years. Since the water drawn from the well is 50 percent reclaimed wastewater, it is the equivalent to 1 L of reclaimed water being consumed daily (Asano, *et al.*, 1992).

Trace Organics Removal. The regulations intend to control the concentration of organics of municipal wastewater origin as well as anthropogenic chemicals that have an impact on health when present in trace amounts. Thus, the dilution requirements and the organics removal specified in Project categories I and IV in Table 4 are to limit average concentration of unregulated organics in extracted groundwater affected by the groundwater recharge operation. The concentration of unregulated and unidentified trace organics is of great concern since other constituents and specific organics are dealt with through the

Table 5. Estimated Log Virus Removal by Wastewater Treatment in Different Project Category[1]

Project category[2]	Treatment requirements	Log virus removal[3]
I	Primary/Secondary/Filtration/ Organics Removal/Disinfection	6.9
II	Primary/Secondary/Filtration/ Disinfection	6.1
III	Primary/Secondary/Disinfection	3.1
IV	Primary/Secondary/Filtration/ Organics Removal/Disinfection	6.3

[1]Adapted from State of California, June 1990.
[2]Refer to Table 4 for the Project category.
[3]Log removal is the negative log of the fraction remaining. Thus, if the fraction remaining is 0.10, it is equivalent to one log removal. Conversely, 99.999% removal or 0.00001 remaining is the equivalent of 5 logs removal.

Table 6. Estimates of Overall Removal of Enteric Viruses in Groundwater Recharge Systems due to the Combined Effects of Treatment Processes, Soil Systems, and Retention in Groundwater[1,2]

Project category	Treatment requirements	Log virus removal
I	Primary/Secondary/Filtration/ Organics Removal/Disinfection	17.0
II	Primary/Secondary/Filtration/ Disinfection	16.2
III	Primary/Secondary/Disinfection	14.1
IV	Primary/Secondary/Filtration/ Organics Removal/Disinfection	12.6

[1]Adapted from State of California, 1990.
[2]Infiltration rate of 7.3 m/day was assumed for virus removal with soils in unsaturated zone.

Table 7. Summary of Available Enteric Virus Data from Unchlorinated Secondary Effluents in California[1]

Agency	Facility	Study Period	Type of Secondary Treatment	No. of Samples	No. of Positive Samples	Min.-Max.[2] Viral Unit /100 L	Estimated Geometric Mean, Viral Units/100 L	Median Value	≤90 % Value
Orange County Water District	County Sanitation Districts of Orange County, Plant No.1	1975-78	Trickling Filter	145	109	<26-26,420	185	3,160	17,562
Orange County Water District	County Sanitation Districts of Orange County, Plant No.1	1978-81	Activated Sludge	105	53	<2->250	2	ND[3]	66
Monterey Regional Water Pollution Control Agency	Castroville	1980-85	Activated Sludge	67	53	<100-73,400	200	53	2,700
County San- itation Districts of Los Angeles County	Pomona	1975	Activated Sludge	60	27	<44-13,659	6	ND[3]	539
Las Virgenes Municipal Water District	Tapia	1980 1988-90	Nitrified Activated Sludge	47	41	<0.26->23	2	2>	23
Total				424	283		2[4]	2[4]	500[4]

[1]After Asano, et al., 1992.
[2]Less than indicated value means the limit of detection, while greater than indicated value means too numerous to count.
[3]ND = Not detected.
[4]Combined result of all activated sludge data.

established maximum contaminant levels and action levels developed by the California Department of Health Services. Approximately 90 percent by weight of the organics comprising the total organic carbon (TOC) in treated municipal wastewater are unidentified (State of California, 1987). One of the health concerns related to the unidentified organics is that an unknown but small fraction of them are mutagenic.

Regulation of the presence of trace amount of organics in reclaimed water can be accomplished by dilution using surface water or groundwater of less contaminated source. When reclaimed water makes up more than 20 percent of the water reaching any extraction well for potable water supply, treatment to remove organics must be provided. Because of lack of an ideal measure for trace amount of organics in reclaimed water as well as in the affected groundwater, total organic carbon (TOC) was chosen, as a surrogate, to represent the unregulated organics of concern. Although TOC is not a measure of specific organic compounds, it is considered to be a suitable measure of gross organics content of reclaimed water as well as groundwater for the purpose of determining organics removal efficiency in practice. However, there is insufficient basis for the establishment of a gross organics standard for the recharge water that protects public health.

Table 8. Maximum Allowable TOC Concentration in Reclaimed Wastewater (Recharge Water) Where Organics Removal to Achieve 1 mg/L TOC in Extracted Well Water is Required (State of California, 1992)

Percent Reclaimed Wastewater in Extracted Water	Maximum Allowable TOC in mg/L in Recharge Water	
	Surface Spreading (Category I)	Direct Injection (Category IV)
0-20	20	5
21-25	16	4
26-30	12	3
31-35	10	3
36-45	8	2
46-50	6	2

The proposed regulations shown in Table 4 require that the groundwater recharge projects by surface spreading resulting in a 20-50 % reclaimed wastewater contribution at any extraction well (Category I), and the recharge project by direct injection resulting in a 0-50 % contribution (Category IV), must provide an organic removal step sufficient to limit the TOC concentration of wastewater origin in extracted water to 1 mg/L. Table 8 shows the maximum TOC concentration that may be allowed in the reclaimed wastewater, for a given percent reclaimed wastewater contribution, to achieve no more than 1 mg/L TOC of wastewater origin in the extracted water.

The numbers in Table 8 assume a 70 % reduction through the unsaturated zone and no TOC removal in the aquifer. The numbers associated with the direct injection were derived by dividing 1 mg/L TOC concentration by the fractional contribution of reclaimed water to native groundwater at the extraction point. Thus, the numbers for the direct injection are 30 % of those for the groundwater recharge by surface spreading. In addition, direct injection projects would have to achieve a 70 % TOC reduction to compensate for the lack of an unsaturated zone in the overall soil-aquifer treatment capability.

The applicability of Table 8 in the field groundwater recharge operations is under study. With well-operated advanced wastewater treatment processes including activated carbon and reverse osmosis, TOC in reclaimed wastewater can be reduced to less than 1 mg/L. At this point, essentially all identifiable organics would be absent in detectable concentrations.

Inorganic Chemicals. Inorganic chemicals, with the exception of nitrogen in its various forms, are adequately under control if all maximum-contaminant limits (MCLs) in the reclaimed wastewater are met. By limiting the concentration of total nitrogen in the reclaimed water, detrimental health effects such as methemoglobinemia can be prevented. In those recharge operations where adequate nitrogen removal cannot be achieved by treatment processes or passage through an unsaturated zone, the criteria provide the alternative method such as well head treatment to reduce the total nitrogen concentration to below the allowable concentration of 10 mg/L as N.

Cost of Pretreatment and Groundwater Recharge

Because requirements for pretreatment for groundwater recharge vary considerably from location to location, a range of pretreatment processes are covered by cost estimates, so that rough preliminary information on costs are available for planning purposes. Culp (1980) summarized pretreatment and recharge costs for five sets of recharge conditions as shown in Table 9. In addition to liquid processing costs, sludge processing costs are included.

The lowest cost system is for secondary treatment in aerated lagoons followed by surface spreading. The next set of conditions is secondary treatment by an activated sludge process followed by surface spreading. The third example is one which closely corresponds to project category I in Table 4. It includes activated sludge process, filtration, and granular activated carbon adsorption (GAC). For direct injection (similar to Project Category IV), it includes activated sludge process, high-lime treatment, recarbonation, flow equalization, filtration, ammonia removal by ion exchange, breakpoint chlorination, GAC adsorption, reverse osmosis, ozonation, and chlorination.

Table 9. Summary of Pretreatment and Groundwater Recharge Costs[a]

Recharge method	Pretreatment system	Unit cost,[b] cents/m^3	
		3.8×10^3 m^3/d	38×10^3 m^3/d
Surface spreading	Secondary treatment in aerated lagoon	19.7	10.2
Surface spreading	Secondary treatment by activated sludge	63.6	29.8
Surface spreading	Activated sludge, filtration, GAC adsorption	100.0	42.2
Surface spreading	Activated sludge, high lime, recarbnation, filtration, GAC adsorption, chlorination	119.2	50.3
Direct injection	Activated sludge, high lime, recarbnation, flow equalization, filtration, NH$_3$ removal by ion exchange, breakpoint chlorination, GAC adsorption, reverse osmosis, ozonation, and chlorination	240.4	107.2

[a] After Culp (1980) with modifications.
[b] Cost figures are adjusted to the 1989 costs using the U.S. Environmental Protection Agency Small City Conventional Treatment Plant and Large City Advanced Treatment Plant Construction Cost Indices, National Average. The costs include sludge processing.

Summary and Conclusions

To increase the natural supply of groundwater, artificial recharge of groundwater basins is becoming increasingly important in groundwater management and particularly in situations where the conjunctive use of surface water and groundwater resources is considered.

Several constraints limit expanding use of reclaimed municipal wastewater for groundwater recharge. The lack of specific criteria and guidelines governing the artificial recharge of groundwater with reclaimed municipal wastewater is one of them; thus, the establishment of statewide policy and regulations for the planning and implementing new groundwater recharge projects is imperative. The rational basis and other background information for the proposed groundwater recharge regulations are presented in this chapter. The regulations will serve as a basis with which future groundwater recharge projects are evaluated. These regulations will be formalized through the agency review process along with the public hearing process, and incorporated into revised Wastewater Reclamation Criteria (State of California, Code of Regulations, Title 22: Environmental Health).

References

Argo DG, Rigby MG (1981) Evaluation of membrane process and their role in wastewaterreclamation, OWRT report no 14-34-0001-8520, vol 3A

Asano T, Wassermann KL (1980) Groundwater recharge operations in California, Journal American Water Works Assoc 72, 7, 380-385

Asano T, Roberts PV (ed) (1980) Wastewater reuse for groundwater recharge, California State Water Resources Control Board, Sacramento CA

Asano T, Sakaji RH (1990) Virus risk analysis in wastewater reclamation and reuse. In: Hahn HH, Klute R (eds) Chemical water and wastewater treatment 483-496, Springer - Verlag Berlin Heidelberg

Asano T (ed) (1985) Artificial recharge of groundwater, Butterworth Publishers, Boston MA

Asano T, Leong LYC, Rigby MG, Sakaji RH (1992) Evaluation of the California wastewater reclamation criteria using enteric virus monitoring data, Proceedings of the Water Quality International '92, Washington DC, May 24-30 1992, IAWPRC London UK

Bennani AC, Lary J, Nrhira A, Razouki L, Bize J, Nivault N (1992) Wastewater treatment of greater Agadir (Morocco): An original solution for protecting the bay of Agadir by using the dune sands, In: Waste water management in coastal areas, March 31-April 2, 1992, Montpellier, CFRP-AGHTM Paris France

Bouwer H (1978) Groundwater hydrology, McGraw-Hill Book Co, New York NY

Bouwer H (1988) Groundwater recharge as a treatment of sewage effluent for unrestricted irrigation, In: Pescod MB, Arar A (eds) Treatment and use of sewage effluent for irrigation. Butterworths, London UK

Chang AC, Page AL (1980) Soil deposition of trace metals during groundwater recharge using surface spreading, In: Asano T (ed) Artificial recharge of groundwater, Butterworth Publishers, Boston, MA

County Sanitation Districts of Los Angeles County (1977) Pomona virus study - final report, State Water Resources Control Board, Sacramento CA

Craun GJ (1979) Waterborne disease - a status report emphasizing outbreaks in groundwater. Groundwater 17, 183

Culp RL (1980) Selecting treatment processes to meet water reuse requirements, In: Asano T and Roberts PV (eds) Wastewater reuse for groundwater recharge, Calif. State Water Resources Control Board, Sacramento CA

Crook J, Asano T, Nellor MH (August 1990) Groundwater recharge with reclaimed water in California. Water Environment & Technology 42-49

Gerba CP, Goyal SM (1985) Pathogen removal from wastewater during groundwater recharge, In: Asano T (ed) Artificial recharge of groundwater, Butterworth Publishers, Boston, MA

Hultquist RH, Sakaji RH, Asano T (1991) Proposed California regulations for groundwater recharge with reclaimed municipal wastewater. American Society of Civil Engineers Environmental Engineering Proceedings, 759-764, 1991 Specialty Conference/EE Div/ASCE, New York NY

McCarty PL et al (1981) Processes affecting the movement and fate of trace organics in the subsurface environment, Environmental Science and Technology, 15, 1, 40

National Research Council (1982) Quality criteria for water reuse, Panel on quality criteria for water reuse, National Academy Press, Washington DC

Nellor MH et al (1985) Health effects of indirect potable water reuse. Journal American Water Works Assoc., 77, 7, 88

Roberts PV (1980) Water reuse for groundwater recharge: an overview, Journal American Water Works Assoc., 72, 7, 375

State of California (1978) Wastewater reclamation criteria, code of regulations, Title 22, Division 4, Environmental Health, Department of Health Services, Sacramento, CA

State of California (November 1987) Report of the scientific advisory panel on groundwater recharge with reclaimed wastewater, Prepared for State Water Resources Control Board, Dept. of Water Resources, and Dept. of Health Services, Sacramento CA

State of California (June 1990). Proposed guidelines for groundwater recharge with reclaimed municipal wastewater, and Background information on proposed guidelines for groundwater recharge with reclaimed municipal wastewater, Interagency water reclamation coordinating committee and the Groundwater recharge committee, Sacramento CA

Todd DK (1980) Groundwater hydrology, 2nd ed. John Wiley and Sons, New York NY

Ward CH et al (1985) (Eds) Ground water quality, John Wiley & Sons, New York NY

DISPERSION OF CONTAMINANTS FROM LANDFILL OPERATIONS

U. Förstner
Department of Environmental Science and Engineering
Technical University of Hamburg-Harburg
Eißendorferstr. 40
D-2100 Hamburg 90
Germany

Contaminated solid materials, which may affect groundwater pollution include municipal solid wastes, sewage sludge, dredged sediments, industrial by-products, wastes from mining and smelting operations, filter residues from waste water treatment and atmospheric emission control, ashes and slags from burning of coal and oil, and from incineration of municipal refuse and sewage sludge. The problem of "contaminated land" was introduced with the accidential detection of large-scale pollution from industrial waste deposits which had been handled inproperly.

As an example, Table 1 lists quantitative figures of the production of waste materials. Municipal solid waste amounts to approximately half a ton per person and year. Sewage sludge is even more. Construction waste including soil movements is more than two tons per person and year, with large differences in quality - most of them unproblematic, a smaller portion, however, highly contaminated. Table 1 also lists industrial and hazardous waste, which is a relative small portion compared with the previous items. However, one has to consider that manufacture of a lower-middle-class car generates approx. 15 tons of wastes, that is 25 times the weight of the car. A considerable proportion of such wastes is related to mining activities, and the quantities of mine tailings are in the same order of magnitude as the actual sediment discharge to oceans.

NATO ASI Series, Vol. G 32
Migration and Fate of Pollutants in Soils and Subsoils
Edited by D. Petruzzelli and F. G. Helfferich
© Springer-Verlag Berlin Heidelberg 1993

Table 1   Quantities of solid waste materials (Baccini and Brunner, 1991; Neumann-Malkau, 1991; and other sources).

	kg/capita/year	
Municipal Solid Waste	130	(Bangladesh)
	520	(U.S.A.)
Sewage Sludge (95% Water)	720	(F.R.G.)
Construction Waste (incl. Soil Movements - ca. 80% - and Road Construction Waste)	2,200	(F.R.G.)
Industrial and Hazardous Waste	80	(F.R.G.)
	million tonnes/year	
Dredged Materials	580	(World)
Incineration Residues	300	(World)
Mine Waste + Tailings	17,800	(World)
(Natural erosion:	26,700)	

Evolution of Barriers Between the Landfill and Groundwater

The temporal evlution of the four major "barriers" in landfill techniques, which can be substantially influenced or rather determined by the engineer, is schematically presented in Figure 1 (Ryser, 1989). Up until the end of the sixties more or less suitable geological sites (barrier 4) were chosen, taking no further preventive measures into consideration. The majority of the landfills now requiring sanitation date back to this time. With the growing awareness that there is a lack of natural water-tight sites, the "envelope" (barrier 3) was introduced at the beginning of the seventies. During the eighties, controlled reactions (barrier 2) by gas collection and utilization gave rise to considerable expectations, which, however, were not all fulfilled; this is particularly valid with respect to the long-term behavior of the landfill (see below). The view that before tipping it is necessary to separate utilizable materials (wood, paper, scrap metals) or substances for further conveyance (batteries etc.) is now widely accepted. At the beginning of the nineties the "barrier" No. 1 represents an essential factor for the control of the contents of the landfill and thereby influence its long-term behavior.

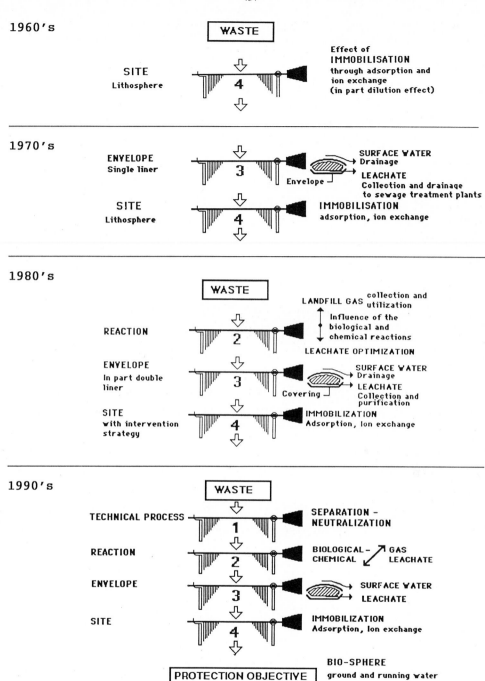

Figure 1 Development of four major barriers for control
of reactor landfill leachates during the 1960's,
1970's, 1980's and 1990's (Ryser, 1989)

Groundwater Pollution from Landfills and Other Sources

With regard to loss of groundwater resources, ascribable to contamination, a United States Library of Congress Report from 1980 lists 1360 well closings in a 30 years span, which can be broken down as follows (Anonymous, 1980): Metal contamination: 619 wells; organic chemical contamination: 242, including 170 from trichloroethylene used to emulsify septic tank grease; pesticide contamination: about 200 wells; industrial (not de-fined) sites: 185 wells; leachate from municipal solid waste landfills: 64 wells; and 26 wells affected by high concentra-tions of chlorides, 23 wells polluted by nitrates. Altogether, there is a significant effect of heavy-metal pollution - about 40% in this study - on the loss of groundwater resources.

In the United States, the national priority list of 1986 (Anonymous, 1986) developed by the EPA exhibits at total of approximately 1000 sites that pose significant environmental or health risks. About 40% of these sites reported metal pro-blems, with lead, chromium, arsenic, and cadmium cited at more than 50 sites (Wilmoth et al., 1991). Only a few industries or activities account for most of the NPL sites with metal prob-lems; these are in particular metal plating, chemical, mining and smelting, battery recycling, wood treating, oil and sol-vent recycle, and nuclear processing industries. Not surpri-singly, landfills account for a high percentage (40%) as chemical wastes were often dumped in municipal landfills.

Comparison of inorganic groundwater constituents up-stream and downstream of 33 waste disposal sites in Germany (Arneth et al., 1989) indicates typical differences in pollu-tant mobilities, which may partly be related to releases dur-ing the acidic phase of the landfill development. High conta-mination factors ("contaminated mean"/"uncontaminated mean"; see Figure 2) have been found for boron, ammonia, and arsenic (this element may pose particular problems during initial phases of landfill operations; Blakey, 1984); heavy metals such as cadmium, chromium, lead, copper and nickel are signi-ficantly enriched in the leachates as well.

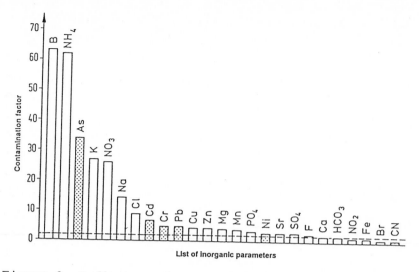

Figure 2   Influence of waste disposal on ground water quality
            from 33 sites in Germany (Arneth et al., 1989)

In the study on the material management within a fictitious
region "Metaland" (Baccini and Brunner, 1991), an estimation
was made by Baccini et al. (1992) on the release of contaminants from landfills subsequent to a 50 years reactive period,
for the case that all barriers will have failed at this time.

Table 2   Effects of landfill leachate on chemical composition
           of a groundwater reservoir in the Metaland model
           study (Baccini et al., 1992). Specific leachate flux
           to a groundwater reservoir of $2 \cdot 10^9$ m^3 was taken as
           0.02 L per kg and year of waste material (net preci-
           pitation: 1000 mm/year).

	C_{org} [mg · L^{-1}]	Cl	Zn	Cd [μg · L^{-1}]	Hg
Mean leachate concentration after 50 year landfill oper.	600	500	600	2	0.1
Mean concentration in the uncontaminated groundwater	0.5	3	5	0.02	0.005
Mean annual increase of concentration in groundwater	0.24	0.2	0.24	0.0008	0.00016
Mean annual increase in %	*50%*	*7%*	*5%*	*4%*	*3%*

The population in Metaland was assumed as being 1 million on an area of 2.500 km², each producing 40 tonnes of waste within a 40 years time span. Table 2 indicates, that a very significant input of TOC (and of ammonia), with an annual increase of 50% compared to the uncontaminated groundwater will occur, and it has been suggested that this input will last over a period of approximately 1000 years.

Critial Processes for Metal Mobilization from Wastes

Regarding the potential release of contaminants from solid waste materials, changes of pH and redox conditions are of prime importance. It can be expected that changes from reducing to oxidizing conditions, which involve transformations of sulfides and a shift to more acid conditions, increase the mobility of typical "B-" or "chalcophilic" elements, such as Hg, Zn, Pb, Cu, and Cd.

The major process affecting the lowering of pH-values (down to pH 2 to 3) is the exposure of pyrite (FeS_2) and other sulfide minerals to atmospheric oxygen and moisture, whereby the sulfidic component is oxidized to sulfate and acidity (H^+-ions) is generated. Bacterial action can assist the oxidation of Fe^{2+}(aq) in the presence of dissolved oxygen.

$$4\,FeS + 9\,O_2 + 10\,H_2O = 4\,Fe(OH)_3 + 4\,SO_4^{2-} + 8\,H^+$$

or

$$4\,FeS_2 + 15\,O_2 + 14\,H_2O = 4\,Fe(OH)_3 + 8\,SO_4^{2-} + 16\,H^+$$
Pyrite

$$2\,NH_4^+ + 3\,O_2 = 2\,NO_2^- + 2\,H_2O + 4\,H^+$$

Acidity is perhaps the most serious long-term threat from metal-bearing wastes. Water seeping from mine refuse has been passing increased metal concentrations into receiving waters for decades. The threat is especially great in waters with little buffer capacity, i.e., in carbonate-poor areas where dissolved-metal pollution can be spread over great distances. The acidity production can develop many years after disposal, when the neutralizing or buffering capacity in a pyrite-containing waste is exceeded.

Primary emissions of high metal concentration occur from waste rocks and tailings, secondary effects on groundwater take place from the ponds. An important and long-term source of metals are the sediments reworked from the floodplain, mainly by repeated oxidation and reduction processes (Moore and Luoma, 1990). High concentration factors were found in inland waters affected by acidic mine effluents (Förstner, 1981).

Long-Term Metal Release from Industrial Solid Waste

Oxidizable sulfur-containing compounds may induce long-term release of metals from certain industrial waste materials. As an example, pigment production wastes, originating from processes involving sulfuric acid, have been studied. Formerly the dilute acid and its accessories were dumped into the North Sea. With a new process the acid can be concentrated; however, there are still residues which have to be deposited. For stabilization, another waste product is added. This material has a high initial pH, which seems to be favorable for the fixation of metals.

A long-term experiment, over 50 days, has established that there are characteristic temporal changes of pH- and redox conditions upon aeration (Figure 3). The redox values increase from minus 200 mV to plus 500 mV, at the same time the pH-values decrease from 10 to 5.4. This is due to the oxidation of small concentrations of sulfur containing compounds. It is also clear from this experiment that, with a standard leaching test, over one hour or one day, such effects would not have been detected.

With an experimental approach, which was first used by Patrick et al. (1973) and Herms and Brümmer (1978), metal mobility of industrial waste materials has been studied in a circulation apparatus by the controlled intensification of significant release parameters such as pH-value, redox-potential and temperature. Here, an ion-exchanger system is used for extracting and analyzing the released metals at an adequate frequency.

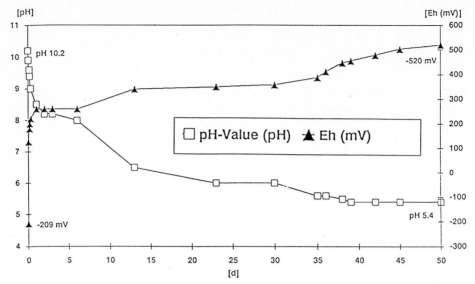

Figure 3   Changes of pH-values and redox-potential in a long-
            term experiment on industrial solid waste material
            (Förstner et al., 1991)

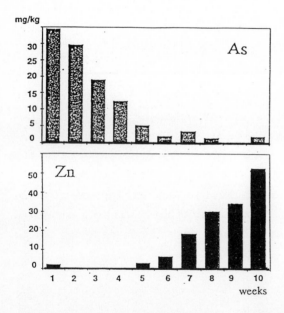

Figure 4

Release of As and
Zn from industrial
solid waste in a
long-term circula-
tion experiment
(from Förstner
et al., 1991)

In a series of experiments, a number of industrial waste materials of different types, intended for co-disposal in borrow
pits, were investigated with this method. In these experiments, special attention was given to the efficiency of individual components with respect to long-term behavior of critical trace elements in such mixed deposits. The kinetics of

element's release from "conditioned" waste material, i.e., treated with high-pH additives, is shown in Figure 4 for As, and Zn. By treatment with pH 5 solutions, mobilization of As is essentially completed after the initial five weeks of the experiment. Mobilization of Zn is strongly enhanced toward the end of the study period. Regarding the latter element, it can be expected that the cumulative percentage of release from the treated material would significantly be enhanced upon continuation of the experiment.

From these data, it could be argued that the long-term effect of the initial high-pH additive on the stability of trace metals in the disposal mixture is less advantageous than previously assumed. Figure 5 presents the cumulative mobilization of zinc from two different waste compositions, which were studied in experiments at pH 5 for 10 weeks. Addition of the high-pH component is reducing release of Zn in the initial stage of the experiment. From the sixth week on, however, the release of zinc is distinctly increasing and is finally surpassing the rates from the untreated industrial waste.

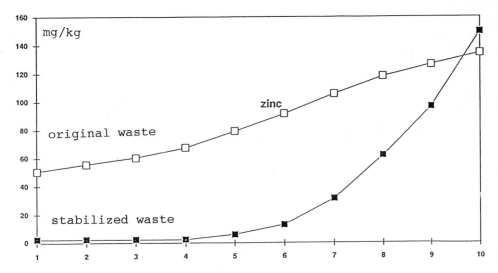

Figure 5   Cumulative mobilization of zinc from original waste (open squares) and subsequent to stabilization by high-pH additive (solid squares)(Förstner et al., 1991)

Long-Term Metal Release from Municipal Solid Waste

In municipal solid waste landfills, initial conditions are characterized by the presence of oxygen and pH-values between 7 and 8. During the subsequent "acidic anaerobic phase", the pH drops to a level as low as 5 because of the formation of organic acids in a more and more reducing milieu; concentrations of organic substances in the leachate are high (Figure 6). In a transition time of 1 to 2 years, the chemistry of landfill changes from acetic to methanogenic conditions. Typically increased concentrations of metals have been found for iron, manganese and zinc in leachates during the acidic decomposition phase compared with the methanogenic phase.

There is not much experience with landfill evolution subsequent to the initial 30 years. What could happen is that the landfill is again oxidized. It has been inferred that oxidation of sulfidic minerals by intruding rainwater may mobilize trace metals. The impact on the underlying groundwater could be even higher if a chromatographic process, involving continuous dissolution and reprecipitation during passage of oxidized water through the deposit, were to preconcentrate critical elements prior to final release with the leachate.

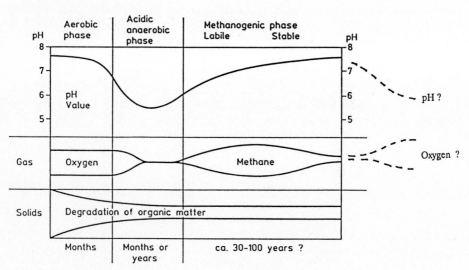

Figure 6   Scheme of chemical evolution of municipal solid waste landfills (from Förstner et al., 1991b)

Experimental investigations by Peiffer (1989) on the long-term
development of sewage sludge materials provide detailed insight into the sequence of processes taking place in the post-
methanogenic stage of such deposits (Figure 7). During an initial phase, anoxic cadmium is bound to sulfides, resulting in
very low metal concentrations in solution (phase 1). Aeration
by addition of dissolved oxygen initiates release of cadmium
from the solid substrate (phase 2); this process is enhanced
by the production of acidity, which lowers the pH from 6.7 to
approximately 6.4 (phase 3). An even stronger effect is observed from the addition of alkaline earth ions (phase 4). The pH
increase, which may be induced from buffering components within the system, leads to a reduction of dissolved cadmium
(phase 5). Formation of new sulfide ions from the degradation
of organic matter brings the concentrations of dissolved cadmium back to its original, extremely low level (phase 6). The
observed pH decrease seems to indicate that zinc and cadmium
are being exchanged for protons, whereas lead and copper, because of their much stronger bonding to the solid substrate,
do not.

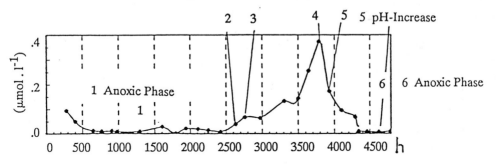

Figure 7   Experiments on cadmium release from solid waste
material (from Peiffer, 1989)

Acid-Producing Potential of Solid Waste Materials

The acid-producing potential not only is related to the oxidation of sulfides, but oxidation of organic matter must be considered as well. Table 3 indicates that the contribution of
protons from organic-N and organic-S in a sample containing

approximately 5% organic carbon is equivalent to the acid-producing potential of 1% FeS_2. Studies of the long-term evolution and diagenesis in a sewage sludge landfill and similar natural sediments (peat, organic soils) by Lichtensteiger et al. (1987) suggest that the transformation of organic material will last for geological time scales (10^3 to 10^7 years).

Table 3   Acid-producing potential of solid matter (from Swift, 1977)

Parameter	Mean Content (%)	Acid Producing Potential (APP)	In 1:10 - Suspension [H$^+$]	pH
Organic Carbon	3,6 %			
Organic-N	0,27 %	0,19 mM/g	0,019	1,71
Organic-S	0,05 %	0,03 mM/g	0,003	2,82
FeS$_2$	1 %	0,33 mM/g	0,033	1,50
Total		*0,55 mM/g*	*0,055*	*1,28*

Pore water data from dredged material from Hamburg Harbor indicate typical differences in the kinetics of proton release from organic and sulfidic sources (Table 4). Recent deposits are characterized by low concentrations of nitrate, cadmium and zinc; when these low-buffered sediments are oxidized during a time period of a few months to years, the concentrations of ammonia and iron in the pore water typically decrease, whereas those of cadmium and zinc increase (with the result that these metals are easily transferred into agricultural crops!).

The different steps are schematically given in Figure 8. Oxidation of sulfides during stage B strongly increases the concentrations of cadmium and zinc in a relative short time.

	Reduced Water	Oxidized Water
Ammonia	125 mg/l	< 3 mg/l
Iron	80 mg/l	< 3 mg/l
Nitrate	≤ 3 mg/l	120 mg/l
Zinc	< 10 μg/l	5000 μg/l
Cadmium	< 0,5 μg/l	80 μg/l

Table 4

Metal mobilization from dredged material after land deposition (Maaß et al., 1985 and other authors)

When acidity is consumed by buffer reactions (phase C), cadmium and zinc concentrations drop, but are still higher than in the original sulfidic system. In phase D, oxidation of organic matter again lowers pH-values and can induce a long-term mobilization of Zn and Cd.

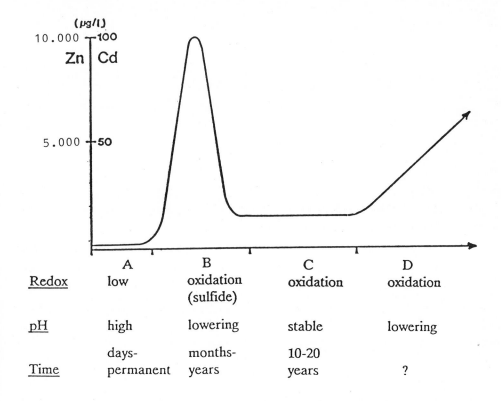

Figure 8   Schematic diagram illustrating different phases of metal release from land-disposed dredged material (Maaß and Miehlich, 1988)

Final Storage of Solid Waste

With regard to the immobilization of contaminants in municipal and industrial waste materials, the term "final storage quality" has been brought into discussion (Baccini, 1989a). Solid residues with final storage quality should have properties very similar to the earth crust (natural sediments, rocks, ores, soil; Table 5). This can be achieved in several ways, e.g., by assortment or thermal, chemical and biological treatment. In most cases this standard is not attained by simple incineration of municipal waste, i.e., by reduction of organic fractions only. There is, in particular, the problem of easily soluble minerals such as sodium chloride. Measures before incineration include the separate collection of (organic) kitchen and garden wastes (containing chlorine and sulfur), which can be transferred into compost; a major decrease of chlorine content, however, would require a significant reduction of PVC in municipal solid waste. After incineration washing of the residues can be performed either with neutral or acidified water. Another possibility is to put the electrostatic precipitator dusts into a thermal process to remove metal such as zinc, cadmium and lead as chlorides at high temperatures ($>1200^\circ$C).

Table 5 Comparison of inventories of chemical components in the two landfill alternatives and in the earth crust (Förstner et al., 1989)

Reactor Landfill	Final Storage	Earth's Crust
Major solid constituents		
Solid "inert" waste	Silicates, oxides	Quarz, Fe-oxide clay, carbonates
Putrefactive waste	[Gypsum, NaCl][a]	(Gypsum, NaCl)
Grease trap waste	(Char)[b]	Kerogenic compounds
Minor solid constituents		
Organic micropollutants	Organic micropollutants	–
Metals in reactive chemical forms	Metal-bearing minerals, mainly oxides	Metals mainly in inert forms
Dissolved constituents		
Protons, electrons	(Protons)	(pH: acid rain)
Organic compounds	(Organic residues)	(Humic acids)
Dissolved salts	[Dissolved salts][a]	(Dissolved salts)

[a] Partial extraction during pretreatment
[b] Minor constituent

Chemical Stability of Bottom Ash Monofills

Long-term release of protons can be expected even from the relatively low organic carbon ashes and slags from municipal waste incineration, as has been suggested by Krebs et al. (1988). Microbial degradation of 1 -2 % of residual organic carbon will produce approximately 1 mol H^+ per kg bottom ash, which is about equivalent to the acid neutralizing capacity of this material (but several orders of magnitude higher than the H^+-input from acid precipitation). Laboratory experiments by Belevi et al. (1992) suggest that non-metal fluxes by leachate (such as chloride, sulfur and DOC fluxes) would adversely impact the environment for years to decades after disposal (Table 6). Heavy metal fluxes by leachate are expected to be compatible with the environment (Table 6); however, additional laboratory and field studies are necessary to assess their be-havior over longer time periods. At present, bottom ash cannot be considered as a material of final storage quality. It should either be disposed into monofills with leachate collec-tion and treatment systems or be treated prior to disposal to achieve final storage quality (Belevi et al., 1992).

Table 6   Chemical behavior of municipal waste incineration bottom ash in monofills (Belevi et al., 1992).

	Initial content [$g·kg^{-1}$]	Fraction mobilized [$g·kg^{-1}$]	%	Leachate conc. [$g·L^{-1}$]	Quality standard [$g·L^{-1}$]	*Time range [years]*
Ca	99	5	5%			
Al	50	3	6%			
Na	36	0.6	2%			
TOC	17	0.7	4%	0.05-5	0.02	*25-250*
K	10	1.2	12%			
Cl	3	2	70%	1 - 10	0.1	*4- 40*
S	2	0.4	20%	0.03-1.5	0.03	*5-250*
				[$mg·L^{-1}$]	[$mg·L^{-1}$]	
Pb	3	0.02	0.7%	<0.03	0.05	
Zn	2	0.007	0.4%	<0.6	1	
Cu	1	0.02	2%	<0.1	0.1	

Chemical Stabilisation of Industrial Waste Material

"Geochemical and biological engineering" emphasize the increasing efforts to use natural resources available at the disposal site for reducing negative environmental effects of all types of waste material, in particular of acid mine wastes (Salomons and Förstner, 1988a,b). Most stabilization techniques aimed at the immobilization of metal-containing wastes are based on additions of cement, water glass (alkali silicate), coal fly ash, lime or gypsum (Calmano, 1988; Goumans et al., 1991).

Experimental studies of the processes taking place with mixed residues from lignite coal incineration indicate favorable effects of incorporation of both chloride and heavy metals in newly formed minerals (Bambauer, 1992; Figure 9).

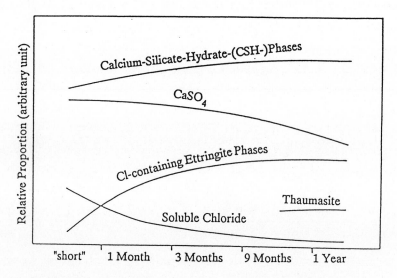

Figure 9 Schematic representation of mineral formations in "stabilisates" from coal residues (Bambauer, 1992)

Ettringite, i.e., calcium-aluminate-monosulfate-hydrate phases, in particular, can act as "storage minerals" for chloride and metal ions. The former may be incorporated at up to 4 kg $CaCl_2$ per m^3 of the mineral mixture. Calcium-silicate-hydrate phases may be formed in a subsequent process, and by filling further pore space these minerals can significantly reduce permeability of the waste body for percolating solutions.

Table 7   Leachability of salts and trace elements in samples
          of "stabilisates" (from Bambauer, 1992)

	NH$_3$ mg/l	Cl mg/l	SO$_4$ mg/l	Zn µg/l	Cu µg/l	Cd µg/l
Initial Concentration	2580	10000	4000	10400	1300	160
Concentration in Eluate	0.5	244	611	21	26	0.11
Passage in %	*0,02*	*2,2*	*15,2*	*0,2*	*2,0*	*0,07*
Guideline FGD-Water	-	-	-	1000	500	50
Limit-Value Class I	50	200	250	1000	100	5

Experimental studies of the leachability of salts and trace
elements from samples of "stabilisates", with a pressure-fil-
tration method, indicate relative high rates of release for
sulfate ions, but not for zinc and cadmium in the eluate
(Table 7; Bambauer, 1992).

Storage under Permanent Anoxic Conditions

Incorporation in naturally formed minerals, which remain stable over geological times, constitutes favorable conditions for
the immobilization of potentially toxic metals in large-volume
waste materials both with respect to environmental safety and
economic considerations. Metal sulfides in particular have
very low solubilities compared with those of the respective
carbonate, phosphate, and oxide compounds (Figure 10).

Marine conditions, which are favorable because of the
higher production of sulfide ions seem in addition to repress
the formation of monomethyl mercury (Craig and Moreton, 1984).
This type of waste deposition under stable anoxic conditions,
with large masses of polluted materials covered with inert
sediment, has become known as "subsediment-deposit".

The first example was planned for highly contaminated
sludges from Stamford Harbour in the Central Long Island Sound
following intense discussions in the U.S. Congress (Morton,
1980, Bokuniewicz, 1982).

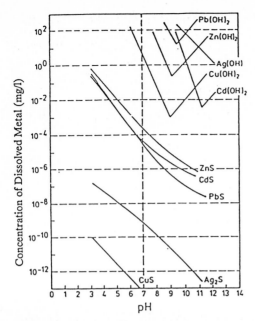

Figure 10

Solubility of
metal sulfides
and metal oxides
(from Ehrenfeld
and Bass, 1983)

Other advantages of near-shore capped mound deposits (Kester
et al., 1983) include the protection of groundwater resources,
since the underlying water is saline, and enhanced degradation
of organic priority pollutants (Kersten, 1988).

The concept of final storage quality is a new challenge
for geochemists. At first, we have to characterize the different solid substrates with respect to the short-, middle- and
long-term mobility of critical constituents. Secondly, in cooperation with biologists we should evaluate the effects of
the different solid substrates and their critical compounds to
organisms. Thirdly, we may give our advise to the engineers
what technologies would be best for reducing the environmental
impact of certain waste materials.

References

Anonymous (1980) Groundwater strategies. Environ Sci Technol
14:1030-1035
Anonymous (1986) National Priorities List Fact Book. HW 7.3,
94 pp. US Environmental Protection Agency, Washington DC
Arneth, J-D, Milde G, Kerndorff H, Schleyer, R (1989) Waste
deposit influences on ground water quality as a tool for
waste type and site selection for final storage quality.
In: The Landfill - Reactor and Final Storage, P Baccini
(ed). Lecture Notes in Earth Sciences 20, pp 399-415
Springer, Berlin

Baccini P (ed)(1989) The landfill - reactor and final storage. Lecture Notes in Earth Sciences 20. Springer Berlin

Baccini P, Brunner PH (1991) Metabolism of the Anthroposphere. Springer-Verlag Berlin-Heidelberg-New York

Baccini P, Belevi H, Lichtensteiger T (1992) Die Deponie in einer ökologisch orientierten Volkswirtschaft. Gaia 1: 34-49

Bambauer HU (1992) Mineralogische Schadstoffimmobilisierung in Deponaten - Beispiel: Rückstände aus Braunkohlenkraftwerken. BWK Umwelt-Special March 1992:S29-S34

Belevi H, Stämpfli DM, Baccini P (1992) Chemical behaviour of municipal solid waste incinerator bottom ash in monofills. Waste Management Research 10: 153-167

Blakey NC (1984) Behavior of arsenical wastes co-disposed with domestic solid wastes. J Water Poll Control Fed 56:69-75

Bokuniewicz HJ (1982) Submarine borrow pits as containments for dredged sediments. In: Kester DR, Ketchum BH, Duedall IW, Parks PK (eds) Dredged Material Disposal in the Ocean, pp 215-227. John Wiley & Sons, New York

Calmano W (1988) Stabilization of dredged mud. In: Salomons W, Förstner U (eds) Environmental Management of Solid Waste: Dredged Materials and Mine Tailings, pp 80-98. Springer-Verlag, Berlin.

Craig PJ, Moreton PA (1984) The role of sulphide in the formation of dimethyl mercury in river and estuary sediments. Mar Pollut Bull 15:406-408

Ehrenfeld J, Bass J. (1983) Handbook for Evaluating Remedial Action Technology Plans. Municipal Environ Res Lab Cincinnati. EPA-600/2-83-076. August 1983

Ehrig H-J (1989). Water and element balances of landfills. In: P Baccini (Ed) The Landfill - Reactor and Final Storage. Lecture Notes in Earth Sciences 20, pp 83-115, Springer, Berlin

Förstner U (1981) Trace metals in fresh waters (with particular reference to mine effluents). In: KH Wolf (ed) Handbook of Strata-Bound and Stratiform Ore Deposits. Vol 9:271-303

Förstner U, Kersten M (1988) Assessment of metal mobility in dredged material and mine waste by pore water chemistry and solid speciation. In: W Salomons, U Förstner (eds) Chemistry and biology of solid waste - dredged material and mine tailings. Springer Berlin, pp 214-237

Förstner U, Calmano W, Kienz W (1991b) Assessment of long-term metal mobility in heat-processing wastes. Water Air Soil Pollut 57-58:319-328

Förstner U, Colombi C, Kistler R (1991) Dumping of wastes. In: Merian E. (ed) Metals and Their Compounds in the Environment. Chapter I.7b, pp 333-355. VCH Verlag, Weinheim

Förstner U, Kersten M, Wienberg R (1989a) Geochemical processes in landfills. In: P Baccini (ed) The landfill - reactor and final storage. Lecture Notes in Earth Sciences 20. Springer Berlin, pp 39-81

Goumans JJJM, Van der Sloot HA, Aalbers ThG (eds)(1991) Waste materials in construction. Studies in Environmental Science 48. Elsevier Amsterdam 672 p

Herms U, Brümmer G (1978) Löslichkeit von Schwermetallen in Siedlungsabfällen und Böden in Abhängigkeit von pH-Wert, Redoxbedingungen und Stoffbestand. Mitt Dt Bodenk Ges 27:23-43

454

Kester DR, Ketchum BH, Duedall IW, Park PK (eds)(1983) Wastes in the ocean. Vol 2: Dredged-material disposal in the ocean. Wiley New York, 299 p

Kersten M (1988) Geochemistry of priority pollutants in anoxic sludges: cadmium, arsenic, methyl mercury, and chlorinated organics. In: W Salomons, U Förstner (eds) Chemistry and biology of solid waste - dredged material and mine tailings. Springer Berlin, pp 170-213

Krebs J, Belevi H, Baccini P (1988) Long-term behavior of bottom ash landfills. Proc 5th Intern Solid Wastes Exhibition and Conf, ISWA 1988, Copenhagen

Lichtensteiger T, Brunner PH, Langmeier M (1988). Klärschlamm in Deponien. EAWAG Project No. 30-681. EC-COST.

Maaß B, Miehlich G, (1988) Die Wirrkung des Redoxpotentials auf die Zusammensetzung der Porenlösung in Hafenschlick-spülfeldern. Mitt Dtsch Bodenkunde Ges 56:289-294

Maaß B, Miehlich G, Gröngröft A (1985) Untersuchungen zur Grundwassergefährdung durch Hafenschlick-Spülfelder. II. Inhaltsstoffe in Spülfeldsedimenten und Porenwässern. Mitt Dtsch Bodenkundl Ges 43/I:253-258.

Malone PG, Jones LW, Larson RJ (1982) Guide to the Disposal of Chemically Stabilized and Solidified Waste. U.S. Environmental Protection Agency, Washington D.C.

Moore JN, Luoma SN (1990) Hazardous wastes from large-scale metal extraction. Environ Sci Technol 24: 1278-1285

Neumann-Malkau P (1991) Anthropogenic mass movement - interfering with geologic cycles? In: Geotechnica Congress Cologne, September 1991, pp 153-154

Patrick WH, Williams BG, Moraghan JT (1973) A simple system for controlling redox potential and pH in soil suspensions. Soil Sci Soc Amer Proc 37:331-332

Peiffer S (1989) Biogeochemische Regulation der Spurenmetallöslichkeit während der anaeroben Zersetzung fester kommuler Abfälle. Dissertation Universität Bayreuth, 197 p

Salomons W, Förstner U (Eds)(1988a) Chemistry and Biology of Solid Waste: Dredged Materials and Mine Tailings. Springer, Berlin

Salomons W, Förstner U (Eds)(1988b) Environmental Management of Solid Waste: Dredged Materials and Mine Tailings. Springer, Berlin

Schoer J, Förstner U. (1987) Abschätzung der Langzeitbelastung von Grundwasser durch die Ablagerung metallhaltiger Feststoffe. Vom Wasser 69:23-32

Swift RS (1977) Soil Organic Matter Studies. IAEA Vienna. pp 275-281

Wilmoth RC, Hubbard SJ, Burckle JO, Martin JF (1991) Production and processing of metals: Their disposal and future risks. In: Merian E (ed) Metals and Their Compounds in the Environment. Chapter I.2, pp 19-65. VCH Verlagsgesellschaft Weinheim

RISK ANALYSIS OF GROUNDWATER CONTAMINATION

Jacques Ganoulis
Division of Hydraulics & Environmental Engineering,
School of Engineering,
Aristotle University of Thessaloniki,
54006 Thessaloniki, Greece

Groundwater contamination is the most critical among various types of pollution that can occur in the water cycle. This is a consequence of the large time scales of the phenomena and the irreversible character of the damage caused. In fact, due to the very slow movement of groundwaters, pollutants can reside for a very long time in the aquifer. As a consequence, groundwater remains polluted for centuries even if the pollutant sources are no longer active. In the same time, because of the complex interaction between pollutants, soil and groundwater, remediation of contaminated subsurface is a very delicate operation. Most of the time total removal and cleaning up of the contaminated soil is necessary or biological techniques should be applied for a long time in order to become efficient.

One very important problem of deterioration of groundwater quality is the increasing salinisation near the soil surface and desertification of millions of hectares of irrigated land around the world. For example in Australia, it has been recognised (Tickell and Humphrys,1984) that a rising of the groundwater table is one of the main causes of waterlogging and salinity increase near the top layer of the soil. As the groundwater moves upward, the salinity is increased by dissolussion of salts in the soil. The rising of the groundwater table is the combined effect of intensive actual irrigation practices together with the disruption of the natural equilibrium between plants, soil and groundwaters. In fact, intensive removal in the past of deep-rooted vegetation, has reduced the natural drainage capacity of the basin and destroyed the natural equilibrium between groundwater recharge and drainage. When the water table rises in a depth less than two meters from the soil surface, salt concentrations are further increased by evaporation and damage to vegetation and soils are very likely.

NATO ASI Series, Vol. G 32
Migration and Fate of Pollutants in Soils and Subsoils
Edited by D. Petruzzelli and F. G. Helfferich
© Springer-Verlag Berlin Heidelberg 1993

The protection of groundwater resources is based on different strategies involving empirical or sophisticated methods. Various traditional strategies for groundwater protection range from the construction of groundwater vulnerability maps and the definition of protection perimeters around pumping wells to the use of sophisticated optimisation multilcriterion-decision-making techniques under risk conditions. A very characteristic example is the definition of adequate waste disposal sites in relation to the risk of groundwater contamination.

The main difficulty in designing groundwater development plans is that groundwater pollution is subject to several types of uncertainties. These are related to the high variability in space and time of the involved hydrogeological, chemical and biological processes. The principal task of the engineering risk analysis is to assess the probability or risk to comply with the groundwater quality standards in the areas to be developed. For example, in groundwater used for irrigation, according to the standards, salinity concentration may not exceed 1000 ppm .

The aim of this chapter is twofold: (1) to present a coherent framework for the application of risk and reliability techniques for groundwater protection, and (2) to illustrate the methodology with a case study in Northern Victoria, Australia.

UNCERTAINTIES IN AQUIFER CONTAMINATION STUDIES

Because of the natural variability in space and time, the main problem for evaluating the risk of groundwater contamination is the fact that physical parameters and variables of the system show random deviations. To this randomness, one must add various other uncertainties due to the scarcity of the information concerning the inputs, the value of the parameters (measurement and sampling uncertainties) and also the imperfection of the models (modelling uncertainties).

Fig.1 shows the variation in space of the porosity in a typical case of an alluvial aquifer. Because of such random variations of physical characteristics, it follows that the output variables are also not deterministic, but also show random variations. For dealing with randomness and uncertainties, risk analysis provides a general framework. The various steps to be undertaken for a comprehensive application of engineering risk analysis to groundwater contamination problems are the following: (1) identification of hazards, (2) risk quantification, (3) consequences of risk, (4) perception of the consequences and (5) risk management. Methods and tools used in groundwater contamination problems are: uncertainty analysis, stochastic simulation and the fuzzy set approach (Ganoulis, 1991b).

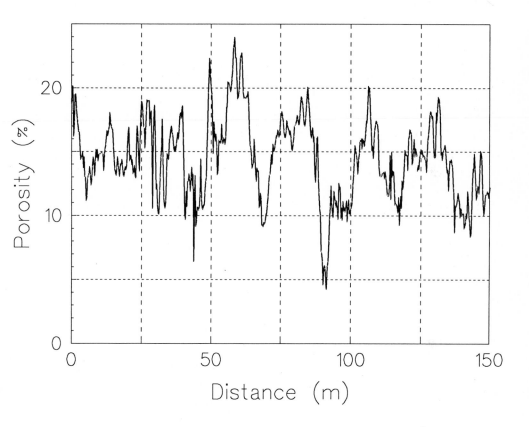

Figure 1. Typical porosity variation in an alluvial aquifer.

A variable can be defined in aquifer systems to characterize groundwater quality. This characteristic variable is called *"load"* or *"exposure,"* l. In groundwater problems the exposure could be the pollutant concentration at given location, estimated with some level of uncertainty. This uncertainty can be encoded with probabilistic methods treating l as a random variable. Alternatively, fuzzy set methods can be used to represent uncertainty in the exposure by expressing l as a fuzzy number (Ganoulis, 1991b). The "*resistance* "or "*capacity*" r could be the maximum groundwater pollutant concentration allowed by the standards. Exposure below this level would not cause adverse effects, as for example cancer. The resistance is also uncertain in many cases. Then probabilistic methods or fuzzy set methods can be used to describe resistance either as a random variable or a fuzzy number. *Failure* is defined as an event when exposure is larger than resistance, l > r. Because of the uncertainties in l and r this event can be expressed as a probability P (l > r). The *consequence of a failure,* in engineering risk analysis, is considered as economic loss, loss of human life or other adverse effects. This consequence L(l) is often expressed as a deterministic function of the exposure. The expression of risk depends on the way uncertainties in the elements of risk analysis are considered. The classical probabilistic formulation generally considers the expected value of risk using the probability density function g (l) of exposure l. In this case the engineering risk ER is

$$ER = \int_{r}^{\infty} L(l)\, g(l)\, dl$$

The use of the risk analysis approach is illustrated for a case study in Australia, where geological and hydrogeological data have been available. The fate of pollutants is simulated on the basis of these data and with computerized mathematical models . The random walk simulation method is applied to analyse the transport and dispersion of pollutants. Spatial and temporal variabilities of the aquifer characteristics can also be taken into account for considering alternative measures for remediation of the groundwater pollution.

RISK ASSESSMENT BY MATHEMATICAL MODELLING

Groundwater Hydrodynamics

Different models and modelling techniques have so far been used in groundwater hydraulics to account mainly for the time variation of the piezometric head in various locations in the aquifer. Lump-type hydrodynamic models have been introduced that use a rather rough grid discretization of the aquifer in space. Models based on finite differences and finite elements have been extensively operate with in the past (Custodio et al. 1988). Chiew and McMahon (1990) have employed the

AQUIFEM-N model in order to take into account in an integrated framework both the surface hydrological processes and the groundwater flow.

With the assumption of a quasi-horizontal regional groundwater flow driven by gradients of the total or hydraulic head H, a mass balance and Darcy's law applied to a semi-confined aquifer of constant porosity give the following partial differential equation

$$\frac{\partial H}{\partial t} = \nabla \cdot (T \nabla \cdot H) + \lambda (H_0 - H) - q \tag{1}$$

where H is the total head , $T = T(x,y)$ is the transmissivity, λ is vertical conductance, H_0 the constant total head in the upper aquifer formation and q is the pumping rate in $m^3/s/m^2$.

Over-relaxation and steady-flow conditions can be used for the numerical integration of Eq.1. Once the velocity fields are computed the risk for groundwater contamination can be quantified by means of a random walk simulation code.

Solute Transport and Random Walks

Depending on the composition of the pollutant (organic, radioactive, microbiological), its interaction with groundwater and soil is expressed as a function of the pollutant concentration. For conservative pollutants, such as saline waters, this interaction is negligeable and for regional groundwater flow, the following mass conservation equation is obtained

$$\frac{\partial C}{\partial t} + \frac{\partial}{\partial x}(uC) + \frac{\partial}{\partial y}(vC) = Dx\left(\frac{\partial^2 C}{\partial x^2}\right) + Dy\left(\frac{\partial^2 C}{\partial y^2}\right) + Dz\left(\frac{\partial^2 C}{\partial z^2}\right) \tag{2}$$

where:
$C(x,y,z,t)$: is the pollutant concentration (M/L^3)
$u(x,y,t)$, $v(x,y,t)$: are the groundwater velocity components (L/T)
Dx, Dy, Dz : are the dispersion coefficients (L^2/T)

In the deterministic approach, the velocity field is obtained by means of hydrodynamic models such as the one described by the Eq. (1). Equation (2) is valid in the 3-D space, if transport in the vertical direction is negligible. For the analytical or numerical integration of Eqs. (1) and (2) the boundary conditions are written for the flow and solute transport. Distinction is made between boundaries of constant or prescribed solute concentration and constant or zero solute flux. In this chapter, x-y coordinates are those of the vertical plane and water intrusion from the upper aquifer formation has been taken into consideration. The dispersion term $\partial^2 C/\partial z^2$, in the direction vertical to the plane, has been neglected.

In fact, equation (2) is a random partial differential equation. The causes of randomness and variability are (i) the random variation of the velocity components (u,v) owing to the spatial variability of the aquifer parameters (porosity, permeability), and (ii) the variation of the dispersion coefficient D as a result of the random fluctuations of the velocity components. In the general case, stochastic simulation and risk analysis techniques can be used to quantify the effect of various uncertainties in the dispersion process (Ganoulis,1985;1986;1991a; 1991b, Ganoulis et al.,1991).

Over the years the advection-dispersion equation (2) has been extensively investigated and numerically approximated by numerous methods. Finite differences and finite elements have been used and produced stable numerical results. However, significant errors are inroduced in all these numerical simulations. These are due to the fact that only a limited number of terms in the Taylor series expansions are taken into account. Explicit algorithms suffer from the so-called numerical diffusion. This is an artificial diffusion related to the truncation errors. It is superimposed on the physical diffusion and leads to an excessive attenuation of the input signals. Implicit finite difference algorithms introduce trailing effects because the initial signals are propagated at velocities that differ from the physical ones.

Several particle-oriented models in hydrological applications have been developed in the past. According to the method of characteristics (Konikow and Bredehoeft, 1978) a large collection of computer-generated particles, each of which is assigned a value of concentration, move along streamlines. Then, these concentration values are corrected to take into account the influence of diffusion by a standard Eulerian finite difference or finite element technique. This method is rather complex and difficult to use and sensitive to the boundary conditions.

It seems that particle methods based on random walks are more flexible and easy to use and lead to relatively accurate results (Ganoulis, 1977, Kinzelbach 1989).

Consider at time t=n Δt a large number of particles N located at the positions

$$r_p^n = (x_p^n, y_p^n) \qquad\qquad p = 1, 2, ..., N \qquad\qquad (3)$$

According to the random walk principle the probability of finding a particle at a given position after time Δt follows a Gaussian law of mean value 0 and variance $s^2 = 2\Delta tD$, where D is the dispersion coefficient. Now the particles are moving from time t=nΔt to time t+Δt=(n+1)Δt according to the relations

$$x_p^{n+1} = x_p^n + u\Delta t + x_1 \qquad\qquad (4)$$

$$y_p^{n+1} = y_p^n + v\Delta t + x_2 \qquad\qquad (5)$$

where u, v are the velocity components of groundwater and x_1, x_2 are random variables following a normal distribution of mean value 0 and variance $s^2 = 2\Delta tD$.

To evaluate probabilities and concentrations of the particles, the area is covered by a regular grid (Fig. 2). Knowledge of the velocity components u, v at the grid points permits particle velocities to be computed by linear interpolation. The probability of reaching a given grid cell and consequently the particle concentrations are evaluated by counting the number of particles which fall within the grid square. The procedure is illustrated in Fig. 2 for three particles initialy located at the same point A. Every particle is moving according to the relations (4) and (5). After 10 time steps the particles occupy three different positions A_1, A_2 and A_3.

3-PART

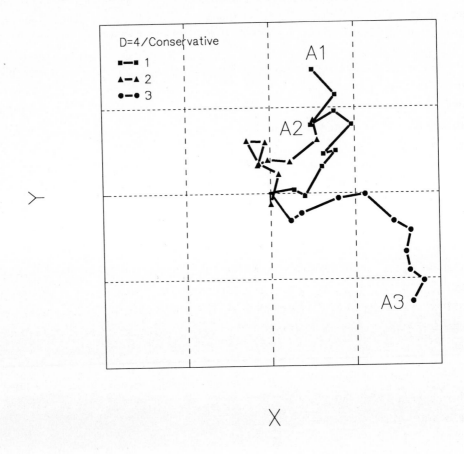

Figure 2 . Random walk of three particles after ten time steps.

This method has some drawbacks: First, to obtain statistically meanigful results at least a thousand particles must used. If applied to multiple sources that continuously emit pollutants the method, calls for a large number of particles to be tracked. Second, because of the statistical origin of the method, the concentration field shows oscillations and averaging in space and time is necessary.

When the deterministic current velocity field is used in equations (4) and (5), the solution obtained tends to that of the deterministic convective-dispersion equation. The decay term $-\lambda C$ can be simulated by eliminating at every time step a number of particles proportional to λ.

Now consider a large number of particles initially located at the same point (point source), but departing at different times $t=n\Delta t$. Each particle is moving during a given travel time $T>t$, where $T=t+m\Delta t=(n+m)\Delta t$. The final position $\mathbf{r}$ $(t+T)$ of each of the particles after time T can be computed by using the randomly varying velocity field $\mathbf{V}(t)$ as follows

$$\mathbf{r}\,(t,t+T) = \int_{t}^{t+T} \mathbf{V}\,(t)\,dt \tag{6}$$

It is obvious that the final position of every particle depends on the initial departing time t. Allocation of a different value of t to each particle, permits different final positions of the particles after time T to be found. Because of the stationarity of the random process, the concentration field and the probability of reaching a given location are independent of t.

On the basis of a relatively long record of time series of velocities V_i, the impact probabilities and consequently the risk at a given location are evaluated numerically. After a travel time $T=(n+m)\Delta t$, equation (6) takes the following form

$$\mathbf{r}\,(t,t+T) = \sum_{n}^{n+m} V_i\,\Delta t \qquad\qquad i=n,\ ....n+m \tag{7}$$

Particles are counted at every location with a help of a grid as in random walk simulations. Pollutant concentrations are proportional to the number of particles located within every square of the grid.

The simulation procedure is possible if the time history of the pumping rate is known from measurements and flowmeter recordings.

APPLICATION TO THE CAMPASPE VALLEY

The study area

The Campaspe River in the north-central Victoria is one of the tributaries of the Murray River. It flows northward from Lake Eppalock to the river Murray, in a basin of approximately 2100 Km^2 (Fig. 3).

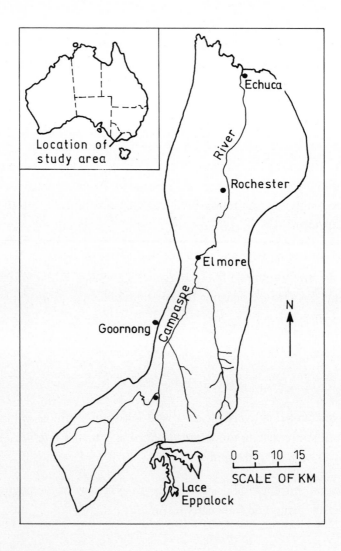

Figure 3. Campaspe river basin.

A typical geological cross-section of the Campaspe River basin is shown in Fig. 4. Two major aquifers are found in the Campaspe Valley: (1) the Shepparton Formation and (2) the Deep Lead. The Shepparton Formation aquifer is a sedimentary geological formation, mainly composed of clays. It vertically extends from the actual ground level to variable depths of about 70m at Rochester and 55m at Elmore. Hydraulically, the Shepparton Formation behaves mainly as a phreatic aquifer in the north and partly as a semi-confined aquifer in the south. Hydraulic conductivities range from 25 to 55 m/day, specific yields from 0.02 to 0.2 and vertical conductances from 0.001 to 0.03 m/year/m.

The Deep Lead extends vertically below the Shepparton Formation to the Paleozoic Bedrock, which, although capable in transmitting water within the fractures, is not active because the water pressures are almost the same as in the Deep Lead. The thickness of the Deep Lead is about 40m at Rochester and 30m at Elmore. It consists of gravel and sands with a hydraulic conductivity up to four times great as that of the Shepparton Formation.

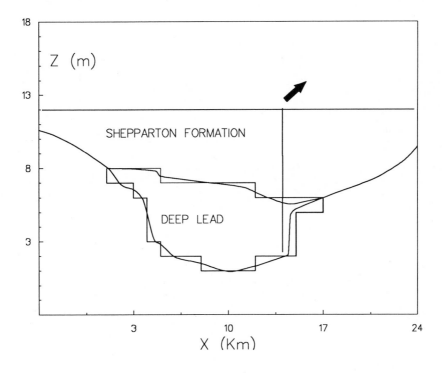

Figure 4. Typical geological cross-section.

Values of hydraulic conductivity in the Deep Lead range from 25 to 200 m/day, with a typical value of 130m/day in the south and lower values in the north. The Deep Lead starts at Axedale, becoming larger and deeper from the south to north. It varies from confined to semi-confined, with storage coefficients ranging from 10^{-4} to 10^{-2}.

The study area is located in the Rochester irrigation district   Three cross-sections have been chosen for mathematical modelling and computer simulations . Most of the computer runs were developed in the cross-section A-C along the Dingee road. In this cross-section the main production bore in the Deep Lead is the RW10032 bore (Houlihan). Data for the salinity variation in this bore cover the period 1982-1990. As indicated in Fig. 5, the data show a reduction of the observed salinity with time. This rather unexpected result seems to be systematic, as shown in Fig.5, where a polynomial regression of the data versus time has been applied. One can see that the rate of decrease of salinity with pumping is lessens with time and there is a tendancy to reach

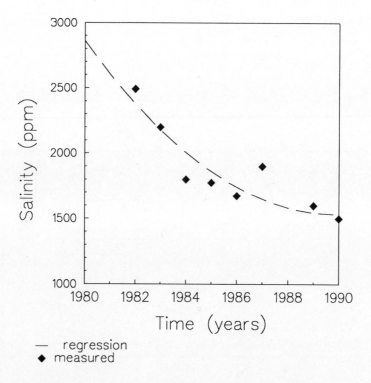

Figure 5. Salinity variation observed in main production bore.

an asymptotic limit. The explanation of this behavior by mathematical modelling has been the principal aim of our study in a general framework of salinity control .

It has been recommended (Reid, 1988) that pumping in the Deep Lead will contribute to lowering the shallow water table in the Shepparton Formation and so to providing a better drainage of the groundwaters in the basin. Furthermore, Williamson (1984) and Nolan and Reid (1989) have studied the salt redistribution resulted by pumping and re-use of groundwaters in the Deep Lead and concluded that further work is needed to obtain the "best policy" for groundwater extraction from the Deep Lead in order to avoid any groundwater quality degradation.

Experience from pumping in the Deep Lead during the past few years shows contradictory results as far as the time variation of the salinity of pumping water is concerned (Reid, 1988): in some cases (Rochester) a systematic improvement of the groundwater salinity with pumping has been monitored, while in some others (Loddon and Campaspe) a degradation of the groundwater quality with pumping has been measured. These different behaviours are mainly related to the initial distribution of the salinity concentration around the pumping wells and the mechanism of redistribution of the salinities during pumping. In all those cases, both advection and dispersion phenomena are important for the salinity redistribution in the vertical as well as in the horizontal planes.

Control and management of the salinity in the basin is a complex process, involving several steps and actions such as evaluation of the present situation, mathematical modelling and definition of various water disposal and treatment strategies. For salinity control plans, pumping in the Deep Lead, use and disposal of groundwaters, water treatment and mixing between waters of different salinity are some of the available options in a multi-objective optimization and decision process. An important component of the whole process is the mathematical modelling and computer simulation of the risk of salinity contamination due to pumping in the Deep Lead.

In the present chapter a mathematical model of relatively local scale is presented for quantification of the risk of solute transport and dispersion near pumping bores in the Deep Lead.At the same time an explanation is given to the apparently contradictory observations concerning the rate of variation of the salinity concentrations with pumping in the Deep Lead.

The model has been developed for a vertical cross section and can easily be extended to three dimensions with similar algorithms.

Results of simulation

The piezometric head variation is shown in Fig.6 for q=10l/s and 100l/s. On the basis of the computed velocity fields, the salinity intrusion with pumping has been simulated by means of the random walk numerical code.

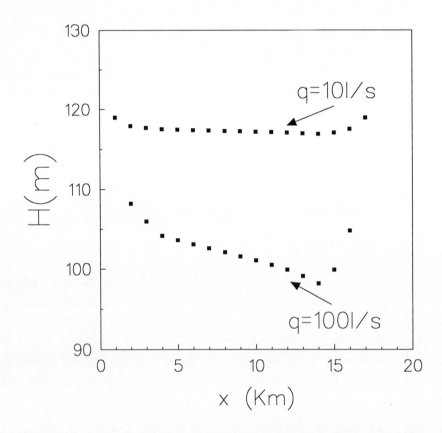

Figure 6. Piezometric head drawdown in Deep Lead for q=10l/s and q=100l/s.

Measurement of salinity contours in the Deep Lead indicate the existence of an almost discontinuous salinity front near the Houlihan well. Fresh water can flow from the vicinity of the well and reduce the initial high salinity of the water in the bore. Based on salinity distribution data from 1986 and 1987 the following initial distribution has been assumed for 1982 (Fig 7): 2550 ppm around the well and 550 ppm in the remaining cross-section.

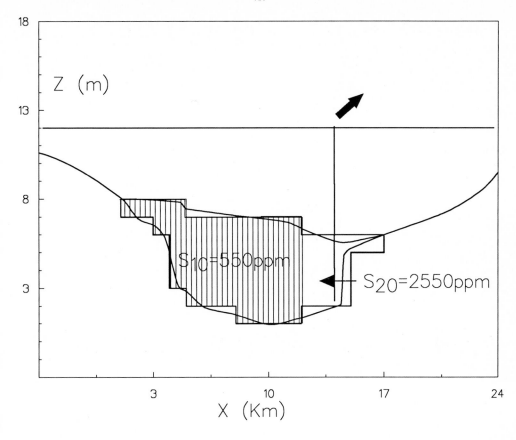

Figure 7. Initial salinity distribution.

The final salinity distribution has been obtained by superposition of the value of 550 ppm (initial difference). Figures 8 and 9 show particle concentrations after one and two years as predicted by the simulation for pumping for six months per year and with $Dx = 0.8 \times 10^{-3} m^2/s$ and $Dy = 10^{-5} m^2/s$. The respective salinity contours are shown in Figs. 10 and 11.

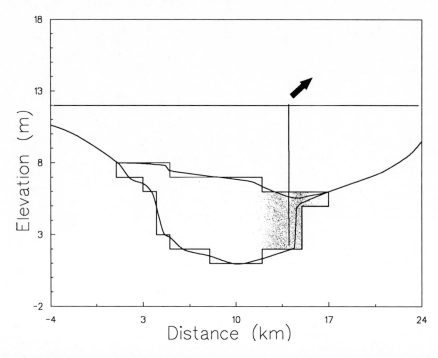

Figure 8. Particle concentrations after t=1 year.

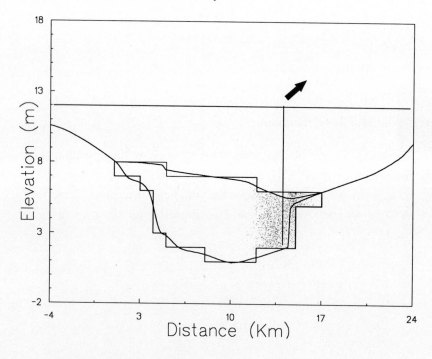

Figure 9. Particle concentrations after t=2 years.

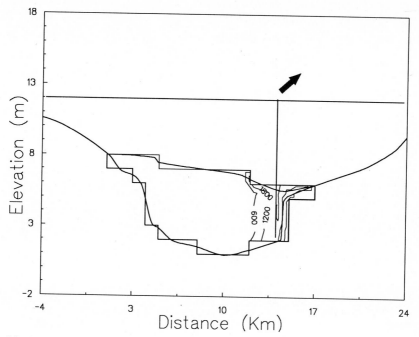

Figure 10. Salinity contours after t=1 year.

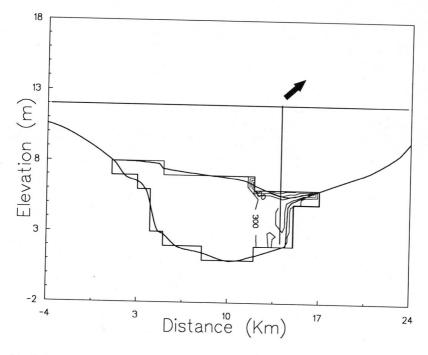

Figure 11. Salinity contours after t=2 years.

The comparison between the simulated results over eight years and the data are presented in Fig. 12 . It is remarkable that so good an agreement has been obtained with realistic values of the dispersion coefficient and the most probable initial condition.

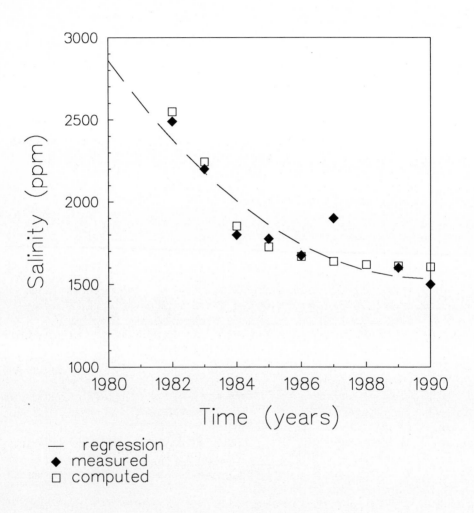

Figure 12. Comparison of measured with simulated results.

DISCUSSION AND CONCLUSIONS

For the assessment of risks of groundwater pollution and the design of groundwater development plans, reliable numerical simulation methods are needed. Contamination risks of groundwater can be quantified and the probability to meet the water quality standards can be found by using Lagrangian algorithms based on random walk simulation.This methodology has been applied in a case study in Victoria, Australia.

It has been recommended that pumping in the Deep Lead will contribute to lowering the shallow water table in the Shepparton Formation and so to provide a better drainage of the groundwaters and reducing the risk of salinisation. In order to understand and explain quantitatively these observations, a computerized mathematical model has been developed, based on random walk simulations. The model has been applied in the RW10032 pumping bore (Houlihan), and the time evolution law of the salinity concentrations has been succefully simulated.

REFERENCES

Bagtzoglou, A.C., A.F.B. Tompson and D.E. Doudherty (1991) Probabilistic Simulation for Reliable Solute Source Identification in Heterogeneous Porous Media, in: Water Resources Engineering Risk Assessment, Ganoulis, J.(ed.), NATO ASI Series, Vol.G29, pp.189-202, Springer-Verlag, Heidelberg

Chiew, F.H.S., McMahon, T.A. (1990) Estimating grounwater recharge using a surface watershed modelling approach J.Hydrol., 114: 285-304

Custodio,E., A. Gurgui , J.P. Lobo Ferreira (Eds.) (1988) Groundwater Flow and Quality Modelling, NATO/ASI Series, Reidel, 843 p.

Ganoulis, J. (1977) Simulation du type Monte-Carlo pour l' étude microscopique de la dispersion en milieu poreux Proc.IAHR Symposium on Hydrodynamic Diffusion and Dispersion in Porous Media, pp.365-373, Pavia, Italy

Ganoulis, J., H. Morel-Seytoux (1985) Application of stochastic methods to the study of aquifer systems. Technical Documents in Hydrology, UNESCO,Paris.

Ganoulis, J. (1986) Sur les échelles spatiales des hétérogenéités en milieu poreux, Hydrogéologie, No.2, 115, BRGM (France).

Ganoulis, J. (1991a) Nitrate contamination of surface and groundwater in Greece In: Nitrate Contamination:Exposure,Consequences and Control, In:.Bogardi & Kuzelka(Eds.), pp.55-64, Springer -Verlag, Heidelberg

Ganoulis, J. (ed.) (1991b) Water Resources Engineering Risk Assessment, NATO ASI Series, Vol G 29, Springer-Verlag, Heidelberg, 552 pp.

Ganoulis, J., L. Duckstein , I. Bogardi (1991) Risk Analysis of Water Quantity and Quality Problems: The Engineering Approach, in: Water Resources Engineering Risk Assessment, Ganoulis, J.(ed.), NATO ASI Series, Vol.G29, pp.3-20, Springer-Verlag, Heidelberg

Kinzelbach, W. (1989) Methods for the Simulation of Pollutant Transport in Ground Water - A Model Comparison", Proceedings of NWWA Conference on Solving Groundwater Problems with Models, Indianapolis

Konikow, L. F., J. D. Bredehoeft (1978) Computer Model of Two-Dimensional Solute Transport and Dispersion in Ground Water, in Techniques of Water Resources Investigations of the United States Geological Survey, Book 7, Chapter C2, United States Government Printing Office, Washington,

Nolan J. , M.Reid (1989) The role of Deep Lead Pumping for Salinity Control and Resource Development within the Shepparton Region, RWC , Invest. Branch Unpublished Report, 1989/11

Reid M. (1988) Campaspe Valley : Deep Lead Groundwater Management Proposal,DIRT Unpublished Report, 1988/27

Tickell, S. J. , Humphrys, W.G. (1984) Groundwater resources and Plain. Dept. of industry, Technology and Resour., Geol. Surv. Rep. 84, 197 pp.

Williamson, R. J. (1984) Campaspe Irrigation District Model, Geological Survey of Victoria,Dept. of Minerals and Energy, Unpublished report 1984/53

Migration of Radionuclides in Natural Media. The Chernobyl' Case

V.S. Soldatov
Institute of Physical Organic Chemistry
Belarus Academy of Sciences
Surganov Str. 13
Minsk 220603
Republic of Belarus.

The present paper is a review of recent works of Russian, Ukrainian and Byelorussian scientists devoted to assessing of Cnernobyl's radiological consequences. Until 1989 most information concerning the accident was classified; this favoured the appearance in the literature of many unreliable and non-professional publications of sensational character. All the restrictions for publication were removed in May 1989. This chapter contains the data taken from the most reliable sources of information. For those who is interested in the problem described in more details, the author recommends the excellent book by Yu.A.Israel and co-authors containing the most complete and systematic information.

General Characteristics of Accident Region

The Chernobyl' nuclear power station is situated on the right bank of the largest tributary of Dnieper river Pripjat, separating the Ukraine and Byelorussia. The border with Russia passes at a distance of approximately 160 km, therefore all the three countries suffered from the accident in the largest degree. The contaminated territory is mainly a lowlands with approximately one-third of cultivated land and one-third of wild forest and marsh. The average density of population is about 20 inhabitants per square kilometre. The largest cities in the contaminated territory are Kiev (~ 3.0 million inhabitants, 125 km distant) and Gomel (0.5 million inhabitants, 150 km distant). The climate in the region of the accident is similar to that in Northern Europe (average temperature in Byelorussia is -7°C in January and 18°C in July ; annual rainfall is 644 mm). The map of the accident region is presented in Fig. 1.

NATO ASI Series, Vol. G 32
Migration and Fate of Pollutants in Soils and Subsoils
Edited by D. Petruzzelli and F. G. Helfferich
© Springer-Verlag Berlin Heidelberg 1993

General Characteristics of the Accident

The fourth reactor of the Chernobyl' power station destroyed at the accident had a power of one MW and was of a water-graphite channel type. It was charged

with a nuclear fuel (Uranium-235) three years before the accident. The total amount of fuel was 190 tonnes. The general picture of the accident was as follows. A heat explosion happened on April 26, 1986. As a result a fire had started and a large amount of radioactive material was released to the atmosphere. The main burning material was graphite. The high temperature in the fire zone was supported by the internal heat emission in the destroyed reactor active zone. The fire lasted ten days during which large amounts of radioactive gases, vapours and aerosols were released to the atmosphere. The most intensive upward moving current of radioactive releases was observed during the first three days after the explosion. The level of radiation at a distance of 10 km from the place of the accident and an altitude 200 m reached 1000 mR/h. The height of the jet reached 1200 m and

Fig. 1. The map of the accident region. The shaded areas correspond to the level of contamination more than 5 Ci/km^2

later stabilized at a level of 200-300 m. The radiation situation was controlled by means of aero-gamma measurements from special helicopters and aeroplanes. The first map of the radioactivity levels was completed on May 1, 1986 (for a region with radius 100 km around the accident). The official map of the whole region of the accident was issued on May 10, 1986. It was used as a basic document for all following actions connected with eliminations of the accident consequences, such as resettling the population, establishing the exclusion zone and the zones of strict control, etc. Simultaneously, the systematic measurements of the radiation situation in approximately half of the European part of the USSR started. The data of aero-gamma measurements were combined with the data of laboratory radiochemical measurements of gamma, beta and alpha activity of the samples of different natural objects: water, bottom sediments, soil, plants, parts of animal bodies etc. Special attention was paid to investigation of humans, settlement areas, houses, industrial and agricultural objects. Several tens of research institutes of the State Committee on Hydro-Meteorologic Service, Soviet, Ukrainian and Byelorussian Academies of Sciences, Academy of Medical Sciences, Agricultural Academy, Ministry of Health, universities, and specialized industrial and military institutes were involved in investigation of the level of radioactive contamination, and the physical and chemical forms of the radionuclides, as well as, their distribution and migration. The ways of safe behaviour, deactivation of the natural and industrial objects, agricultural animals, food, etc. were also extensively investigated. Thousands of scientists have been working (and are still working) on these problems.

Amount and Radionuclide Composition of Chernobyl' Release

The total amount of the radioactive matter released into the environment in the Chernobyl' accident was evaluated as 77 kg of pure radionuclides with an activity of about 150 MCi. Table 1 gives quantities and half-life times of the most important radionuclides in the release.

It is worth noting, that the confidence limits for the quantities of Cs, and Te isotopes are much large than for the other cases and are probably about 30%. With the figures given in the Table, the ratio between ^{131}I and ^{137}Cs is 12.5 which is in agreement with the experimental data. The data in the Table 1 were taken from the information of the USSR Governmental Commission prepared for the IAEA Meeting in August 1986 in Vienna. The total amount of the fuel discharged during first 10 days of the accident was estimated as 4% of the total fuel inventory in the Chernobyl'-4 reactor at the beginning of the accident. In these estimates only off-site fuel was taken into account. Later these estimate was repeatedly refined, and the figure of 3.5% with an error margin of ±0.5% can be accepted as accurate. Practically all radioactive noble gases and Tritium were discharged into the atmosphere.

Characteristics of External Irradiation

From the radiology stand-point, the most important radionuclides are those causing the largest damage to human health. The external irradiation dose is entirely caused by gamma radiation and can be estimated for different periods of time from the amounts of the discharged radionuclides and their half-life time values.

Table 1. Quantity of the radionuclides with half life time more than two days in the active zone of the Chernobyl'-4 reactor before the accident and the amount of the discharged radionuclides

Radio-nuclide	Half life days, years	Inventory amount		Discharge		
		MCi	10^{18}Bq	MCi	10^{16}Bq	%
^{239}Np	2.35 d	1300	48.1	41.6	154	3.2
^{99}Mo	2.73	130	4.81	3.0	11.1	2.3
^{132}Te	3.27	120	4.44	40	148	33
^{131}I	8.04	90	3.33	30	111	33
^{140}Ba	12.6	130	4.81	7.3	27.0	5.6
^{141}Ce	32.5	130	4.81	3.0	11.1	2.3
^{103}Ru	39	130	4.81	3.8	14.1	2.9
^{89}Sr	51	52	1.92	2.1	7.8	4.0
^{91}Sr	58.5	40	1.48	1.3	4.8	3.2
^{95}Zr	64	130	4.81	4.2	15.6	3.2
^{110}Ag	250	0.13	0.48	0.0042	0.0016	3.2
^{144}Ce	284	90	3.33	2.5	9.3	2.8
^{106}Ru	368	52	1.92	1.5	5.6	2.9
^{134}Cs	2.07 y	4.0	0.148	1.3	4.8	33
^{135}Sb	2.7	0.5	0.0185	0.016	0.059	3.2
^{90}Sr	29.12	5.2	0.192	0.2	0.74	4.0
^{137}Cs	30.17	7.2	0.266	2.4	8.9	33
^{238}Pu	87.75	0.027	0.0010	0.00081	0.003	3.0
^{239}Pu	20,460	0.020	0.00075	0.00061	0.002	3.0
^{240}Pu	6,537	0.027	0.0010	0.00081	0.003	3.0
^{241}Pu	14.4	4.7	0.175	0.142	0.525	3.0

This estimation is based on a model considering an infinite plane covered with a mixture of radionuclides. An exposure dose rate at the distance **Z** from the surface caused by a radionuclide **i** is given by the equation

$$R_i(t) = A_i^0 * K_i * K_L^i * exp(-\lambda_i * t) \quad (1)$$

where $R_i(t)$ is exposure dose rate at the distance **Z** from the surface, A_i^0 is a constant, K_i is a constant (exposure doze rate at contamination density equal to 1 [e.g. (mR/h)/(kBq/m^2) or (mR/h)/(Ci/km^2)], K_L^i-is the shielding coefficient of the radiation by a layer of the soil, λ_i is a constant of radioactive decay. This equation has a simple solution for an impermeable surface (K_L^i=1). Since

$$R_i(t) = dD_i/dt \qquad (2)$$

and

$$\lambda_i = 0.693/T^i_{1/2} \qquad (3)$$

where dD_i is a dose of external radiation, accumulated at the time interval dt and $T^i_{1/2}$ is a half-life time of an i-radionuclide. Then

$$D_i = A_i^o * K_i * \int_{t_1}^{t_2} \exp(-0.693\ t/T^i_{1/2}) * dt \qquad (4)$$

and

$$D_i = 1.44 * A_i^o * K_i * T^i_{1/2} \qquad (5)$$

for $t_1 = 0$ and $t_2 = \infty$.

It follows from the Eqns. 4 and 5 that an exposure dose in absence of vertical migration of radionuclides is proportional to the surface contamination, the K_i value and the half life time $T^i_{1/2}$. Table 2 gives estimates of the contributions of different isotopes to the integral exposure dose for   t equal to 4 months and 100 years. It is seen that during the first period most of the exposure dose is due to Te, I, Ba, Ru, Zr and Cs isotopes, with ^{134}Cs and ^{137}Cs contributing 15.7%. In the "cesium spots" their fraction is much higher (39.1%). At the longer period of exposure the role of Cesium increases reaching 76 to 93% of the total dose.

It is interesting to evaluate the exposure doses which were formed in Belarus due to release of ^{131}I and ^{137}Cs, the two most dangerous gamma emitters. According to the 1988 report of the UN Scientific Committee on the Effect of Radiation the ratio of the initial collective exposure doses of ^{131}I and ^{137}Cs was equal to 16 for the territory of Belarus:

$$A^o_{I\text{-}131} / A^o_{Cs\text{-}137} = 16$$

The other constants necessary to apply Eqn. 5 are: $K_{I\text{-}131} = 7.3$; $K_{Cs\text{-}137} = 10.7$ $[(\mu R/h)/(Ci/km^2)]$; $T_{1/2(I\text{-}131)} = 193$ h ; $T_{1/2(Cs\text{-}137)} = 264,552$ h. Substitution of these values into Eqn. 5 gives

$$D_{Cs\text{-}137} / D_{I\text{-}131} = 131$$

This estimation shows that, in spite of the fact that the initial activity of the ^{131}I was 16 times higher than that of ^{137}Cs, the total exposure dose of ^{131}I is negligibly small compared to that of ^{137}Cs.

Table 2. Contributions of different isotopes into exposure dose of external -irradiation in Byelorussia for the period 4 months and 100 years after the accident

	$100 \ D_i / \Sigma D_i$,%			
Isotope	sector "North"		"Cesium spots" in Belarus	
	4 months	100 years	4 months	100 years
^{239}Np	0.9	0.2	0	0
^{99}Mo	0.2	0	0	0
^{132}Te	22.1	26.7	5.1	2.5
^{131}I	6.6	10.5	1.5	1
^{140}Ba	11.3	6.5	2.6	0.6
^{141}Ce	0.8	0	0.2	0
^{103}Ru	10.2	11.6	2.6	1.2
^{95}Zr	27.8	1.4	8.3	0.2
^{144}Ce	0.7	0	0.6	0
^{106}Ru	3.7	4.0	3.4	1.6
^{134}Cs	9.9	22.1	16.0	15.5
^{137}Cs	5.8	17.0	59.5	77.4

The consideration above explains why ^{137}Cs is considered as the most ecologically hazardous isotope. The radioactive contamination of the districts distant from Chernobyl' is almost entirely due to the presence of ^{137}Cs. The area of the territory polluted and the number of people living in this area is given in Table 3.

Table 3. Characteristics of the scale of the ^{137}Cs contamination

Density of ^{137}Cs contamination, Ci/km	Expos. dose* rate, mR/h	area, thousands km^2	populat., thousands
1 -5	0.010-0.080	103	1489
5-15	0.080-0.160	18	281
15-40	0.160-0.450	7.2	79
>40	0.45-2.2	3.1	3

* The data for Belarus

A certain amount of ^{137}Cs was carried away from the USSR territory. The highest level of contamination, 3 Ci/km, was registered in Sweden. The areas with a relatively high level of contamination (1-3 Ci/km) diminished by about two orders of magnitude one year after the accident. The amount of ^{137}Cs brought to different countries by the radioactive air masses and the fall-out is characterized in Table 4.

Table 4. Contents of Cesium-137 in the fall-outs in different countries after Chernobyl' accident

Country	Cont., 10^{14} Bq	Country	Cont, 10^{14} Bq
U.S.S.R	407	France	8.3
Poland	92	Hungary	7.9
Rumania	67	Checo-	
Yugoslavia	61	Slovakia	5.9
Sweden	34	D.R.China	5.4
Bulgaria	27	U.K.	4.4
Germany	22	Albania	3.9
Finland	19	U.S.A.	2.8
Turkey	18	Canada	2.5
Austria	11	Ireland	2.5
Italy	11	Switzerland	2.0
Norway	11	Netherlands	0.68

α - and β- Emitters

The contribution of α - and β- emitters into external irradiation is negligibly small. Nevertheless, their danger for humans can be high due to their ability to accumulate in a certain organs of human body causing their intensive irradiation. The most dangerous source of α -radiation are ^{239}Pu and ^{240}Pu, and that of β- radiation is ^{90}Sr. The main danger of ^{90}Sr is due to its accumulation in bone tissues causing heavy marrow irradiation and increasing the risk of blood diseases. Plutonium isotopes are present as fine dust or smoke in microparticles and are most dangerous when present in breathing air. Entering the lungs they stick to the tissues increasing danger of lung cancer. At the same time, plutonium entering a body through the digestive system is less dangerous since it is almost not retained in the organism. Fortunately, the volatility of the above nuclides is not high and the contamination level in the both cases is lower than the standard established outside the 30-km zone (3 Ci/km^2 for ^{90}Sr, 0.1 Ci/km^2 for Pu - isotopes).

Physical and Chemical Forms of Radionuclides

The explosive release of the nuclear fuel and construction material has produced a large amount of fine particles and lead to a high level of radioactive deposits in the vicinity of the reactor. The following fire resulted in abundant release of microparticles of the dispersed burning materials and their elevation with the upward-moving thermal current to the height of several hundreds meters.

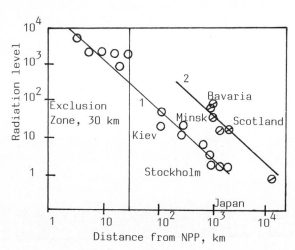

Fig. 2. Level of the external irradiation related to the natural radioactive background assumed equal to 0.01 mR/h as a function of distance 7-10 days after the accident. 1 - dry weather, 2 - periodic rains. (Israel Yu A, 1990)

Condensation of the vapours of the radioactive matter on the different natural microparticles already present in the air and the following aggregation produced a large variety of different types of microparticles. Different materials such as boron, rare earth metals, lead, sand, etc. were thrown into the burning reactor. All this resulted in a large variation in the particle composition: uranium oxides, zirconium, iron, silicon dioxide and silicates, carbon and bitumen particles etc. Most of the radioactive matter was released into the environment in the form of micro-particles which are mainly responsible for the ground contamination. Therefore, the local contaminations are strongly influenced by the meteorological conditions, especially by rainfall, this resulted in a heterogeneous ("spotty") character of the contamination distribution. These features of the accident provide a marked difference with the character of the radioactive environment contamination due to the atmospheric atomic bomb tests. In the latter case, the bomb materials evaporated in the explosion and were gradually released from the high stratospheric levels over periods of years after the bomb tests. The composition of the radionuclides is also different. The Chernobyl' case is characterized by brief massive deposition simply dependent on the distance from the place of the accident (Fig.2).

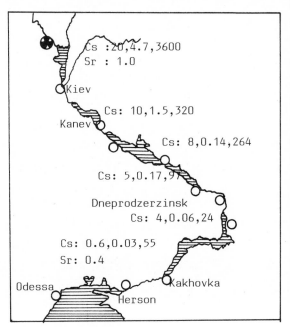

Fig. 3. Contamination of the Dnieper water reservoirs. The figures mean concentration of the isotope in water $(Ci/l*10^{-11})$, in the bottom slit (Ci/km^2) and the total amount (Ci).

Radioactive particles, known as "hot particles", have sizes ranging between fractions of and several tens of micrometers and activities between 0.1 and 100 Bq/particle. About half of the total fall-out is situated in the area within a radius of ten km from the accident site. The number of hot particles has a magnitude from 10^5 to 10^3 per square meter at distances 40 to 250 km from the place of the accident. Most of them remain in the surface layer of the ground about 1 cm thick. In the case of sandy soils they can penetrate down to 5 cm. In the peatbog soils they were detected at a depth of 30 cm. The main way of horizontal transfer of the radionuclides is their transport by winds and surface water streams. Nevertheless, regular examination of the contaminated territory from air and by laboratory analysis showed evidence of stability of the contamination isolines. This shows little importance of horizontal migration processes. The following important example supports this conclusion. Practically all ^{137}Cs (1.1 MCi) is situated in the area drained by the Dnieper and its tributaries. In spite of that, only 2.465 Ci (0.2%) of ^{137}Cs was released into Kiev water reservoir during the period from June 1986 to May 1989. The release of ^{90}Sr in that period was 1600 Ci, while its total amount was about 0.25 MCi. Contamination of the Dnieper water reservoirs decreases rapidly with the distance from Chernobyl', as seen from Fig. 3. Concentration of ^{90}Sr drops much less than that of ^{137}Cs which is due to higher mobility of Sr in water streams. The bottom sediment activity ranged between $6*10^{-5}$ and $1*10^{-7}$ Ci/kg and is about 10 times lower than the activity of the soil in their vicinity. The ratio ^{90}Sr/^{144}Ce in the bottom silt of Kiev water reservoir is one or two order of magnitude lower than in the fuel fall-out. This shows that the main part of ^{90}Sr has already been washed out to the Black Sea. The annual input into the Black Sea in 1988-1989 was about 75 Ci and the total amount of ^{90}Sr in Kiev water reservoir is not more than 100 Ci.

It is necessary to note that natural waters contamination is due to the radioactive suspended matter. The water itself after filtration through a microfiltration membrane never reached MPC levels for the radionuclides. Nevertheless, nowadays the tendency toward increasing the solubility of Cs and Sr is observed - from 0.01 and 1 % in 1986 to 0.2 and 25 % respectively in 1989.

Vertical Migration of Radionuclides

Vertical migration of radionuclides in soils was a subject of extensive investigation since it controls the dynamics of external irradiation and the contamination of agricultural production and groundwater in the contaminated territories.

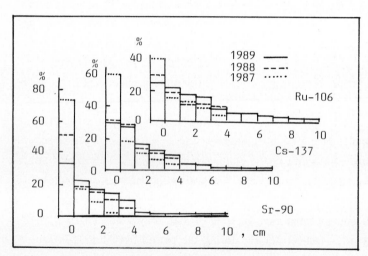

In the summer of 1986 the landscape-geochemical field network was established with the aim of long term observation of radionuclide migration processes in different landscapes and types of soils. In each of the observation networks the vertical migration of ^{134}Cs, ^{106}Ru, ^{144}Ce, ^{110}Ag and ^{90}Sr was studied by periodic (one or two times a year) soil sampling and laboratory analysis from different depths.

Fig. 4. Examples of diagrams of vertical distribution of the radionuclides. Sod-podzolic soil. (Petriaev E P, 1990)

Migration of Cs was studied by measuring quantities of ^{134}Cs to avoid disturbances from pre-Chernobyl ^{137}Cs. The typical vertical profiles of radionuclide distribution are presented in Fig. 4.

The mathematical analysis of these dependences done by A.A.Kasimovski has shown that the best fit between the experimental and theoretical data for the dependence of specific activity of the soil (Bq/g) as a function of the soil layer depth x (g/cm^2) is given by the superposition of an exponential and a Gaussian function:

$$(x,t) = (a_1/L)\exp(-x/L) + (a_2/\beta\sqrt{Dt})\exp(x-vt)^2/(4Dt) \quad (6)$$

where L is the "migration depth" parameter, g/cm^2; D is the diffusion coefficient, g^2/cm^4*s; v is the rate of mass transfer, g/cm^2*s; β is the normalizing coefficient; and a_1 and a_2 are fractions of the specific activity, related to each of the migration processes corresponding to the exponential or Gaussian distribution. The first term corresponds to a slow transfer of radionuclides chemically bound with soil particles. Their fraction is increasing in the series ^{90}Sr < ^{106}Ru < ^{144}Ce < ^{137}Cs. The second term describes "fast" migration. The series of migration rates is almost opposite to the previous one. Only a small fraction of radionuclides participates in this process. Nevertheless, it is most important in contamination of deep layers of the soil. At present more than 90% of the total amount of the radionuclides is contained in the top 3 cm layer of undisturbed turf. Table 5 gives an example of the parameters of the model computed for different types of soils.

Table 5. Parameters of the model for migration of radionuclides. The measurements are done in June 1987 (examples). (Israel Yu A, 1990)

SOIL	Radio nuclide	Exponent		Gaussian		
		L	a_1/a, %	Dt	vt	a_2/a, %
Sod-alluvial low gleyed	^{134}Cs ^{106}Ru ^{90}Sr	1.1 1.3 0.8	85 74 70	10 19	4.5 0.9	15 26 30
Alluvial sod-gleyed	^{134}Cs ^{106}Ru ^{144}Ce ^{90}Sr	0.52 0.60 0.56 0.6	99 97 96 75	10 6.4 6.3	13.5 13.4 8.5	1 3 4 25
Sod-podzolic cultivated	^{134}Cs ^{106}Ru ^{144}Ce ^{90}Sr	0.94 0.94 1.2 0.95	94 94 96 70	6 6 6	6 6 6	6 6 4 30

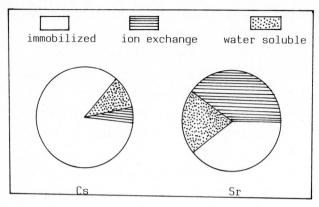

Fig. 5. Ratio between different states of Cs and Sr radionuclides in a surface layer of sod-podzolic sandy soil. The sample is taken in a distance 40 km from the accident place in 1989. (Petriaev E P, 1991)

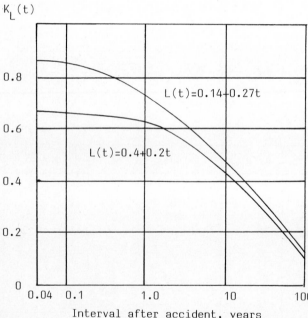

Fig. 6. Shielding coefficient as a function of time. (Israel Yu A, 1990)

The parameter L characterizes penetration of the main part of the radionuclides into the soil and has little dependence upon the type of radionuclide but varies significantly with the type of soil. It linearly increases with time

$$L(t) = p + qt \qquad (7)$$

Constants **p** and **q** vary within the limits 0.14 to 0.77 and 0.22 to 0.41, respectively, depending upon the type of soil.

One factor complicating radionuclide migration is that a radionuclide in the soil is not represented by a one migrating substance but forms several different species: soluble ionic substances, complexes with humic and fulvic acids, inorganic precipitates, ion exchange compounds, etc. Their conversion into the soluble (mobile) form can be due to chemical, physical or biological processes and is strongly influenced by the initial state of the radionuclide, organomineral composition of the soil, pH, and aeration conditions. It is known, for instance, that in presence of humic acids the rate of destruction of hot particles can increase by two orders of magnitude. Even storage in an air dry state can change the ratio of soluble to non-soluble forms of radionuclides. It was noticed also that during a year storage of samples of sod-podzolic soil the quantity of water-soluble Cs increased 2.5 times and ^{90}Sr increased 4 times. In peat-boggy soil these figures are 5 and 4, respectively, and in podzolized soil 5 and 1.2, respectively. At the same time, the solubility of Ce and Ru isotopes decreases. The ratio between different forms of radionuclides in the soil (water soluble, acid soluble, ion exchange, immobilized etc.) can vary over a wide range and is dependent upon external conditions. Nevertheless, some regularities can be summarized. The immobilized form is always predominant, and the water soluble part is always small. The soluble and ion exchange part of Sr is always larger than that of Cs. This is probably due to the ability of Cs to undergo isomorphous substitution for **K** in alumosilicates. A typical example of different forms for the ratio of Cs and Sr is given in Fig. 5.

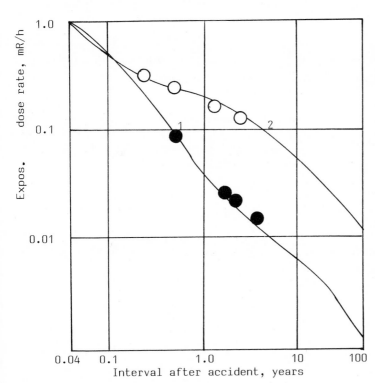

Fig. 7. Exposure dose rate of external gamma radiation in different regions as a function of time. Normalized data. R is taken equal 1 mR/h 15 days after the accident. 1- Kiev, 2- cesium spots. The points are experimental data. (Israel Yu A, 1990)

Due to an enormous complexity of the problem and a lack of information, mathematical modelling of radionuclide migration has a formal character. Nevertheless it allows to make predictions of the radiologic situation. The knowledge of parameters of vertical radionuclide distribution allows determination of K_L^i (in Equation 1) as a function of the migration depth (L), which, in turn, is a function of time. Fig. 6 gives the shielding coefficient for different rates of vertical migration (given in the figure in the form of a linear equation **L=L(t)**).

In Fig. 7 the dependence of the dose rate which accounts for radioactive decay and vertical migration of all types of the radionuclides is given for two particular places.

The data presented show that accounting for vertical migration of the radionuclides leads to correct prediction of radiologic situation in the environment.

ACKNOWLEDGEMENT. I am grateful to Dr. M.V. Mal'ko, Dr. I.V. Rolevich and Dr. V.P. Kasperchik for their help and valuable consulting at preparation of the present paper.

References

Alexahin R M, Krishev I I, Fesenko S V, Sanzharova N I (1990) Radioecologic Problems of Nuclear Energetics (in Russian). Atomnaya Energia 68: 320-328

Belyaev S, Borovoi A, Volkov V, Gagarinski A (1991) Chernobyl' Five Years After, Nuclear Europe Worldscan No 3-4: 22-2

Boris'uk L G, Gavril'uk V I, Dotsenko I S a.o. (1990) Vesti AN BSSR Ser.Phys.-Energ. Nauk No 4: 38-41

Burkart W, Crompton N E A (1991) Assessing Chernobyl's Radiological Consequences, Nuclear Europe Worldscan No 3-4: 27-30

Kuzmenko M I (1990) Radioecologic Studies of Water Reservoirs of the Ukrainian SSR (in Russian). Hydrobiol. Journ. 26: 86-99

Il'in L A, Pavlovskiy O A (1988) Radiologic Consequences of the Accident at Chernobyl' NPP (in Russian) Atomnaya Energia 65:119-129

Israel Yu A, Vakulovskiy S M, Vetrov V A, Petrov V N, Rovinskiy F Ya, Stukin E D (1990) Chernobyl':Radioactive Contamination of Natural Media, (in Russian) Gidrometizdat,Leningrad

Nikitchenko I N (1990) Some Results of Investigation of Radioactive Contamination of Agricultural Areas in Byelorussia (in Russian). Vesti AN BSSR Ser. Phys.-Energ. Nauk No 4: 23-30

Petriaev E P, Ovs'iannikova S V, Sokolik G A a.o. (1990) Change in Radionuclides State in the Contaminated Territories of Byelorussia (in Russian). Vesti AN BSSR Ser.Phys.-Energ. Nauk No 6: 78-83

Petriajev E P, Sokolik G A, Leonidova S L a.o. (1990) Distribution of "Hot" Particles in Southern Regions of Byelorussia (in Russian) Vesti AN BSSR Ser.Phys.-Energ. Nauk No 4: 42-49

Petriajev E P, Sololik G A, Ivanova T G a.o. (1990) Estimate and Forecast of Contamination of Soils in Southern Districts of Byelorussia (in Russian).Vesti AN BSSR Ser . Phys.-Energ. Nauk No 4: 73-77

Petriaev E P, Ovs'iannikova S V, Rubenchik a.o. (1991) State of Radionuclides of Chernobyl' Fall-out in Soils of Byelorussia (in Russian) Vesti AN BSSR Ser Phys.-Energ. Nauk No 4: 48-55

Silant'ev A N, Shkuratova I G, Bobovnikova Ts I (1989) Vertical Migration of Radionuclides of the Chernobyl's Fall-out in the Soils (in Russian). Atomnaya Energia 86: 194-197

Sobotovich E V (1990) Contamination of Natural Waters by Technogenic Radionuclides from Chernobyl' Release (in Russian). Water Resources No 6: 39-46

Sobotovich E V, Dolin V V (1990) Migration Mechanism of the "Hot" particles Radionuclides in Soils, Surface and Ground Waters" (in Russian). No 6: 51-55

Subject Index

NATO ASI Series G

NATO ASI Series G